Natural Resource Management and Policy

Volume 56

Series Editors

David Zilberman, College of Natural Resources, University of California, Berkeley, CA, USA

Renan Goetz, Department of Economics, University of Girona, Girona, Spain

Alberto Garrido, ETS, Technical University of Madrid, Madrid, Spain

There is a growing awareness to the role that natural resources, such as water, land, forests and environmental amenities, play in our lives. There are many competing uses for natural resources, and society is challenged to manage them for improving social well-being. Furthermore, there may be dire consequences to natural resources mismanagement. Renewable resources, such as water, land and the environment are linked, and decisions made with regard to one may affect the others. Policy and management of natural resources now require interdisciplinary approaches including natural and social sciences to correctly address our society preferences. This series provides a collection of works containing most recent findings on economics, management and policy of renewable biological resources, such as water, land, crop protection, sustainable agriculture, technology, and environmental health. It incorporates modern thinking and techniques of economics and management. Books in this series will incorporate knowledge and models of natural phenomena with economics and managerial decision frameworks to assess alternative options for managing natural resources and environment.

Keijiro Otsuka · Yukichi Mano · Kazushi Takahashi
Editors

Rice Green Revolution
in Sub-Saharan Africa

 Springer

Editors
Keijiro Otsuka
Graduate School of Economics
Kobe University
Kobe, Japan

Yukichi Mano
Graduate School of Economics
Hitotsubashi University
Tokyo, Japan

Kazushi Takahashi
Graduate School of Policy Studies
National Graduate Institute for Policy
Studies
Tokyo, Japan

ISSN 0929-127X   ISSN 2511-8560 (electronic)
Natural Resource Management and Policy
ISBN 978-981-19-8045-9   ISBN 978-981-19-8046-6 (eBook)
https://doi.org/10.1007/978-981-19-8046-6

This Springer imprint is published by the registered company Springer Nature Singapore Pte Ltd.
The registered company address is: 152 Beach Road, #21-01/04 Gateway East, Singapore 189721, Singapore

Preface

The ultimate objective of development economics is to provide a useful development strategy or prescription for the policies that must be implemented to realize economic development in low- and middle-income countries. This edited volume discusses how to realize a Green Revolution in sub-Saharan Africa (SSA), where agricultural production has stagnated for an extended period. We focus particularly on rice as the most promising crop in SSA because of the potential for high technology transferability from tropical Asia, which itself realized a rice Green Revolution nearly a half-century ago (Otsuka and Larson 2013, 2016). Therefore, an important question is how to transfer Asian rice technology most effectively to SSA. A related question concerns what types of supplementary policies are needed to realize and accelerate rice sector development in SSA. Our research team has conducted a number of case studies to address these questions over ten years. This book represents the synthesis of our research efforts.

Our answer to the first question is simple but fundamental: the provision of rice cultivation training is an effective means of transferring Asian rice Green Revolution technology, thereby dramatically enhancing the productivity of rice farming in SSA. A common mistake is the neglect of the knowledge that the rice Green Revolution is intensive not only in the use of improved seeds and inorganic fertilizer but, more importantly, in cultivation management. Thus, the rice Green Revolution is possible in SSA if rice farmers are trained to learn proper cultivation practices through extension. This is the central message of this book.

We also found that mechanization, particularly the introduction of tractors, is necessary for many areas in SSA where oxen are not available for plowing due to trypanosomiasis caused by the tsetse fly, so manual labor is often used for land preparation. Thus, tractors can be used as a substitute for manual labor in SSA. This enables intensive land preparation complementary to subsequent rice cultivation activities, leading to significant improvements in efficiency and effectiveness. This is in sharp contrast to tropical Asia, where a tractor is a substitute for draft animals, with little land productivity differences between the two. Another finding is that rice farmers and millers are often indifferent to the quality of rice, leading to the production of low-quality rice in SSA, which cannot compete with imported Asian rice. Thus, the government must provide information on improving the quality of rice to farmers, traders, and rice millers. We also confirmed the importance of irrigation in enhancing the rice production capacity in SSA. In particular, we argue that the investment returns for irrigation facilities are likely to be high if proper rice cultivation practices are disseminated to rice farmers to improve the productivity of irrigated rice farming.

This book is a joint product of a group of development economists and the Japan International Cooperation Agency (JICA) Ogata Sadako Research Institute for Peace and Development. We are particularly grateful for the comments and encouragement of its staff: Hitoshi Fujiie, Rinko Jogo, Nobuko Kayashima, Ryota Kubo, Koji Makino, Etsuko Masuko, Sachiko Mitsumori, Hironobu Murakami, Yoshie Sasabe, Maiko Takeuchi, Minoru Yamada, and Naotaka Yamaguchi. We are also indebted to a large number of JICA staff, including Shinjiro Amameishi, Shuichi Asanuma, Tomohiro Azegami, Kojiro Fujino, Yuichi Matsushita, and Fumihiko Suzuki. Last but not least, we would like to point out that this research has been stimulated by the two most prominent rice cultivation experts and practitioners devoted to the improvement of rice farming in SSA: Motonori Tomitaka and Tatsushi Tsuboi. We wish to thank all of them wholeheartedly.

Kobe, Japan	Keijiro Otsuka
Tokyo, Japan	Yukichi Mano
Tokyo, Japan	Kazushi Takahashi
June 13, 2022	

References

Otsuka K, Larson DF (2013) An African Green Revolution: finding ways to boost productivity on small farms. Springer, Dordrecht

Otsuka K, Larson DF (2016) In pursuit of an African Green Revolution: views from rice and maize farmers' fields. Springer, Dordrecht

Contents

Editors and Contributors

About the Editors

Keijiro Otsuka is a professor of development economics at the Graduate School of Economics, Kobe University and a chief senior researcher at the Institute of Developing Economies in Chiba, Japan since 2016. He received a Ph.D. in economics from the University of Chicago in 1979. He majors in Green Revolution, land tenure and land tenancy, natural resource management, poverty reduction, and industrial development in Asia and Sub-Saharan Africa.

Yukichi Mano is a professor at Hitotsubashi University, Japan, and is a fellow at Tokyo Center for Economic Research (TCER). He received a Ph.D. in Economics from the University of Chicago in 2007. His scholarly interests include agricultural technology adoption, horticulture, and high-value crop production, business, and management training (KAIZEN), human capital investment, migration and remittance, and universal health coverage in Asia and Sub-Saharan Africa.

Kazushi Takahashi is a professor at the National Graduate Institute for Policy Studies (GRIPS) and is a director of the Global Governance Program at GRIPS, Japan. He received a Ph.D. in Development Economics from GRIPS. His scholarly interests include agricultural technology adoption, rural poverty dynamics, microfinance, human capital investment, and aid effectiveness in Asia and Sub-Saharan African countries.

Contributors

Joseph A. Awuni Department of Economics, School of Applied Economics and Management Sciences, University for Development Studies, Tamale, Northern Region, Ghana

Kei Kajisa School of International Politics, Economics and Communication, Aoyama Gakuin University, Shibuya-ku, Tokyo, Japan

Yoko Kijima Graduate School of Policy Studies, National Graduate Institute for Policy Studies (GRIPS), Minato-ku, Tokyo, Japan

Masao Kikuchi Center for Environment, Health and Field Sciences, Chiba University, Kashiwanoha, Kashiwa, Chiba, Japan

Eustadius Francis Magezi Department of Agricultural Economics, Graduate School of Agricultural Science, Tohoku University, Miyagi, Japan

Yukichi Mano Graduate School of Economics, Hitotsubashi University, Kunitachi-shi, Tokyo, Japan

Douglas J. Merrey Gainesville, FL, USA

Yuko Nakano Faculty of Humanities and Social Sciences, University of Tsukuba, Tsukuba, Ibaraki, Japan

Timothy N. Njagi Tegemeo Institute of Agricultural Policy and Development, Tegemeo Institute, Egerton University, Nairobi, Kenya

Tatsuya Ogura Department of Agricultural and Resource Economics, Graduate School of Agricultural and Life Sciences, University of Tokyo, Bunkyo-ku, Tokyo, Japan

Keijiro Otsuka Graduate School of Economics, Kobe University, Kobe, Hyogo, Japan

Takeshi Sakurai Department of Agricultural and Resource Economics, Graduate School of Agricultural and Life Sciences, University of Tokyo, Bunkyo-ku, Tokyo, Japan

Kazushi Takahashi Graduate School of Policy Studies, National Graduate Institute for Policy Studies (GRIPS), Minato-ku, Tokyo, Japan

Hiroyuki Takeshima International Food Policy Research Institute (IFPRI), Development Strategy and Governance Division, District of Colombia, WA, USA

Trang Thu Vu Research Department, Asian Development Bank Institute (ADBI), Chiyoda-ku, Tokyo, Japan

Part I
Introduction

Chapter 1
The Rice Green Revolution in Sub-Saharan Africa: Issues and Opportunities

Keijiro Otsuka, Yukichi Mano, and Kazushi Takahashi

Abstract The time is ripe to pursue a Green Revolution in rice in sub-Saharan Africa (SSA) as a means of promoting food security and poverty reduction. This is partly because rice is an up-and-coming crop in this region, and partly because, as will be demonstrated in this volume, we have now accumulated deep knowledge about rice cultivation in SSA. With the aim of generating relevant policy recommendations, this book attempts to show what needs to be done to realize a rice Green Revolution in SSA. It is based on more than ten years of empirical inquiries into rice production by our research team in selected countries, namely Mozambique, Tanzania, Uganda, Kenya, Ghana, and Senegal, along with the newly added case of Cote d'Ivoire. This chapter explains why rice is important, provides a conceptual framework for realizing a rice Green Revolution, and proposes several major hypotheses to be tested in this book.

K. Otsuka (✉)
Graduate School of Economics, Kobe University, 2-1 Rokkodai-cho, Nada-ku, Fourth Academic Building, 5Th Floor, Room 504, Kobe 657-8501, Hyogo, Japan
e-mail: otsuka@econ.kobe-u.ac.jp

Y. Mano
Graduate School of Economics, Hitotsubashi University, 2-1 Naka, Isono Building Room 324, Kunitachi-shi 186-8601, Tokyo, Japan
e-mail: yukichi.mano@r.hit-u.ac.jp

K. Takahashi
Graduate School of Policy Studies, National Graduate Institute for Policy Studies (GRIPS), 7-22-1, Roppongi, Room 1211, Minato-ku 106-8677, Tokyo, Japan
e-mail: kaz-takahashi@grips.ac.jp

© JICA Ogata Sadako Research Institute for Peace and Development 2023
K. Otsuka et al. (eds.), *Rice Green Revolution in Sub-Saharan Africa*, Natural Resource Management and Policy 56, https://doi.org/10.1007/978-981-19-8046-6_1

1.1 Introduction

While rice production doubled from 2008 to 2018 in sub-Saharan Africa (SSA),[1] rice consumption in this region has increased more rapidly, resulting in the growth of rice imports from Asia. The need for a Rice Green Revolution is now more of a critical issue than ever in SSA because, as will be shown in this chapter, rice is now one of the most important staple crops in the region. Moreover, there is enormous potential for SSA to improve the productivity of lowland rice farming (Balasubramanian et al. 2007).[2] For these reasons, a rice Green Revolution should be pursued urgently in sub-Saharan Africa to promote food security and poverty reduction.

It must be pointed out at the outset that the Green Revolution is not simply a "seed-fertilizer" revolution, as was pointed out a half-century ago by Johnston and Cownie (1969), but management-intensive (Otsuka and Larson 2013, 2016; Otsuka and Muraoka 2017). That is, the adoption of improved seeds and fertilizer can have significant and sustainable impacts on crop yields and profitability if water, soil, weeds, and rice plants are all effectively managed using improved cultivation practices.[3] Thus, we define the Green Revolution as the development and diffusion of high-yielding varieties coupled with the application of improved cultivation practices.

Nevertheless, policymakers and researchers have often assumed that the Green Revolution requires only the use of modern inputs, such as modern varieties (MV) and chemical fertilizers (Gollin et al. 2021; Carter et al. 2021), ignoring the role of improved cultivation practices widely adopted during the Asian rice Green Revolution, such as transplanting (in rows in a timely manner), bunding, and land leveling[4] (David and Otsuka 1994; Abe and Wakatsuki 2011; Rashid et al. 2013; Ragasa and Chapoto 2017). This misunderstanding explains, at least partly, why SSA has failed to realize the rice Green Revolution before now, even though the revolution began in tropical Asia (i.e., South and Southeast Asia) more than a half-century ago and resulted in dramatic increases in rice and wheat production in the 1970s and 1980s.[5]

This book attempts to show what exactly should be done to realize a rice Green Revolution in SSA based on more than ten years of empirical inquiries into rice

---

[1] Doubling rice production from 2008 to 2018 was the target of the Coalition for African Rice Development organized jointly by the Japan International Cooperation Agency and the Alliance for a Green Revolution in Africa.

[2] We advocate increases in the productivity of lowland rice cultivation, but not upland rice, such as NERICA, because of the absence of significant productivity effects of upland rice technology except in a few countries in SSA (for example, see Kijima et al. (2008, 2011)).

[3] We believe that this is true not only for rice but also for other crops (Otsuka and Muraoka 2017; Takahashi et al. 2020).

[4] Water is stored evenly by bunding and leveling, which kills weeds and facilitates even growth of rice plants.

[5] According to informal interviews with rice experts who were engaged in rice research and extension in the 1960s and 1970s, transplanting, leveling, and bunding were roughly practiced by rice farmers in Asia before the Green Revolution.

production by our research team in selected countries in this region, namely Mozambique, Tanzania, Uganda, Kenya, Ghana, and Senegal, in addition to the new addition of Cote d'Ivoire. Since we are interested in poverty reduction and food security, we focus on smallholder rice farming in SSA (Larson et al. 2014, 2016). Following Schultz (1964), we assume that smallholders are rational and willing to adopt new profitable technologies if they are available. This book, which is based on a number of down-to-earth case studies conducted in several African countries, amply attests to the relevance of the Schultz thesis.

This book is the sequel to the two earlier books on the same subject: K. Otsuka and D. F. Larson eds., 2013, *An African Green Revolution: Finding Ways to Boost Productivity on Small Farms*; and 2016, *In Pursuit of an African Green Revolution: Views from Rice and Maize Farmers' Fields*. The main message of the first book was that rice is likely to be the most promising cereal crop in SSA because of the high transferability of well-established Asian rice Green Revolution technology, whereas the central message of the second book was that rice cultivation training programs are effective in dramatically increasing rice yields in SSA. This third book has broader coverage in terms of topics, study periods, and locations. In fact, this book deals not only with the impact of the rice cultivation training programs in both rainfed and irrigated areas in the short and long term but also with the impacts of mechanization, irrigation, and improved rice milling technology.

This chapter reviews the production trends of major cereal crops in the past in SSA in Sect. 1.2 and examines the prospect of rice production for the future in Sect. 1.3. The conceptual framework is provided in Sect. 1.4, along with the formulation of testable empirical hypotheses. Section 1.5 sets out the structure of the book.

1.2 Trends in the Past

1.2.1 Population Pressure and the Increasing Importance of Crop Yield

In SSA, the rural population has continued to grow and cultivable lands have become scarcer over time (Holden and Otsuka 2014). Earlier, farmers converted uncultivated forest and grazing land on the hills into upland crop fields. However, they gradually converted unused marshy land into lowland paddy fields because of the increasingly limited availability of uncultivated land in the hills. However, in general, the land–labor ratio, measured by arable land per capita in the rural population in Fig. 1.1, continued to decline in SSA. While it is true that the land–labor ratio has always been much lower in tropical Asia than in SSA, the level in Asia in 1961 was no different from that of SSA in 2018. This fact indicates that the population pressure on limited land on the eve of the Green Revolution in Asia was as severe as the current population pressure in SSA.

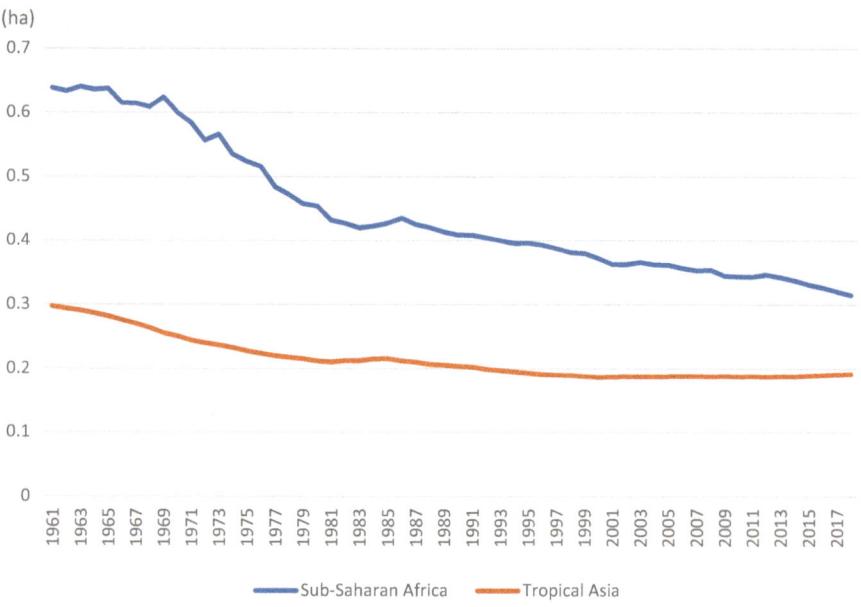

Fig. 1.1 Changes in arable land per capita in rural areas in tropical Asia and SSA (*Source* FAOSTAT 2019)

The theory of agricultural intensification proposed by Boserup (1965) and the theory of induced institutional and technological innovations formulated by Hayami and Ruttan (1985) provide a basis for inducing an intensification of the farming system and land-saving technological change. This is likely to result in increasing land productivity as a way of responding to increasing population pressure. Table 1.1 shows annual growth rates of cereal production per capita in rural areas, harvested area per capita in rural areas, and output per harvested area in SSA by decade. Output is measured by a simple sum of the weights of maize, rice, wheat, millet, and sorghum. Since yams and other root crops are also widely grown in SSA, measuring food production only by cereal crops is likely to be inaccurate.

Keeping such reservations in mind, let us examine the broad trends shown in Table 1.1. It seems clear that cereal output per capita in rural areas did not increase in the 1970s to 1990s,[6] whereas harvested area per capita in rural areas generally declined during the same period. An important observation is that output per capita in rural areas steadily increased in the 2000s and 2010s, owing to the increasing growth rate of output per harvested area, which can be considered a proxy for land productivity. Also, harvested area per capita in rural areas slightly increased after the turn of the century, presumably because of the increasing use of arable land for

---

[6] It may be noticed in Table 1.1 that output per harvested area grew substantially in the 1960s and 1970s. This is due to the relatively fast growth of maize yield in these early periods (see Fig. 1.3), which was caused by the dissemination of a newly developed hybrid maize in Eastern and Southern Africa (Smale et al. 2013).

Table 1.1 Annual growth rates of cereal output per capita (rural areas), harvested area per capita (rural areas), and grain output per harvested area in SSA (%)

Period	Output per capita (rural areas)	Harvested area per capita (rural areas)	Output per harvested area
1961–71	0.8	−0.5	1.3
1971–81	0.0	−3.4	3.4
1981–91	−0.4	1.4	−1.7
1991–2001	−0.5	−1.4	0.9
2001–11	1.6	0.3	1.3
2011–18	2.2	0.6	1.5
Ratio: 2018/1961	1.35	0.72	1.87

Source FAOSTAT (2019)
Amount of production is the sum of the weights of maize, rice, wheat, millet, and sorghum production, labor is proxied by rural population, and land is measured by the total harvested area of the five grains

cereal production.[7] According to the last row in Table 1.1, output per capita in rural areas increased by 35%, harvested area per capita in rural areas declined by 28%, and output per harvested area increased by 87% from 1961 to 2018. Thus, it is clear that the declining land–labor ratio was more than compensated for by increasing land productivity so as to boost labor productivity. Interestingly, this is similar to the development paths of South and East Asia in the past (Otsuka and Fan 2021).

The fact that land productivity has been increasing in recent years in SSA indicates that economic forces have been at work to promote the intensification of farming systems to increase crop yields in the face of the increasing scarcity of agricultural land.

1.2.2 Yield Trend of Cereal Crops in SSA

It is a mistake to assume that yield of cereal crops has been completely stagnant in SSA. As is shown in Fig. 1.2, it is true that cereal crop yield per hectare of harvested land had been either stagnant or grew only slowly from 1961 to around the turn of the century. The yield continued to stagnate in the case of sorghum and even declined in the case of millet in this century. As shown by Otsuka and Muraoka (2017), the yield of these crops did not appreciably increase in tropical Asia either, and hence, a yield gap between the two major regions did not emerge. Thus, there is little opportunity for SSA to learn technology from tropical Asia in millet and sorghum production.

---

[7] Arable land per capita in rural areas continuously declined according to Fig. 1.1, but harvested area per capita in rural areas slightly increased during the last two periods according to Table 1.1. These observations indicate that an increasingly larger portion of arable land has been cultivated. For example, harvested area was 35% of arable land in 1980 but had reached 51% by 2018.

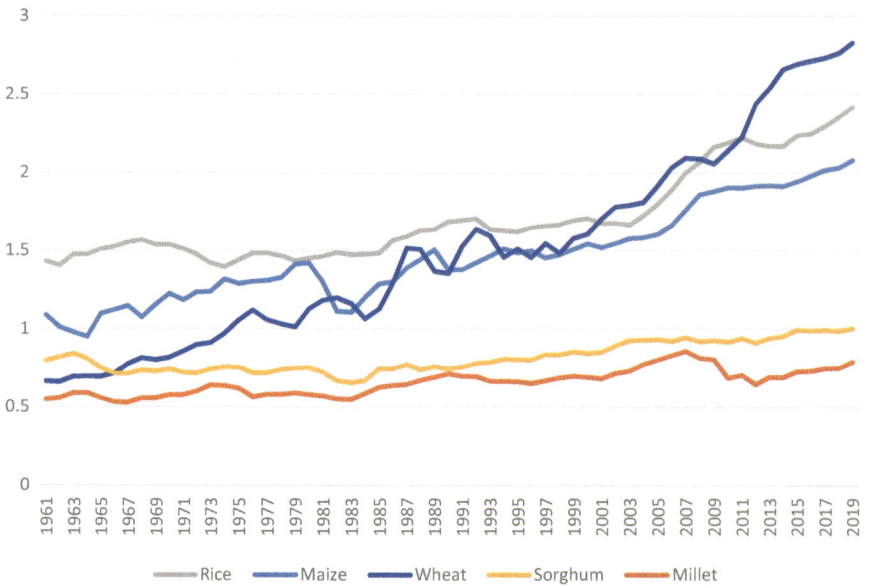

Fig. 1.2 Changes in average yield of major cereals in SSA (ton/ha), 3-year moving averages (*Source* United States Department of Agriculture 2021)

Therefore, any expectation that the yield of these crops will increase significantly in SSA in the near future is unfounded.

The average wheat yield in SSA continuously increased, starting with 0.7 tons per hectare in 1961 and reaching nearly 3 tons per hectare in 2019. The yield of 3 tons is comparable to the average in other regions (Shiferaw et al. 2013). South Africa has the largest wheat harvested area, and its yield reached 3.6 tons per hectare in the late 2010s, which is much higher than in India. Wheat is grown by fully mechanized large-scale irrigated commercial farms in South Africa, leading to a wheat Green Revolution in this country. The wheat yield of smallholders in Ethiopia also grew, even though it was lower than 2.8 tons per hectare in the late 2010s.

It is noticeable that the yield of rice and maize increased from roughly 1.7 tons to 2.4 tons per hectare and from 1.5 tons to 2.1 tons per hectare from the turn of the century to the present. Whether this is an onset of the Green Revolution or temporary growth is a critical question. Although we are not sure if the initial yield growth of these crops will lead to sustainable growth in coming years, the thrust of this edited volume is that rice yield can continue to grow if proper government interventions, to be identified in this volume, are made to facilitate the rice Green Revolution.

How is the differential yield growth observed in Fig. 1.2 related to changes in harvested areas shown in Fig. 1.3? Several interesting phenomena can be noted. First, while the maize harvest area has increased sharply in recent years, the harvested area of millet began declining around 2010, and that of sorghum remained unchanged. These changes are consistent with the increasing trend of maize yield and the declining trend of millet yield. Second, the wheat harvested area has been small

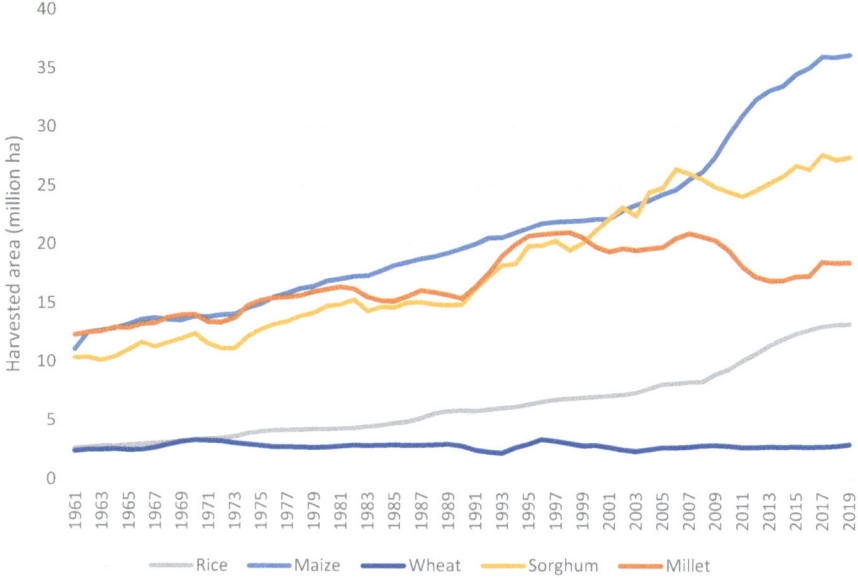

Fig. 1.3 Changes in aggregate harvested area in SSA by major cereal crop, 3-year moving averages (*Source* United States Department of Agriculture 2021)

and has remained essentially constant. This indicates a limited area suitable for wheat cultivation in SSA, i.e., the cold climate in southern and highland parts of the continent. There is also no room for expanding irrigated large-scale commercial wheat farms in South Africa. Thus, we can hardly expect that wheat will be a major staple crop produced in SSA. Third, the paddy harvested area slowly increased up to 2005 and then expanded sharply, so that recently, it approached the harvested area of millet. There seems to be no doubt that rice has become a major staple crop in SSA.

1.2.3   The Increasing Importance of Rice

Table 1.2 shows the estimated total consumption,[8] consumption per capita, total production, and net imports by major cereal crops in 1980 and 2018. It is clear that maize is the most important crop in terms of consumption and production. Both maize consumption and production more than tripled over the last 38-year period. Maize was marginally exported, but most countries in SSA were largely self-sufficient. Consumption per capita of maize increased by 16% from 1980 to 2018. Total and per capita consumption of milled rice increased much more dramatically; total consumption increased by 5.65 times, and per capita consumption almost doubled over a mere 38-year period.

---

[8] Consumption is estimated by adding production and net imports, without regard to changes in stock.

Table 1.2 Consumption, production, and import of major cereals in SSA

Item	1980				2018			
	Estimated consumption (million tons)[a]	Estimated consumption per capita (kg)	Production (million tons)[b]	Net imports (million ton)	Estimated consumption (million tons)[a]	Estimated consumption per capita (kg)	Production (million tons)[b]	Net imports (million ton)
Maize	23.93	65.55	25.38	−1.45	78.89	75.96	79.36	−0.47
Milled Rice	6.45	17.67	4.13	2.33	35.92	34.58	20.89	15.03
Wheat	5.38	14.72	2.62	2.75	25.05	24.12	7.50	17.55
Millet	7.02	19.22	6.96	0.06	13.44	12.94	13.43	0.01
Sorghum	10.08	27.61	10.25	−0.17	24.69	23.78	24.13	0.57

Source FAOSTAT (2019)

[a]Consumption is estimated by adding domestic production and net import

[b]Milled rice equivalent is used

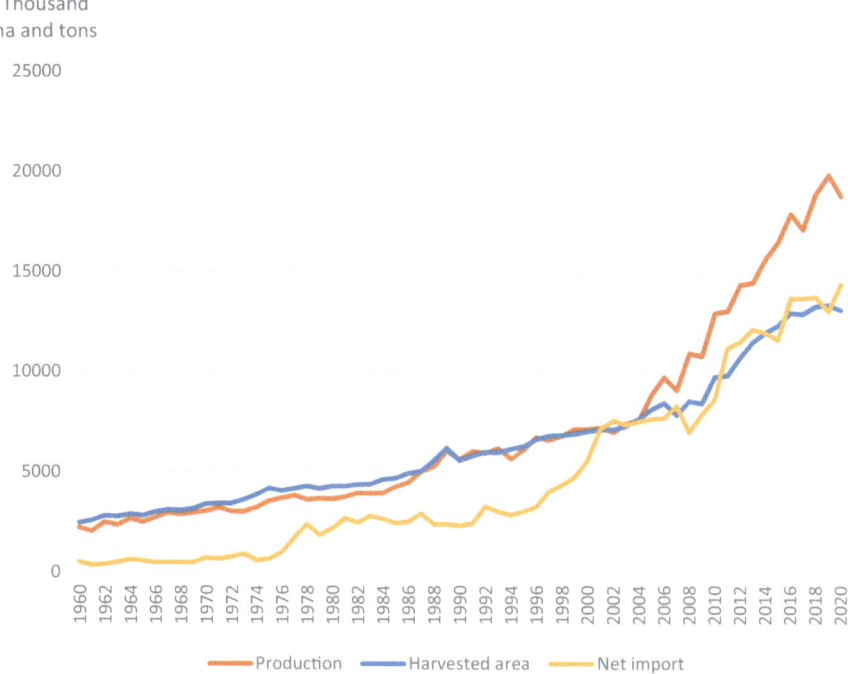

Fig. 1.4 Changes in rice production, harvested area, and net imports in SSA[9] (*Source* United States Department of Agriculture 2021)

Although milled rice production increased five times, net imports increased more sharply and accounted for 42% of total consumption in 2018. Evidently, the importance of rice in SSA has increased significantly in recent years. Wheat consumption increased significantly, although less dramatically than rice. Its production also increased but more slowly than consumption, resulting in huge net imports in 2018. Since the area suitable for wheat production is limited in SSA, we can hardly expect rapid wheat production growth and reduced growth of wheat imports. Although the production of millet and sorghum increased slowly, SSA is largely self-sufficient in these crops because of the declining per capita consumption. The importance of millet and sorghum declined.

As is shown in Fig. 1.4, rice production, expressed by the milled rice equivalence, has steadily increased for the last 60 years. The growth in rice production is primarily accounted for by an increase in the harvest area, as the trends of the two curves look similar, particularly until 2005. The gap between the two curves widened in the 2010s, suggesting that the area expansion no longer sustained production growth, thereby increasing rice imports from Asia. Thus, the increasing trend of rice imports cannot be reversed unless substantial yield growth is achieved. Considering the increasing importance of rice in production, consumption, and trade, it is an opportune time for African countries to pursue a rice Green Revolution.

1.3 Prospects for Future

1.3.1 Emerging Yield Gap

In order to examine the prospect of rice farming for the future, it is instructive to examine long-term paddy yield trends in the past and the yield gap between the most advanced regions and others and compare yield in SSA with that in Asia. We assume that if there is a yield gap between the advanced regions and others, and between SSA and Asia, there is room for less advanced regions in SSA to catch up with more advanced regions, including Asia. It is also instructive to compare the case of rice with maize to predict future development.

Figure 1.5, which shows the average paddy yield in SSA and India as well as the top five countries in SSA,[10] provides valuable information. The average yield in India was roughly 1.7 tons per hectare in the early 1960s before the Green Revolution began, roughly the same as the average yield in SSA in the 1990s. Yield in India has been growing since then, whereas yield in SSA gradually increased this century. This increasing yield trend in SSA can be attributed, at least in part, to the Green Revolution in selected areas, judging from the fact that the average yield of the top five countries began increasing sharply from the early 1990s and approached the level in India in the 2010s. Their average yield of nearly 3.8 tons per hectare in 2019 is comparable to the average yield of about 4.1 tons per hectare in tropical Asia in the same year. There seems to be no doubt that the rice Green Revolution has successfully taken place in selected areas in SSA, particularly in irrigated areas.

It is also important to emphasize that the yield gap between the top five countries in SSA and the regional average has widened since 1990. This increasing gap indicates that the vast yield gap as of 2019 resulted from efforts to improve yield in the top five countries rather than the innate difference in yield between the top five and the other countries in the region. Consequently, there seems to be the possibility that less advanced regions may be able to catch up with advanced regions in SSA.

For comparative purposes, Fig. 1.6 shows the yield trend of maize in SSA and India. First, there was practically no yield gap between SSA and India from the 1960s to the mid-1980s, which indicates the small difference in agro-climate between SSA and India. Second, the maize yield in India began increasing in the late 1980s due to the Green Revolution, resulting in a yield gap of roughly one ton per hectare

---

[9] Production is milled rice equivalent.

[10] India is chosen because its agro-climate is relatively similar to that of SSA, compared with other rice growing countries in Asia. The top five countries are Kenya, Niger, Benin, Senegal, and Mali. They are top five countries in terms of the average paddy yield from 2001 to 2020. While Kenya and Niger were almost fully irrigated, other three countries were not.

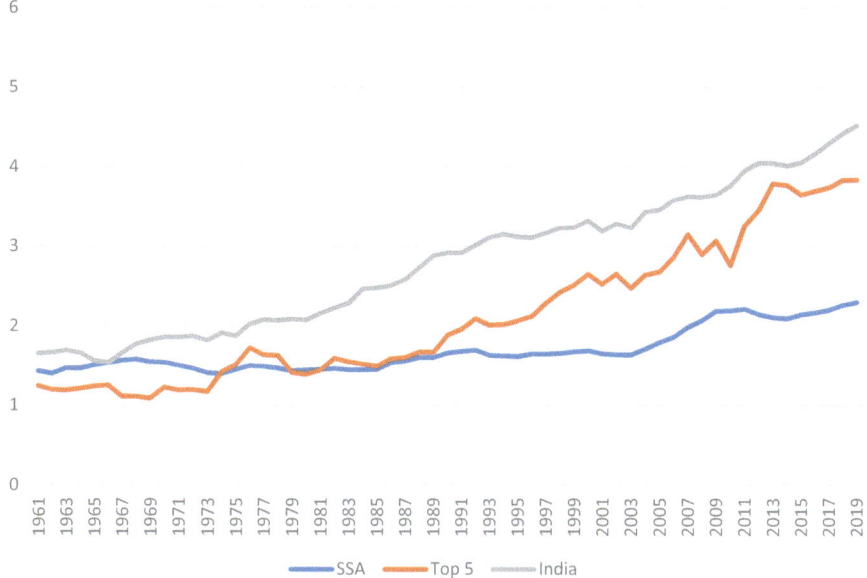

Fig. 1.5 Changes in average paddy yield in SSA, top five countries,[11] and India (ton/ha), 3-year moving averages (*Source* United States Department of Agriculture 2021)

between India and SSA in recent years. This gap is substantially smaller than for rice, which is more than two tons per hectare. Third, the yield gap between the top five countries in SSA and India has disappeared in recent years,[12] indicating that the maize Green Revolution took place to a significant extent in advanced regions. Fourth, the yield gap between the top five countries and the average is comparatively small, which indicates that there is little room for less advanced regions to catch up with advanced regions in maize farming in SSA. Because improved maize varieties are highly location-specific (Smale et al. 2013), catching up may be more difficult in maize than rice production.

We do not mean to argue that priority should be placed solely on the development of the rice sector, ignoring the maize sector. Considering the utmost importance of maize in SSA, the development of the maize sector must receive high priority. Our analysis strongly suggests that the establishment of productive technology must first be pursued in the case of maize, as argued by Otsuka and Muraoka (2017). In contrast, the dissemination of already established advanced technology to wider areas is an urgent issue in rice. This provides the basis for the main argument of this book: rice cultivation training is a vitally important entry point to the rice Green Revolution in SSA.

---

[11] Top five countries are Kenya, Niger, Senegal, Benin, and Mali.

[12] The top five countries are Ethiopia, Zambia, Uganda, Mali, and Cote d'Ivoire. They are the top five in terms of the average yield from 2001 to 2020. We excluded South Africa because of the dominance of large commercial farms.

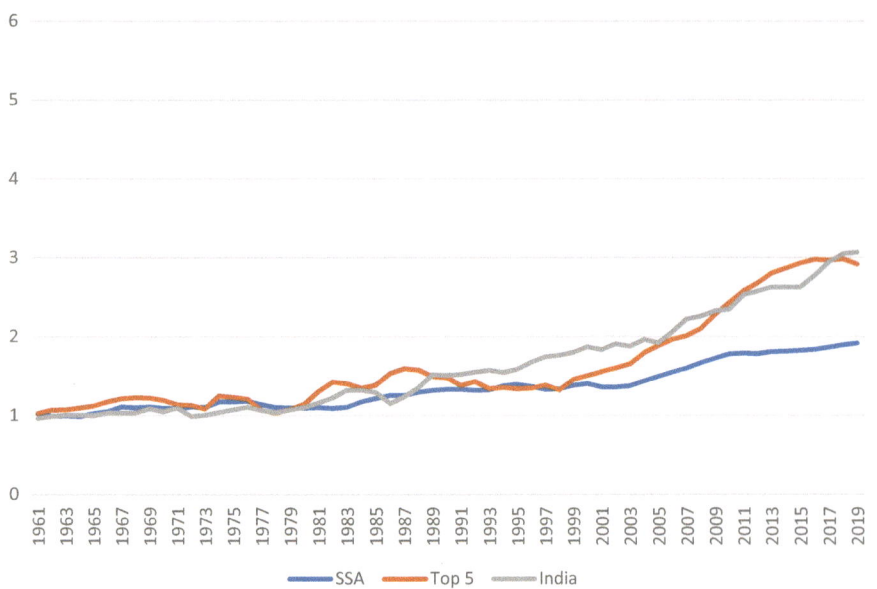

Fig. 1.6 Changes in average maize yield in SSA, top five countries,[13] and India (ton/ha), 3-year moving averages (*Source* United States Department of Agriculture 2021)

1.3.2 Changes in the Real Prices of Rice and Maize

Figure 1.7 shows that global rice production has continued to increase over the past several decades. Since more than 90% of rice is produced in Asia, the Asian rice Green Revolution significantly contributed to increasing rice production globally. While it is difficult to tell exactly when it took place, it seems that it has had substantial and sustainable impacts on rice production since the late 1960s. This was partly because it took a few decades to disseminate new technologies. Moreover, improved varieties, such as pest- and disease-resistant and drought- and flood-tolerant varieties, were successively developed (Evenson and Gollin 2003; Janaiah et al. 2005; Estudillo and Otsuka 2006; Emerick et al. 2016), which are likely to have contributed to sustainable growth in rice production in Asia. This book is concerned primarily with how to launch a rice Green Revolution in SSA, which is expected to increase the productivity of rice farming for a few decades, but not how to sustain it beyond that period. In order to sustain the Green Revolution over long periods in this region, continuous development and diffusion of improved rice production technologies and management methods will be required.

As a result of the Asian Green Revolution, real rice prices continued to decline due to increasing rice production until the "food crisis" in 2008. The real rice price at around the year 2000 was merely one-third of the level around 1970. This sharp

---

[13] Top five countries are Ethiopia, Zambia, Uganda, Mali, and Cote d'Ivoire.

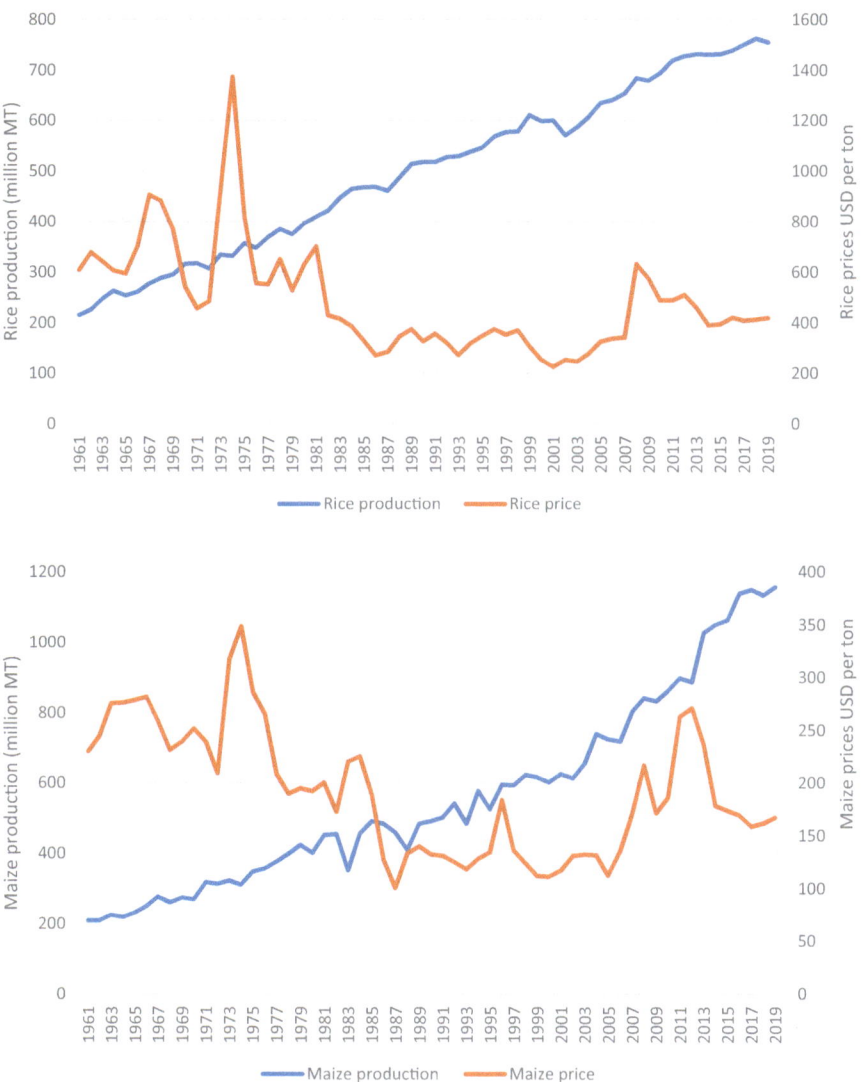

Fig. 1.7 World production and real prices for rice and maize in 2010 US dollars (*Source* FAOSTAT 2019; World Bank Pink Sheet 2020)

reduction in real rice prices indicates that the primary beneficiaries of the Green Revolution were rice consumers, including those in SSA. Rice farmers in SSA suffered from lower rice prices without receiving any benefit from improved productivity of rice farming, as well as rice farmers in Asia who failed to adopt improved technology due to unfavorable production environments (David and Otsuka 1994). These lower rice prices may have reduced incentives to develop and disseminate improved technologies in rice production in SSA.

The lower panel of Fig. 1.7 indicates that the real price of maize also declined from the 1960s to around 2005 because of the increasing maize yield and production. The real prices of rice and maize jumped in 1974 and 2008 despite only a slight reduction in global production due to speculation with incorrect expectations. The real maize prices have remained comparatively high in the 2010s.

If real rice prices decline or remain relatively low, reliance on cheap imports, rather than boosting production, can be a possible option for SSA. There are a couple of reasons that cereal prices, particularly rice prices, will increase. The first is climate change, which will increase cereal prices by 12–18% by 2050 (Rosegrant et al. 2021). Extreme weather events may also occur more frequently than before, likely leading to a temporary shortfall in production. Another potential threat is the increasing cost of food production in high-performing Asian countries, which depend on labor-intensive production methods due to small farm size. Unless farm size expansion and labor-saving mechanization occur, there is a fear that Asia will become a major importer of cereals that will drive up cereal prices globally (Otsuka et al. 2016a, b; Yamauchi et al. 2021).

Although it is challenging to predict cereal prices in the future, it seems safe to assume that they will not decline in the coming years. Thus, it makes sense to devote serious efforts to boosting rice and maize production in SSA by realizing the Green Revolution. It must be pointed out that, since SSA accounts for a small share of global rice production, the Green Revolution in SSA, even if successful, will not affect the international price of rice to a significant extent.

1.4   Conceptual Framework and Hypotheses

1.4.1   A Conceptual Framework

The rice Green Revolution involves the intensification of rice cultivation, which is defined as the intensive use of modern inputs, such as improved rice seeds and inorganize fertilizer, combined with improved cultivation practices. Paddy yield and profitability of rice farming are enhanced by such intensification. Some improved cultivation practices, such as proper seed preparation, fertilizer application, and pest and weed control, are also expected to improve paddy quality (JICA 2021). Although intensification usually means increased use of inputs, we would like to include appropriate harvesting, drying, and storing activities as a part of the intensification process, to the extent that these activities increase the value of output and profitability of rice farming. Figure 1.8 illustrates our conceptual framework, which explains how to achieve the rice Green Revolution through the intensification of rice farming.

We hypothesize that the chief development strategy ought to be a rice cultivation training program because it has proven to be highly effective in improving the productivity of rice farming, based on previous studies (Kijima et al. 2012; deGraft-Johnson et al. 2014). We can hardly expect a significant and sustainable

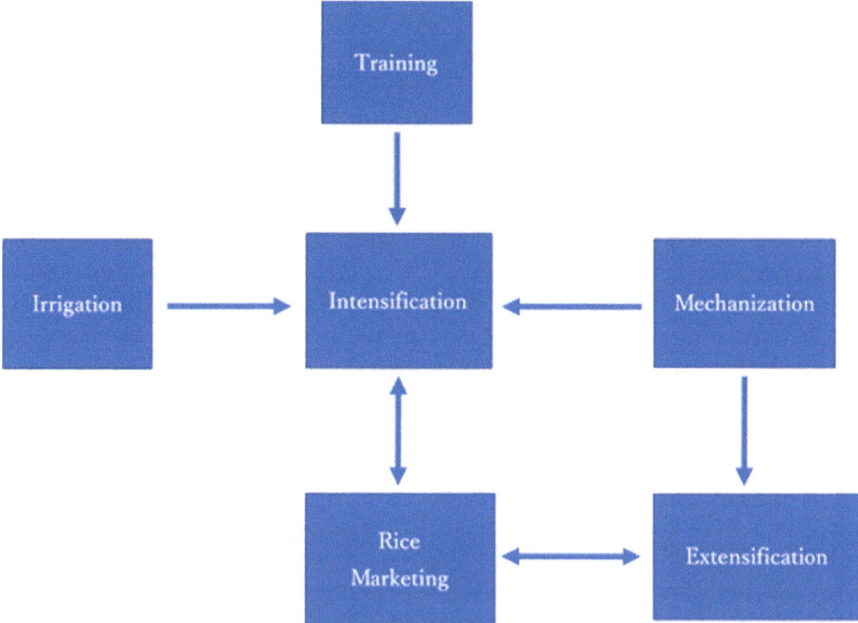

Fig. 1.8 A Conceptual framework to achieve a rice Green Revolution through intensification. *Note* Intensification of rice cultivation is defined as the intensive use of modern inputs, such as improved rice seeds and inorganize fertilizer, combined with improved cultivation practices, including appropriate harvesting, drying, and storing activities

improvement of rice productivity without improving cultivation practices, such as bund construction, leveling, weeding, water and pest control, and proper timing and spacing of transplanting, as is the case in Asia. Because these cultivation practices are knowledge-intensive, it requires training farmers to acquire an accurate understanding of these practices and ensure proper implementation to realize their full yield potential.

This emphasis on rice cultivation training does not imply that other factors, such as investments in mechanization and irrigation, are unimportant.[14] On the contrary, these investments play critical roles in facilitating the intensification of rice cultivation. We postulate that mechanization, particularly the introduction of power tillers (two-wheel hand tractors), facilitates proper land preparation, which prevents the overgrowth of weeds by keeping even water depth, thereby inducing the intensive adoption of complementary cultivation practices, such as proper spacing of transplanting. Four-wheel tractors may not be appropriate for intensification because it is difficult to maneuver heavy machines in small muddy paddy fields and move across

---

[14] We do not consider the lack of credit availability to be a primary constraint on the rice Green Revolution judging from case studies on credit in Kenya (Njagi et al. 2016) and Tanzania (Nakano and Magezi 2020). Although this book does not analyze the role of input markets, it can be a critical factor affecting the intensification of rice farming.

paddy fields without agricultural roads. The use of power tillers and four-wheel tractors is also expected to lead to the extensification of rice farming, i.e., cultivation of formerly unused land and conversion of non-rice land to paddy land, owing to the efficiency in land preparation, including construction and plowing of new paddy fields.

Like Asia, irrigated areas tend to be more intensively cultivated throughout the year, as the availability of irrigation water increases the benefit of intensification, and they are thus far more productive than rainfed areas in SSA (David and Otsuka 1994; Balasubramanian et al. 2007; Kajisa and Payongayong 2011; Nakano and Otsuka 2011; Cassman and Grasssini 2013; Nakano et al. 2013). Because irrigated paddy area accounts for a minor fraction of the total paddy area in SSA,[15] whether a rice cultivation training program effectively improves productivity in a rainfed area is a critical question to be investigated in this book. We may expect that, since bunding and leveling improve water control, their adoption may enhance productivity in rainfed farming.

It is often argued that milled rice produced in SSA cannot compete with higher-quality imported Asian rice (Diako et al. 2010; Demont 2013; Demont et al. 2017). We believe that the correct timing of harvesting and proper drying of harvested paddy and, probably more importantly, proper milling machines and proper grading of milled rice are critically important in the quality improvement of milled rice. Reardon et al. (2014) report that the introduction of improved milling machines triggered revolutionary changes in the quality of milled rice in Asia. Improved milling technology, especially removing stones and other impurities, combined with the grading of rice and differential pricing of paddy based on the quality, may stimulate proper seed-bed preparation, fertilizer application, and pest and weed control as well as the adoption of appropriate harvesting and post-harvesting activities to produce and deliver high-quality paddy (Ibrahim et al. 2020; Kapalata and Sakurai 2020; Ogura et al. 2020). Production and delivery of high-quality paddy, which is regarded as a part of intensification in this study, will, in turn, affect the rice marketing activities.

1.4.2 Hypotheses

Following our introduction of the conceptual framework above, we would like to postulate four key hypotheses to be tested in this book. While the impact of the rice training program on the intensification of rice farming is expected to be significant, one may wonder if it will differ between the shorter and the longer run, between rainfed and irrigated areas, and between training participants and non-participants. We therefore propose the following hypothesis:

Hypothesis 1: Rice cultivation training programs have a significant impact on the intensification of rice farming, thereby improving paddy yield and the profitability of

---

[15] According to Balasubramanian et al. (2007), the proportion of irrigated rice land is about 15%, which is much lower than the estimated irrigation ratio of roughly 50% in tropical Asia.

rice farming, (a) in both the shorter and longer run, (b) in both rainfed and irrigated areas, and (c) not only for training participants but also non-participants through information spillovers.

The purpose of mechanization is commonly assumed to save labor and draft animals in Asia (e.g., Hayami and Ruttan 1985). In SSA, where land preparation depends mainly on manual labor, however, the introduction of power tillers and four-wheel tractors makes it possible to carry out land preparation more thoroughly and speedily. The effects of the adoption of power tillers and four-wheel tractors could be different because power tillers can be more easily maneuvered in muddy paddy fields and moved across fields without destroying bunds. Thus, it will likely be profitable to use power tillers for both intensification and extensification and four-wheel tractors particularly for extensification. Regarding the impact of tractorization, it seems worth testing the following hypothesis:

Hypothesis 2: The adoption of power tillers contributes to both the intensification and extensification of rice farming, whereas that of four-wheel tractors contributes primarily to the extensification.

There is no question that the presence of irrigation facilitates the intensification of rice farming. While it is relatively easy to assess the impact of irrigation on the productivity of rice farming on farmers' fields, it is more difficult to assess the rate of return on investment in irrigation facilities because of the difficulty in collecting the data on the cost of construction and maintenance and in predicting future benefits which are affected by increased production efficiency and future rice prices. In our view, the extent of increased production efficiency critically depends on the provision of rice cultivation training programs in conjunction with the irrigation projects. The estimation becomes more difficult if we attempt to include general equilibrium effects on the development of related activities, such as rice milling and trading, and the delivery service of inputs. Thus, our tentative hypothesis on the impact of irrigation investment is as follows:

Hypothesis 3: While investment in a large-scale irrigation scheme may have modest expected returns, its economic viability increases if proper rice cultivation training on modern inputs and improved management practices is provided and also if the general equilibrium effect is taken into account.

The quality of milled rice depends on rice varieties, the timing of harvesting, drying of paddies, and the quality of rice milling machines, e.g., the use of destoners and color sorters. Thus, rice millers can play a key role in improving the quality of rice because they can introduce improved machines to produce high-quality milled rice and because they can affect the quality of paddy grown by farmers by introducing grading and differential pricing.[16] Traders may also introduce grading and differential pricing of milled rice if markets become conscious about milled rice quality. Therefore, it seems reasonable to postulate the following hypothesis:

---

[16] Although rice millers may have incentives to offer information about proper harvesting, drying, and storing to procure high-quality paddy land, it may be prohibitively costly for them to ensure delivery of such paddy fields from those farmers who receive their instructions.

Hypothesis 4: Rice millers play the role of innovators in the improvement of the quality of rice by installing improved milling machines and enhancing farmers' incentives to produce high-quality paddy. Rice traders are also expected to provide incentives to farmers to produce high-quality paddy by offering quality-based pricing when markets become conscious about rice quality.

1.5   Structure of the Book

This book consists of six parts: Part I is concerned with an overview of extensification, intensification, and the Green Revolution in sub-Saharan Africa; Part II examines the impact of rice cultivation training programs; Part III explores the effect of mechanization, irrigation, and improvement of rice quality and milling technologies; Part IV summarizes findings of this study and proposes strategies to achieve a full-fledged rice Green Revolution in SSA.

After this introductory chapter, we review the literature on the role of agricultural extension in Chap. 2 to show the broad context in which the various case studies of the impact of management training on rice farming were undertaken. Since we believe that the rice cultivation training program is the fundamental activity required to realize a rice Green Revolution, four chapters are devoted to assessing the program. Applying a randomized controlled trial, Chap. 3 confirms the sustainable impacts of the training program and dissemination of improved cultivation practices from training participants to non-participants in an irrigated area in Cote d'Ivoire. Chapter 4 demonstrates the significant impact of the training in rainfed areas and irrigated areas in Tanzania. Furthermore, it shows the sustainable impact of training and the technology being successfully disseminated by training participants to non-participants. Chapter 5 shows the evidence of the sustainable impacts of a rice cultivation training program and the passing on of improved cultivation practices from training participants to non-participants in a rainfed area in Uganda. Chapter 6 provides evidence that the rice training program effectively enhances the productivity of rice farming even in an unfavorable rainfed area in Mozambique, where farming without applying improved seeds and chemical fertilizer is practiced.

In Part III, we begin by reviewing the literature on the impact of agricultural mechanization and irrigation on rice farming intensification in Chap. 7. The effect of tractorization in Cote d'Ivoire is dealt with in Chap. 8 and in Tanzania in Chap. 9. Both chapters demonstrate that the adoption of power tillers promotes both the intensification and extensification of rice farming. Furthermore, Chap. 9 finds that adopting four-wheel tractors has had little impact on the intensification, whereas the adoption of draught animals significantly enhances the intensification.

As explained previously, estimating the rate of return on large-scale irrigation investment is not easy, as it involves the complex calculation of associated costs and benefits. Furthermore, it is difficult to assess the general equilibrium effects of irrigation investment on related activities, such as rice milling and trading. Yet, Chap. 10 attempts to estimate the rate of return on the Mwea Irrigation Scheme in Kenya by asking a hypothetical question of what the rate of return would be if investment in irrigation facilities were made now. According to the estimation results, the rate of return is unlikely to be high unless rice prices are assumed to be relatively high. Moreover, because of the possible complementarity between improved cultivation practices and the availability of irrigation water, well-designed training on appropriate rice cultivation may significantly enhance the rate of return to large-scale irrigation investments. Chapter 11 assesses the efficiency of small- to medium-scale irrigation management in Senegal and finds that relatively large-scale irrigation is more efficiently managed than smaller ones.

Both Chap. 12, in which we attempt to assess the efficiency of introducing improved rice milling machines in Mwea in Kenya, and Chap. 13, where we attempt to assess the impacts of providing market information and rice miller's pricing policies to farmers on the marketing and production behavior in Ghana, support the hypothesis that rice millers play a key role in improving the quality of rice. The implication is that rice millers ought to be a focal point of rice value chain upgrading rather than a holistic approach to improve the entire value chain without a clear focus.

The concluding chapter, Chap. 14, proposes a strategy for a full-fledged rice Green Revolution in SSA based on in-depth case studies reported in Chaps. 3–6 and 8–13. Specifically, we conclude that there is no reason to doubt the success of the widespread rice Green Revolution in SSA as long as rice cultivation training programs, coupled with the introduction of power tillers and irrigation investment, are offered as a way of promoting the intensification of rice farming.

References

Abe SS, Wakatsuki T (2011) Sawah ecotechnology—a trigger for rice green revolution in sub-Saharan Africa. Outlook on Agric 40(3):221–227

Balasubramanian V, Sie M, Hijmans RJ, Otsuka K (2007) Increasing rice production in sub-Saharan Africa: challenges and opportunities. Adv Agron 94(1):55–133

Boserup E (1965) The conditions of agricultural growth: the economics of agrarian change under population pressure. Allen and Unwin, London

Carter M, Laajaj R, Yang D (2021) Subsidies and the African Green Revolution: direct effects and social network spillovers of randomized input subsidies in Mozambique. Am Econ J Appl Econ 13(2):206–229

Cassman KG, Grassini P (2013) Can there be a Green Revolution in sub-Saharan Africa without large expansion of irrigated crop production? Glob Food Sec 2(3):203–209

David CC, Otsuka K (1994) Modern rice technology and income distribution in Asia. Lynne Rienner, Boulder

deGraft-Johnson M, Suzuki A, Sakurai T, Otsuka K (2014) On the transferability of the Asian rice Green Revolution to rainfed areas in sub-Saharan Africa: an assessment of technology intervention in Northern Ghana. Agric Econ 45(5):555–570

Demont M (2013) Reversing urban bias in African rice markets: a review of 19 national rice development strategies. Glob Food Sec 2(3):172–181. https://doi.org/10.1016/j.gfs.2013.07.001

Demont M, Fiamohe R, Kinkpe AT (2017) Comparative advantage in demand and the development of rice value chains in West Africa. World Dev 96:578–590. https://doi.org/10.1016/j.worlddev.2017.04.004

Diako C, Sakyi-Dawson E, Bediako-Amoa B, Saalia FK, Manful JT (2010) Consumer perceptions, knowledge, and preferences for aromatic rice types in Ghana. Nature and Science 8: 12–19. http:// 197.255.68.203/handle/123456789/1233

Emerick K, de Janvry A, Sadoulet E, Dar MH (2016) Technological innovations, downward risk, and the modernization of agriculture. Am Econ Rev 106(6):1537–1561

Estudillo JP, Otsuka K (2006) Lessons from three decades of Green Revolution in the Philippines. Dev Econ 44(2):123–148

Evenson RE, Gollin D (2003) Crop variety improvement and its effect on productivity: the impact of international agricultural research. CABI, Wallingford UK

Food and Agriculture Organization of the United Nations (2019) FAOSTAT. Rome

Gollin D, Hansen CW, Wingender A (2021) Two blades of grass: the impact of the Green Revolution. J Political Econ 129(8):2344–2384

Hayami Y, Ruttan VW (1985) Agricultural development: an international perspective. Johns Hopkins University Press, Baltimore

Holden ST, Otsuka K (2014) The role of land tenure reforms and land markets in the context of population growth and land use intensification in Africa. Food Policy 48(1):88–97

Ibrahim LA, Sakurai T, Tachibana T (2020) Local rice market development in Ghana: experimental sales of standardized premium quality rice to retailers. Japanese J Agric Econ 22:118–122

Janaiah A, Hossain M, Otsuka K (2005) Is the productivity impact of the Green Revolution in rice vanishing? Econ Pol Wkly 40(53):5596–5600

Japan International Cooperation Agency, JICA (2021) JICA Technical Manual for Rice Cultivation in Africa—CARD Implementation Review 2008–20. https://riceforafrica.net/knowledge/jica

Johnston BF, Cownie J (1969) The seed-fertilizer revolution and labor absorption. Am Econ Rev 59(4):569–582

Kajisa K, Payongayong E (2011) Potential of and constraints to the rice Green Revolution in Mozambique: a case study of the Chokwe irrigation scheme. Food Policy 36(5):615–626

Kapalata D, Sakurai T (2020) Adoption of quality-improving rice milling technologies and its impacts on millers' performance in Morogoro Region, Tanzania. Japanese J Agric Econ 22:101–105

Kijima Y, Otsuka K, Sserunkuuma D (2008) Assessing the impact of NERICA on income and poverty in central and western Uganda. Agric Econ 38(3):327–337

Kijima Y, Otsuka K, Sserunkuuma D (2011) An inquiry into constraints on a Green Revolution in sub-Saharan Africa: the case of NERICA rice in Uganda. World Dev 39(1):77–86

Kijima Y, Ito Y, Otsuka K (2012) Assessing the impact of training on lowland rice productivity in an African setting: evidence from Uganda. World Dev 40(8):1610–1618

Larson DF, Otsuka K, Matsumoto T, Kilic T (2014) Should African rural development strategies depend on small farms? An exploration of the inverse productivity hypothesis. Agric Econ 45(3):355–367

Larson DF, Muraoka R, Otsuka K (2016) Why African rural development strategies must depend on small farms. Glob Food Sec 10:39–51

Nakano Y, Bamba I, Diagne A, Otsuka K, Kajisa K (2013) The possibility of a rice Green Revolution in large-scale irrigation schemes in sub-Saharan Africa. In: Otsuka K, Larson DF (eds) An African Green Revolution: finding ways to boost productivity on small farms. Springer, Dordrecht

Nakano Y, Magezi EF (2020) The impact of microcredit on agricultural technology adoption and productivity: evidence from randomized control trial in Tanzania. World Dev 133:104997

Nakano Y, Otsuka K (2011) Determinants of household contributions to collective irrigation management: the case of the Doho rice scheme in Uganda. Environ Dev Econ 16(5):527–551

Njagi T, Mano Y, Otsuka K (2016) Role of access to credit in rice production in sub-Saharan Africa: the case of Mwea irrigation scheme in Kenya. J Afr Econ 25(2):300–321

Ogura T, Awuni JA, Sakurai T (2020) The impact of quality-based pricing scheme on local paddy transactions in the northern region of Ghana. Japanese J Agric Econ 22:147–151

Otsuka K, Fan S (2021) Agricultural development: new perspectives in a changing world. International Food Policy Research Institute, Washington, DC

Otsuka K, Larson DF (2013) An African Green Revolution: finding ways to boost productivity on small farms. Springer, Dordrecht

Otsuka K, Larson DF (2016) In pursuit of an African Green Revolution: views from rice and maize farmers' fields. Springer, Dordrecht

Otsuka K, Liu Y, Yamauchi F (2016a) The future of small farms in Asia. Dev Policy Rev 34(3):441–461

Otsuka K, Liu Y, Yamauchi F (2016b) Growing advantage of large farms in Asia and its implications for global food security. Glob Food Sec 11:5–10

Otsuka K, Muraoka R (2017) A Green Revolution for sub-Saharan Africa: past failures and future prospects. J Afr Econ 26(S1):i73–i98

Ragasa C, Chapoto A (2017) Limits to Green Revolution in rice in Africa: the case of Ghana. Land Use Policy 66:304–321

Rashid S, Dorosh PA, Malek M, Lemma S (2013) Modern input promotion in sub-Saharan Africa: insights from Asian Green Revolution. Agric Econ 44:705–721

Reardon T, Chen KZ, Minten B, Adriano L, Dao TA, Wang J, Gupta SD (2014) The quiet revolution in Asia's rice value chains. Ann N Y Acad Sci 1331:106–118

Rosegrant MW, Wiebe K, Susler TB, Mason-D'Croz D, Willenbockel D (2021) Climate change and agricultural development. In: Otsuka K, Fan S (eds) Agricultural development: new perspectives in a changing world. International Food Policy Research Institute, Washington, DC

Schultz TW (1964) Transforming traditional agriculture. Yale University Press, New Haven, CT

Shiferaw B, Smale M, Braun HJ, Duveiller E, Reynolds M, Muricho G (2013) Crops that feed the world 10. Past successes and future challenges to the role played by wheat in global food security. Food Security 5:291–317

Smale M, Byerlee D, Jayne T (2013) Maize revolution in sub-Saharan Africa. In: Otsuka K, Larson DF (eds) An African Green Revolution: finding ways to boost productivity on small farms. Springer, Dordrecht

Takahashi K, Muraoka R, Otsuka K (2020) Technology adoption, impact, and extension in developing country agriculture: a review of the recent literature. Agric Econ 51(1):31–45

United States Department of Agriculture, Foreign Agricultural Service (2021) PSD Data Sets

World Bank (2020) World Bank Pink Sheet, Washington, DC

Yamauchi F, Huang J, Otsuka K (2021) Changing farm size and agricultural development in East Asia. In: Otsuka K, Fan S (eds) Agricultural development: new perspectives in a changing world. International Food Policy Research Institute, Washington, DC

Keijiro Otsuka is a professor of development economics at the Graduate School of Economics, Kobe University and a chief senior researcher at the Institute of Developing Economies in Chiba, Japan since 2016. He received a Ph.D. in economics from the University of Chicago in 1979. He majors in Green Revolution, land tenure and land tenancy, natural resource management, poverty reduction, and industrial development in Asia and sub-Saharan Africa.

Yukichi Mano is a professor at Hitotsubashi University, Japan, and is a fellow at Tokyo Center for Economic Research (TCER). He received a Ph.D. in Economics from the University of Chicago in 2007. His scholarly interests include agricultural technology adoption, horticulture, and high-value crop production, business and management training (KAIZEN), human capital investment, migration and remittance, and universal health coverage in Asia and sub-Saharan Africa.

Kazushi Takahashi is a professor at the National Graduate Institute for Policy Studies (GRIPS) and is a director of the Global Governance Program at GRIPS, Japan. He received a Ph.D. in Development Economics from GRIPS. His scholarly interests include agricultural technology adoption, rural poverty dynamics, microfinance, human capital investment, and aid effectiveness in Asia and sub-Saharan African countries.

Part II
Training as a Central Development Strategy

Chapter 2
The Role of Extension in the Green Revolution

Kazushi Takahashi and Keijiro Otsuka

Abstract Given that the rice Green Revolution involves adopting management-intensive production practices, the role of extension is critically important to its realization in sub-Saharan Africa (SSA). This chapter reviews the existing literature on the role of extension services, examines their effectiveness, and identifies challenges in advancing the research. After briefly examining the evolution of extension models, we discuss the importance and difficulty of obtaining credible estimates of the impact of extension because of endogeneity concerns. We then review the empirical evidence from existing studies, focusing primarily on the rice sector in SSA. We find that most studies show positive and significant impacts of extension on rice intensification, income, and profits, particularly when training includes management-intensive cultivation practices. Thus, the provision of rice cultivation training ought to be a vital entry point to the full-fledged Green Revolution in SSA. We also argue the importance of using income and profits as outcomes and consider spillover effects to fully understand the benefits of extension. Finally, we discuss some remaining research issues, including who should become farmer trainers to facilitate the information spillover.

2.1 Introduction

To achieve the Green Revolution in sub-Saharan Africa (SSA), lowland rice production is found to be particularly promising because of the direct transferability of cultivation technologies from tropical Asia (Otsuka and Larson 2013, 2016). The previous chapter indicates that two types of technologies have made the Asian rice

K. Takahashi (✉)
Graduate School of Policy Studies, National Graduate Institute for Policy Studies (GRIPS), 7-22-1, Roppongi, Room 1211, Minato-ku 106-8677, Tokyo, Japan
e-mail: kaz-takahashi@grips.ac.jp

K. Otsuka
Graduate School of Economics, Kobe University, 2-1 Rokkodai-Cho, Nada-ku, Fourth Academic Building, 5Th Floor, Room 504, Kobe 657-8501, Hyogo, Japan
e-mail: otsuka@econ.kobe-u.ac.jp

© JICA Ogata Sadako Research Institute for Peace and Development 2023
K. Otsuka et al. (eds.), *Rice Green Revolution in Sub-Saharan Africa*, Natural Resource Management and Policy 56, https://doi.org/10.1007/978-981-19-8046-6_2

Green Revolution possible. The first one is modern inputs, such as improved seed varieties and inorganic fertilizers. Since the essence of these technologies is embedded in the tangible inputs per se, farmers can at least reap some benefits from applying them to the field. In order to realize the full benefits, farmers need to further learn proper methods such as the optimal input mix and the appropriate timing for its use. The second technology consists of cultivation practices. In the case of rice, this includes seed selection, nursery set-up, bund construction, leveling of paddy fields, and straight-row transplanting, among other practices. Appropriate water control is of particular importance for lowland rice cultivation, an approach widely practiced in tropical Asia even before the onset of the Green Revolution but not in SSA. Because these technologies are intangible and knowledge-intensive, farmers should not only be aware of their existence and expected returns but also understand how to adopt them properly.

Since information on management practices has the characteristics of a public good, such as non-excludability and non-rivalry, public-sector agricultural extension agents are expected to play crucial roles in disseminating management-intensive technologies. This is especially important at the initial stage when dynamic disequilibrium caused by the introduction of new information is most significant, and farmers have limited opportunities to learn from each other (Anderson and Feder 2007; Takahashi et al. 2020). However, it would be prohibitively costly to directly train all individual farmers to increase their awareness and knowledge of new technologies through extension services. Moreover, a public extension system may include some inefficiencies in the delivery of information on the one hand, and farmers may need multiple sources of information to build sufficient confidence to shift to new technology on the other (Centola 2010; Beaman et al. 2021). Thus, there is renewed interest in the literature on the role and effectiveness of farmer-to-farmer extension systems for the wider diffusion of new agricultural technologies, including rice Green Revolution technologies in SSA.

The primary objective of this chapter is to review existing studies on the role of extension services in disseminating new agricultural technologies and improving farmers' welfare, with an emphasis on the rice sector, in order to set the stage for our empirical studies on the impact of rice cultivation training programs to be reported in Chaps. 3–6. After reviewing the evolution of actual extension models in Sect. 2.2, we discuss the pitfalls of evaluating the impact of extension and choosing appropriate outcome variables in Sect. 2.3. We then review the recent empirical evidence on the impacts of rice cultivation training programs in Sect. 2.4 and those of farmer-to-farmer extension systems, paying particular attention to the issue of information spillovers in Sect. 2.5. We confirm from the literature review that the rice cultivation training program has significant impacts on the adoption of new technologies by the training participants as well as improving their welfare. However, we find that the issue of who should become the contact farmers to facilitate farmer-to-farmer extension has been inadequately explored empirically. Finally, in Sect. 2.6, we discuss the remaining research issues, some of which are addressed in the following chapters.

2.2   Role of Extension in Disseminating New Technology

The development of new agricultural technologies can provide a key to improving agricultural production and productivity. It is an activity often carried out by national or international agricultural research organizations. Extension agents play a major intermediary role in transforming new agricultural technologies and knowledge developed at research organizations into innovative practices of farmers through agricultural training (Niu and Ragasa 2018). The ultimate goals of the extension services and training are to improve agricultural productivity, farm income, and food security by helping farmers to make better production decisions and thus improve the welfare of farmers and consumers. However, not all farmers have sufficient capacities and risk-tolerance to accept new technologies, especially in the initial phase when the benefit and know-how of the technology may be less clear. According to Rogers' diffusion theory (Rogers 1995), the adoption process of an innovation follows an S-shaped growth curve in which the cumulative number of adopters is on the vertical axis and the length of time is on the horizontal axis. This is because there are a small number of innovators who are not afraid of risk and are willing to change the status quo by adopting new technology at the initial stage, and then the number of followers increases through contact with adopters at a growing rate until the number of non-adopters decreases.[1] For this reason, in disseminating agricultural technology, it is often observed that a small set of lead farmers or contact farmers, who have higher capacity and willingness to take risks, are targeted as the initial recipients of technology.

Traditional extension services used government-driven, top-down approaches to new information delivery. A typical example is the Training and Visit (T&V) system, introduced by the World Bank in 1974 and implemented in more than 70 developing countries. It collapsed in 1998 due to the high cost of the system (Anderson and Feder 2007). This system is characterized by a single hierarchical line of command with several levels of field and supervisory staff. Extension agents regularly (generally bi-weekly) visit and deliver packaged messages to contact farmers who are expected to teach the technology learned to other farmers. However, this system shows some weaknesses hampering its effectiveness, such as low staff morale and financial unsustainability due to high costs for personnel and other operations. Moreover, the selection of contact farmers was biased toward the resource-rich farmers who were not necessarily the most capable nor representative of most farmers in the community. Regular visits were often unproductive because new information could not be provided every time extension agents visited, resulting in substantial inefficiency.

Indeed, while an earlier study on the impact of the T&V system in Kenya finds high returns (Bindlish and Evenson 1993), a later study that uses the same data set with a more careful estimation fails to find a significant impact on farming efficiency and productivity (Gautam and Anderson 1999). Gautam (2000) attributes the lack of

---

[1] In his classical article, Griliches (1957) demonstrates that timing and speed of diffusion depends on the profitability of new technology.

consistent impact to inefficient delivery mechanisms, including mistargeting farmers and contents. The mixed evidence on return to the extension has raised skepticism among policymakers about the effectiveness of investments in agricultural extension. Furthermore, in accordance with the structural adjustment programs in the 1980s and 1990s, agricultural extension has gradually been downsized in many developing countries.

From the late 1980s, there has been a tendency toward decentralized, demand-driven, inclusive extension approaches that focus on farmers' learning through participation. Farmer field schools (FFS), first developed as a way to introduce a knowledge-intensive integrated pest management (IPM) system in Asia during the Green Revolution period, have become a prominent participatory and learner-centered approach to improve the relevance and efficiency of agricultural extension. In a typical FFS, a group of 20–25 farmers receives 9–12 half-day sessions of experimentation and non-formal training and consultation during a single crop-growing season (Anderson and Feder 2007). Through group interactions, participating farmers can deepen their understanding by communicating with other attendees about the right way to apply the new technology. Some farmers are then selected to receive additional training so as to be qualified as farmer trainers, who then take up training responsibilities for other non-participating farmers (Anderson and Feder 2007).

Although FFS aims to overcome the shortcomings of traditional, top-down extension systems, its effectiveness is still inconclusive. An earlier study in Indonesia finds negligible impacts of FFS on rice yield and pesticide use by graduates and their neighbors (Feder et al. 2004), while another article in the Peruvian Andres finds a positive and significant impact on the IPM knowledge related to tomato production (Godtland et al. 2004). Davis et al. (2012) examine the impact of FFS in three east African countries and find mixed evidence that participation in FFS increases agricultural productivity and income in Kenya and Tanzania, but not in Uganda. A systematic literature review on FFS by Waddington et al. (2014) indicates the positive impact on intermediate outcomes, such as knowledge and adoption of improved practices, as well as final outcomes including agricultural production and farmers' income—see also Antholt (1998) for a review of T&V and FFS systems. However, the authors also caution that studies offering credible estimates of impacts are still scarce and are at best limited to short-term evaluations. In consequence, further research has been conducted to assess the rigorous impact of extension activities in more recent years.

2.3 Evaluating the Impacts of Extension

2.3.1 Estimation Framework

One possible reason for the mixed results of agricultural extension is the difficulty in identifying its impacts. When evaluating the impact of extension, a basic estimation model takes the following form:

$$Y_{ij} = \beta_0 + \beta_1 D_{ij} + X_{ij}\gamma + e_{ij} \tag{2.1}$$

where Y_{ij} is the outcome of a household/plot i in village j, including farmers' knowledge, adoption of new technologies, input use, yield, and profit; D_{ij} is a dummy variable taking the value of 1 if a household receives training/extension services and 0 otherwise; X_{ij} is a set of covariates, such as land size, household, and its head's demographic characteristics; and e_{ij} is the error term. The parameter of interest is β_1, which is supposed to capture the average impact of the training program, holding other things equal. However, establishing a causal relationship between D_i and Y_{ij} is a formidable task due to endogeneity concerns, wherein the error term is correlated with the training dummy, i.e., $\text{cov}(D_{ij}, e_{ij}) \neq 0$.

An endogeneity problem arises when farmers participating in training have different observable and unobservable characteristics from non-participants. For example, extension agents will purposely select farmers who are richer and more capable, more willing to take risks, and more motivated, or farmers with such characteristics will be more likely to participate in training when everybody is allowed to attend the training. As long as such participation rules are obvious and factors affecting the participation decision are included in the covariates of X_{ij}, we can obtain unbiased estimates of $\widehat{\beta_1}$. However, it is difficult, if not impossible, to generate the relevant data on generally unobservable characteristics, such as motivation and risk-tolerance, relegating them to the error term.

Another source of the endogeneity problem is program placement. The placement bias arises if extension agents purposely select target villages for training, which have systematically different characteristics from non-training villages. This selection will happen if extension agents target high-potential villages to demonstrate the benefits of new technologies or target low-performing villages to reduce poverty. In either case, if the different characteristics correlated with D_{ij} are not fully controlled in the regression, the estimated coefficient will be biased.

Several econometric techniques have been developed to overcome the endogeneity problems, such as the instrumental variable (IV) technique, propensity score matching (PSM), inverse probability weighting (IPW), endogenous switching regression (ESR), difference in difference (DID), and doubly robust (DR) estimation. Each of these quasi-experimental evaluation techniques has pros and cons with different underlying assumptions. For example, using the PSM with DID approach, Davis et al. (2012) show causality between the training and outcomes by controlling for observable and time-invariant factors affecting participation in training. However, their study has been criticized for the non-random placement of FFS (Buehren et al. 2017).

Thus, an increasing number of recent studies rely on experimental methods using randomized controlled trials (RCT) where eligibility to participate in training is randomized at the household or village level. Because of randomization, it is ensured, if well-executed, that no systematic relationship between D_{ij} and e_{ij} exists for both observable and unobservable characteristics, yielding the unbiased estimates of $\widehat{\beta_1}$. Using an RCT framework, Kondylis et al. (2017) compare the impact of two different extension systems on adopting sustainable land management in Mozambique. To obtain unbiased estimates, they randomly assigned a typical extension system in which extension agents train contact farmers and an alternative system in which contact farmers directly receive three days of additional central training. Their results show that contact farmers who receive direct training are more likely to adopt the new technology. However, contact farmers' adoption of the technology does not have significant impacts on its dissemination to other farmers, even though the technology shows some positive returns for contact farmers because of increased yield and decreased labor costs.

2.3.2 Choice of Outcomes

In evaluating the impact of agricultural extension, intermediary outcomes, such as knowledge improvement and technology adoption, are often used, which are important and of interest in themselves. However, it is also desirable to identify the extent to which farmers' welfare improves, in terms of income and profit (obtained by subtracting the opportunity cost of family labor and other family-owned resources from income). Two reviews of the literature by Otsuka et al. (2016) and Takahashi et al. (2020) point out that most existing studies on agriculture concentrate on only intermediate outcomes, such as knowledge and adoption of technology, and yield at best. Calculating the imputed value of unpaid family labor is particularly difficult for estimating profits because of the difficulty in estimating the shadow value of family labor. When the prevailing market wage rate is used to impute the value of family labor, profits often become negative, particularly in the context of SSA, indicating that the imputation method may be inappropriate (Takahashi et al. 2020).

Despite such difficulties, addressing incomes and profits are crucial for at least two reasons. First, since improved management practice is care-intensive, it generally requires more labor input than conventional practices. Thus, the improved yield does not necessarily guarantee higher incomes or profits for farmers adopting new technologies. Indeed, Takahashi and Barrett (2014) show that a set of improved rice management practices, called the System of Rice Intensification (SRI), has yield advantages of about 64%, but because of reallocation of labor from non-farm activities to rice production due to the increased labor use, total household income has not significantly improved for farmers adopting SRI in Indonesia. Similar findings are made in the case of contract farming in Madagascar (Bellemare 2018). It is thus essential to understand the net income gains from adopting management practices by carefully considering the opportunity cost of family labor and trade-off across various economic activities.

Second, farmers generally adopt technology not because of yield advantage but because of advantage in profits. Numerous examples exist where new agricultural technologies offer higher yield potentials but not profit, and in consequence, they are not diffused widely. The aforementioned SRI is one such example. Conversely, a technology with no yield advantage can be adopted widely because of positive profits due to reduced input costs (Michler et al. 2019). Understanding the profitability of technology is crucial to better identification of whether the limited diffusion of technologies is due to external or internal constraints that prevent farmers from adopting them or due to the lack of perceived benefits farmers can reap (Macours 2019). Because resource allocations, including family labor, change in response to the introduction of new technologies, it is crucial to analyze either change in income from the whole activities or change in profit in the activities under question by correctly assessing the opportunity cost of family labor.

2.4 Empirical Evidence on the Training Impacts in the Rice Sector

2.4.1 Impacts in Irrigated Conditions

As shown previously, much of the literature on the impact of agricultural extension/training shows a mixed picture about its effectiveness. However, accumulated evidence on the impacts of training in the rice sector consistently shows the positive and significant impacts on technology adoption and productivity in many developing countries, especially in SSA, when management practices are included in the training component. This may be because improved rice farming is management-intensive, so training is likely to enhance the profitability of rice farming by improving rice cultivation practices.

One of the largest field experiments has been conducted in Bangladesh, where more than 180 villages and 5,000 farmers are involved. The allocation of training of SRI was first randomized at the village level and then at the household level within a treated village (Barrett et al. 2021). Training intensification, in terms of the number of trainings conducted and the proportion of farmers eligible to participate in training, is intentionally varied between villages in order to investigate the heterogeneous impact of training. After two years of experiments, the study finds that the probability of adopting SRI increases significantly for the treated farmers relative to control farmers who are not eligible to receive training. The magnitude is larger for villages with more prolonged exposure to training (i.e., training is repeated twice in consecutive years), indicating that agricultural training is a good entry point to disseminate new knowledge- and care-intensive agricultural technologies. Moreover, the treated farmers as well as non-treated farmers in treated villages achieve higher yields, rice income, and profits relative to the control farmers in non-treated villages. The fact that non-treated farmers in treated villages perform better than pure control farmers indicates some spillover effects of training. They further show that, once training is given, there is no additional increase in yield, income, or profits even if training is provided repeatedly in successive years.

In the case of SSA, Kijima et al. (2012) estimate the short-term impact of culti-vation training programs in irrigated areas of Uganda. The training program imple-mented by the Japan International Cooperation Agency (JICA) introduced improved rice management practices for land preparation, such as bund construction, leveling, and puddling, as well as better transplanting methods, such as transplanting of young seedlings[2] with appropriate spacing between seedlings. Based on the IPW method, Kijima et al. (2012) show that training increases the intensification of rice produc-tion. As a result, rice yield and income of participant farmers significantly increase relative to non-participating farmers.

Similarly, Conteh et al. (2013) examine the ex-post impact of rice training in irrigated areas of Sierra Leone. Sierra Leone Agricultural Research Institute (SLARI) first offers training for the district agricultural extension agents and contact farmers, and together, they subsequently provide field training to other farmers. In the training program, Asian-type management practices, such as bunding, leveling, and straight-row planting, are taught along with the recommendation to use modern varieties (MVs) and inorganic fertilizers. Based on PSM, the results show that the training participation generally increases the adoption of improved management practices and profits from rice farming, although different matching algorithms yield somewhat mixed results. The author concludes that one of the main constraints on the growth in agricultural output, particularly rice cultivation, is the lack of efficient extension programs, with some caveats that the evidence is not entirely decisive.

The focus of the studies in the rice sector in SSA tends to be skewed toward irrigated areas. Nevertheless, given that rainfed ecology is dominant in this region, it is essential to enhance our understanding of rainfed rice cultivation. Few studies have compared the relative impacts of training in irrigated and rainfed conditions in SSA. One of the rare exceptions is Mgendi et al. (2021), who compare the differen-tial training impacts of agricultural technologies between irrigated and rainfed areas in Tanzania. In the study area, a Chinese-funded program, the Agricultural Tech-nology Demonstration Center (ATDC), promotes improved management practices, including land harrowing and leveling, transplanting by standardized spacing, and basal fertilization, among others. Using IPW, the study finds that ATDC programs and other non-ATDC training programs facilitate the adoption of improved manage-ment practices. The rice yield of trained farmers significantly increases relative to that of non-trained farmers in irrigated areas. However, there is no statistical differ-ence in yield between trained and non-trained farmers in rainfed areas. Based on this finding, the authors emphasize that not only agricultural training programs but also supportive factors, such as irrigation facilities, should be provided for enhancing the adoption of agricultural technology and rice yield.[3] The importance of supporting

---

[2] In Uganda, transplanting of around four-week-old seedlings is common. JICA experts recommend reducing it to around three weeks.

[3] The agro-ecological conditions of rainfed areas are diverse, ranging from favorable area free from drought and flood to drought and flood prone environments. Thus, the finding that training is not effective in rainfed areas may not be generalizable to other communities.

facilities, especially irrigation, is also discussed in Kajisa and Payongayong (2011) and Ragasa and Chapoto (2017).

The experience of SSA thus seems to resemble that of tropical Asia, where the rice Green Revolution has been successful in irrigated and favorable rainfed areas because of the direct effects (whereby the favorable environment facilitates the intensive use of land) and the indirect effects through the complementarity between improved cultivation practices and the use of MVs and fertilizer (David and Otsuka 1994).

2.4.2 *Impacts in Rainfed Conditions*

Supporting facilities should indeed be developed to realize the Green Revolution in SSA, as we will discuss below. Nonetheless, some studies show the positive impacts of management training even without irrigation. A spectacular example of training impacts in rainfed conditions is found in the Kilombero Valley in Tanzania (Alem et al. 2015; Nakano et al. 2018a). This is a favorable rainfed area because additional water flows from nearby mountain ranges, and the soil is fertile. One of the unique features of this study site is that the large private rice plantation offered management training to farmers nearby to facilitate proper seed selection and dibbling with wide spacing between seedlings, along with the use of MVs and chemical fertilizer. Although the disseminated technologies deviate from the original principles, it resembles a package recommended by SRI, and thus Nakano et al. (2018a) call it "modified SRI." The estimation results by Alem et al. (2015) based on ESR and those of Nakano et al. (2018a) based on PSM both show that training effectively increases the adoption of packaged technologies and promotes rice intensification, which significantly contributes to higher rice yield. Nakano et al. (2018a) further show that paddy yield was reported to be as high as 4.7 tons per hectare on plots where the trainees adopted modified SRI technologies. This is higher than the highest paddy yield under rainfed conditions in tropical Asia to the best of our knowledge. On the other hand, the yield was at most three tons per hectare without adopting modified SRI, even if the same trainees cultivated the plot. Positive impacts on yield lead to the significant improvement of profits for modified SRI adopters.

While not as remarkable, statistically significant positive impacts on yield and income are also observed in the government-led management training programs in rainfed areas of northern Ghana, which are not as favorable for rice production due to the paucity of rainfall (deGraft-Johnson et al. 2014). Chapters 5 and 6 in this volume further examine the impact of training on the productivity of rice farming in rainfed areas in Uganda and Mozambique. The case of Mozambique is of particular interest because the prevailing rice production technology is primitive, and the production environments are relatively unfavorable.

2.4.3 Contract Farming

The involvement of the private sector is also found in the case of contract farming in Benin, in which private firms provide not only improved seeds and fertilizer on credit but also improved knowledge of production practices[4] in return for the delivery of a certain amount of output at predetermined prices. Due to the downsized public-sector extension systems, investigating whether private-sector contract farming is conducive to improving rice yields, profitability, and farmers' incomes is a critical issue in realizing the rice Green Revolution in SSA. While side-selling becomes an issue and thus contract farming in staple foods is considered unviable (Otsuka et al. 2016), an empirical study by Maertens and Velde (2017) shows that contract farming can contribute to the expansion of rice area, increased production and commercialization, increased price, and thus increased income in Benin.

Furthermore, Arouna et al. (2021) use an RCT to identify which attributes of contract farming are productive in Benin: fixed price to reduce uncertainty, production management training to mitigate information friction, and advanced provision of inputs to relax credit constraints. Because of feasibility constraints, they randomly offer a contract that (1) provides a price guarantee, (2) combines extension training with the price guarantee, or (3) provides input loans in addition to the extension training and price guarantee. Their findings suggest that each contract scheme significantly improves rice yield and income relative to the control group, with no significant additional impacts from additional items (i.e., extension training and input loans). Their result indicates that the simplest contract to fix price can have a significant impact on rice yield and household welfare, although it is hard to understand why rice yield and household welfare improved, even when training in cultivation practices was not offered.

Counter-evidence for the punchline results of Arouna et al. (2021) is provided by Olounlade et al. (2020), where farmers who participate in contract farming can have a smaller rice income than non-participants in Benin. The authors attribute this to the rigid contract scheme, where contract farmers cannot sell products to other buyers even when the market price is more favorable than agreed contract prices. Thus, the importance of offering a fixed-price contract remains ambiguous, even when contract farming improves rice productivity through management training and input credits. We also question if it pays the private sector to offer management training if knowledge of management practices is a local public good, and hence, information spillovers are rampant. Failure of contract farming in rice production is also reported in Cote d'Ivoire (Mano et al. 2020 and Chap. 8 in this volume) and Mozambique (Veldwisch 2015; van der Struijk 2013). Indeed, Veldwisch (2015) and van der Struijk (2013) report that rice contract framing in the Chokwe irrigation scheme, Mozambique, ran well in the first few years but eventually collapsed due to contract enforcement problems, including rampant side-selling to a competitive rice miller.

---

[4] It is unclear, however, what cultivation practices were introduced by contract farming in Benin.

2.5 Empirical Evidence of Farmer-to-Farmer Extension

2.5.1 The Existence of Spillover Effects

Many studies in SSA show significantly positive impacts of training in cultivation practices in the rice sector. However, still, less is known about the magnitude of information spillovers from participant farmers to non-participant farmers. Farmer-to-farmer extension was expected to play an essential role in most extension systems, including T&V and FFS approaches. A fundamental difficulty in identifying such spillover effects is a reflection problem originally suggested by Manski (1993): people in the same community behave similarly not only because they learn from each other but also because they have similar socio-economic characteristics and face similar production environments. Although identifying the pure learning effect separated from other confounders is thus challenging, ignoring spillover effects may cause serious empirical and practical problems, at least in the following two aspects.

First, if non-participant neighbors receive positive impacts from training participants, the estimation of training impacts is generally biased downward. To know this, let us assume that the average benefit of the training impact is expressed as R (>0), and the average spillover benefit as S (>0) with the counterfactual outcome of F in the absence of training. Outcomes can be knowledge acquisition, technology adoption, increased yield, income, and profit. Now suppose that there are three groups of people who live in a training village and participate in training (group A), live in the training village but do not participate in training (group B), or live in a non-training village and do not participate in training (group C). Suppose further that spillover exists only within a village, not across villages. Thus, the average outcome of group C is F, while that of group A is F + R, and that of group B is F + S. It is evident that the true direct training impact R is underestimated by the magnitude of S if we compare the outcomes of group A and group B by ignoring spillover. Under the existence of spillover, we will have a biased estimate of R – S and generate misperceptions unless we select a comparison group of non-participants from non-training communities where there is no spillover. Otherwise, it violates one of the critical assumptions—the stable unit treatment value assumption (SUTVA)—according to which the observation of one unit should not be affected by the treatment of other units (Takahashi et al. 2019).

Second, if spillover effects are not properly measured, we will underestimate the cost-effectiveness of the training program, leading to under-investment in profitable opportunities. This is because public investment can be justified if social benefits exceed social costs. The former includes both direct and indirect (spillover) effects (i.e., R + S) for the participant and non-participant farmers. Thus, in evaluating cost-effectiveness, one needs to know the extent to which the farmers' welfare improves, again signifying the importance of identifying the impacts on income and profits.

It must also be emphasized that the absence of the spillover effects implies that new production knowledge generated by training is not a public good. Consequently, there would be no basis to support the public-sector extension system. Thus, the assessment of the spillover effect is a fundamental issue of new technology dissemination. In sum, the existence of spillover effects can be a threat to identification because of the violation of SUTVA but is raison d'être for public-sector extension services.

2.5.2   Evidence of Spillover Effects[5]

One issue that has been intensively discussed in the recent literature is who should be the farmer trainer to maximize the spillover effects or fully facilitate social learning. According to Takahashi et al. (2020), the literature typically considers three types of potential farmer trainers: (a) those who are at the center of the information network; (b) those who are innovative, eager to take risks associated with the adoption of new technologies, and often knowledgeable and productive; and (c) those who have socio-economic and farm characteristics similar to the majority of farmers in a community. In general, (a) and (b) are not distinguishable, and often T&V systems target those contact/lead farmers, while participatory approaches, including FFS, involve more (c)-type farmers.

The advantages of selecting types (a) and (b), which can be called heterophilous networks, are that the number of technology adopters would increase faster because lead farmers have broader connections with others, that information would be more accurately disseminated because they are more knowledgeable, and that demonstration effects in terms of productivity and profitability improvement would be more significant because they are more productive. Moreover, such innovators would effectively integrate new knowledge into local practices when adaptation is required in the local context. The role of an innovative farmer trainer is essential, particularly when peer farmers want to free-ride and reduce the uncertainty associated with the adoption of new technologies by strategically delaying their adoption (Foster and Rosenzweig 1995; Bandiera and Rasul 2006; Maertens 2017).

On the other hand, if the gains of new technology are heterogeneous, reflecting the diversity in growing conditions and farmer characteristics, there is no guarantee that lead farmers' success can be replicated by other farmers (Suri 2011; Magnan et al. 2015; Tjernstrom 2017; Shikuku 2019). In other words, better performance cannot be solely attributed to the advantages embedded in technology but to observable and unobservable characteristics of lead farmers and their agricultural plots (Barrett et al. 2004). In this case, it would be easier for followers to learn from farmer trainers who are endowed with similar but not outstanding characteristics, which can be defined as a homophily network.

The empirical findings of these competing views are mixed. Maertens (2017) provides evidence of the superiority of heterophilous networks, indicating that lead farmers play a more significant role than ordinary farmers in social learning. Dillon et al. (2018) suggest that the network-based targeting approach is more effective than a random selection of lead farmers in encouraging broader adoption. Beaman et al.

---

[5] This subsection draws heavily on Takahashi et al. (2020).

(2021) also find similar results. In contrast, Beaman and Dillon (2018) and Lee et al. (2019) show that the network-based targeting approach is not necessarily superior to a random selection of entry points to improve the knowledge of other farmers.

Matuschke and Qaim (2009) and Weimann (1994) show the superiority of homophilous networks. They contend that the vertical flow of information from the lead farmer to peer farmers often fails, and successful information exchange is more likely based on horizontal, socially proximate relationships. Based on an RCT in Malawi, BenYishay and Mobarak (2019) demonstrate that ordinary farmers perform better than knowledgeable and productive farmers as farmer trainers in terms of efforts and resultant technology diffusion, especially when ordinary farmers are incentivized by monetary reward. The study by Takahashi et al. (2019) using RCT in Cote d'Ivoire (Chap. 3 in this volume) also shows that randomly selected training participants, who are representative of the majority in a community, effectively work as a catalyst to disseminate information to non-training participants. Similar spillover effects are observed in Uganda, where the adoption of new technologies by non-participants in cultivation training gradually increased (see Chap. 5).

Somewhere between these two views is the perspective provided by Feder and Savanstano (2006). These authors find that farmers are more inclined to learn from peer farmers who are slightly superior to them, but not excessively so. Referring to Rogers (1995), Fisher et al. (2018) also share this view and discuss the differential roles of homophily and heterophily networks as follows:

> Homophilous and heterophilous networks have distinct and complementary roles in the diffusion of innovations. Heterophilous networks, such as that between lead and follower farmers, are more important in triggering awareness of a new technology, because new ideas most often enter a system through individuals who have higher status and are more innovative. Homophilous networks are, however, more useful than heterophilous ties in persuading potential adopters of the merits of the innovation [...] If farmer-to-farmer extension is to have a greater role in encouraging follower farmers to adopt innovations it may be necessary to identify lead farmers that are capable and motivated to train other farmers but not too socially distant from the target population of farmers in terms of personal characteristics and innovativeness...... (pp. 321–322).

As discussed by Fisher et al. (2018), if lead farmers are more suitable as entry points to introduce and decode new, unfamiliar technological information, but ordinary farmers are more suitable to the wider diffusion of tested technology, compromising these two views would be potentially the most effective approach. That would entail an approach whereby lead contact farmers receive initial training and provide training in new technologies to socially proximate farmers who are slightly more capable than ordinary farmers. Then, the latter set of farmers teaches the technologies to other farmers comprising the majority of the community. This stepwise approach is precisely the case presented by Nakano et al. (2018b and Chap. 4 in this volume) in Tanzania.

2.6 Conclusion

This chapter reviews the literature on the impact of extension services in developing countries. Many existing studies show mixed evidence on the impact partly because of the difficulty in solving endogeneity problems. Another potential problem is the choice of technological knowledge disseminated by extension services. Needless to say, if the choice is not based on empirical observations, there is a risk of choosing inappropriate knowledge. This book advocates the training of improved cultivation practices of rice farming based partly on intensive discussions with agricultural experts who work in SSA and partly on the positive results of our earlier empirical studies assessing the significant impacts of training of rice cultivation practices in Ghana and Uganda (deGraft-Johnson et al. 2014; Kijima et al. 2012). By now, a growing number of studies have found credible evidence of the positive impact of training in cultivation practices on the adoption of improved technology, yield, income, and profits in the rice sector of SSA. One of the purposes of this volume is to firmly establish the proposition that the provision of rice cultivation training programs can provide a vital entry point to the full-fledged Green Revolution in SSA, not only in irrigated areas but also in relatively favorable rainfed areas.

When evaluating the impacts of training, one crucial issue is whether to use income or profits as outcomes. We argue that unless the net income gain is identified, we cannot know the welfare impact on farmers, which should be an essential consideration for farmers to adopt or discontinue the new technology. Furthermore, such income gains should be calculated by considering the spillover effects. Otherwise, we may underestimate the cost-effectiveness of the training programs. However, identifying the true training impacts and spillover effects is quite complex under the existence of information spillover because it violates the SUTVA. One of the unique attempts is provided by Chap. 3 in this volume, where the authors use an RCT for random assignment of training at the household level within a village and ask treated farmers not to disseminate the taught technology to their neighbors during the experimental phase. After finding the positive impact of training for the treated farmers, they relax this restriction and find that control farmers catch up with treated farmers because of information spillovers.

When spillover exists, one of the hot issues in the literature is who should be the farmer trainers. While there are contrasting views of the superiority of homophilous and heterophilous networks, a compromise approach is discussed by Feder and Savanstano (2006) and Fisher et al. (2018) and empirically tested by Chap. 4 in this volume. In this approach, 20 lead farmers were initially selected and trained for better management practices. The trained lead farmers were responsible for training five other farmers close to them, who were then encouraged to diffuse the improved practices to other ordinary farmers. The study found that the yield gap between the lead and intermediary farmers, as well as other ordinary farmers, widened immediately after the training because the former had adopted new practices faster than the latter. However, intermediary farmers soon caught up with the lead farmers, and ordinary farmers belatedly caught up with the lead and intermediate farmers. Eventually, the yield gap among these groups gradually declined.

Overall, we may hardly expect a significant and sustainable improvement in rice productivity without any improvement in cultivation practices through training programs. This book (Chaps. 3–6) provides new evidence of the impact of training by agricultural extension in the intensification of rice production in Tanzania, Cote d'Ivoire, Mozambique, and Uganda. We confirm that positive impacts of training were realized even without any improvement in irrigation, marketing, or credit programs. However, we do not argue that the provision of training is a sufficient condition to realize a full-fledged Green Revolution in SSA. The literature review suggests that complementary facilities, such as irrigation and agricultural machinery, are also critical. We thus discuss the role of irrigation facilities in the intensification of rice farming in Chaps. 10 and 11 and that of agricultural machinery in Chaps. 8 and 9.

References

Alem Y, Eggert H, Ruhinduka R (2015) Improving welfare through climate-friendly agriculture: the case of the system of rice intensification. Environ Resour Econ 62:243–263

Anderson JR, Feder G (2007) Agricultural extension. In: Evenson RE, Pingali P (eds) Handbook of agricultural economics, Chapter 44, vol 3. Agricultural Development: Farmers. Farm Production and Farm Markets. Elsevier, Amsterdam, pp 2343–2378

Antholt CH (1998) Agricultural extension in the twenty-first century. In: Eicher CK, Staatz JM (eds) International agricultural development, 3rd edn. Johns Hopkins University Press, pp 354–369

Arouna A, Michler JD, Lokossou JC (2021) Contract farming and rural transformation: evidence from a field experiment in Benin. J Dev Econ 151:102626

Bandiera O, Rasul I (2006) Social networks and technology adoption in Northern Mozambique. Econ J 116(514):869–902

Barrett CB, Islam A, Malek MA, Pakrashi D, Ruthbah U (2021) Experimental evidence on adoption and impact of the system of rice intensification. Am J Agric Econ: forthcoming

Barrett CB, Moser CM, McHugh OV, Barison J (2004) Better technology, better plots, or better farmers? Identifying changes in productivity and risk among Malagasy rice farmers. Am J Agric Econ 86(4):869–888

Beaman LA, BenYishay A, Magruder J, Mobarak AM (2021) Can network theory-based targeting increase technology adoption? Am Econ Rev 111(6):1918–1943

Beaman LA, Dillon A (2018) Diffusion of agricultural information within social networks: evidence on gender inequalities from Mali. J Dev Econ 133:147–161

Bellemare MF (2018) Contract farming: opportunity cost and trade-offs. Agric Econ 49:279–288

BenYishay A, Mobarak AM (2019) Social learning and incentives for experimentation and communication. Rev Econ Stud 86:976–1009

Bindlish V, Evenson R (1993) Evaluation of the performance of T&V extension in Kenya. World Bank Technical Paper No. 208. World Bank, Washington, DC

Buehren N, Goldstein M, Molina E, Vaillant (2017) The impact of strengthening agricultural extension services: evidence from Ethiopia. Policy Research Working Paper 8169, World Bank

Centola D (2010) The spread of behavior in an online social network experiment. Science 329(5996):1194–1197

Conteh AMH, Yan X, Mvodo MES (2013) Evaluating the effect of farmers' training on rice production in Sierra Leone: a case study of rice cultivation in lowland ecology. Int J Humanities Soc Sci 7(7):1926–1933

David CC, Otsuka K (1994) Modern rice technology and income distribution in Asia. Lynne Rienner, Boulder

Davis K, Nkonya E, Kato E, Mekonnen DA, Odendo M, Miiro R, Nkuba J (2012) Impact of farmer field schools on agricultural productivity and poverty in east Africa. World Dev 40(2):402–413

deGraft-Johnson M, Suzuki A, Sarurai T, Otsuka K (2014) On the transferability of the Asian rice green revolution to rainfed areas in sub-Saharan Africa: an assessment of technology intervention in Northern Ghana. Agric Econ 45:555–570

Dillon A, Porter M, Ouedraogo A (2018) Social network targeting of agricultural technology: Adoption, input substitution and yield effects. Selected Paper prepared for presentation at the agricultural & applied economics association annual meeting, Washington, DC, August 5–7

Feder G, Murgai R, Quizon JB (2004) Sending farmers back to school: the impact of farmer field schools in Indonesia. Rev Agric Econ 26:45–62

Feder G, Savastano S (2006) The role of opinion leaders in the diffusion of new knowledge: the case of integrated pest management. World Dev 34(7):1287–1300

Fisher M, Holden ST, Thierfelder C, Katengeza SP (2018) Awareness and adoption of conservation agriculture in Malawi: what difference can farmer-to-farmer extension make? Int J Agric Sustain 16(3):310–325

Foster AD, Rosenzweig MR (1995) Learning by doing and learning from others: human capital and technical change in agriculture. J Political Econ 103(6):1176–1209

Gautam M (2000) Agricultural extension: the Kenya experience: an impact evaluation. Operations Evaluation Department, World Bank, Washington, DC

Gautam M, Anderson JR (1999) Reconsidering the evidence on returns to T&V extension in Kenya (Vol. 2098). World Bank Publications

Godtland EM, Sadoulet E, de Janvry A, Murgai R, Ortiz O (2004) The impact of farmer field schools on knowledge and productivity: a study of potato farmers in the Peruvian Andes. Econ Dev Cult Change 53:63–92

Griliches Z (1957) Hybrid corn: an exploration in the economics of technical change. Econometrica 25:501–522

Kajisa K, Payongayong E (2011) Potential of and constraints to the rice Green Revolution in Mozambique: a case study of the Chokwe irrigation scheme. Food Policy 35:615–626

Kijima Y, Ito Y, Otsuka K (2012) Assessing the impact of training on lowland rice productivity in an African setting: evidence from Uganda. World Dev 40(8):1610–1618

Kondylis F, Mueller V, Zhu J (2017) Seeing is believing? Evidence from an extension network experiment. J Dev Econ 125:1–20

Lee G, Suzuki A, Nam VH (2019) Effect of network-based targeting on the diffusion of good aquaculture practices among shrimp producers in Vietnam. Word Dev 124:104641

Macours K (2019) Farmers' demand and the traits and diffusion of agricultural innovations in developing countries. Annu Rev of Resour Econ 11:483–499

Maertens A (2017) Who cares what others think (or do)? Social learning and social pressures in cotton farming in India. Am J Agric Econ 99(4):988–1007

Maertens M, Velde KV (2017) Contract-farming in staple food chains: the case of rice in Benin. World Dev 95:73–87

Magnan N, Spielman DJ, Lybbert TJ, Gulati K (2015) Leveling with friends: social networks and Indian farmers' demand for a technology with heterogeneous benefits. J Dev Econ 116:223–251

Mano Y, Takahashi K, Otsuka K (2020) Mechanization in land preparation and agricultural intensification: the case of rice farming in the Cote d'Ivoire. Agric Econ 51:899–908

Manski CF (1993) Identification of endogenous social effects: the reflection problem. Rev Econ Stud 60(3):531–542

Matuschke I, Qaim M (2009) The impact of social networks on hybrid seed adoption in India. Agric Econ 40(5):493–505

Mgendi G, Mao S, Qiao F (2021) Is a training program sufficient to improve the smallholder farmers' productivity in Africa? Empirical evidence from a Chinese agricultural technology demonstration center in Tanzania. Sustainability 13:1527

Michler JD, Tjernström E, Verkaart S, Mausch K (2019) Money matters: the role of yields and profits in agricultural technology adoption. Am J Agric Econ 101:710–731

Nakano Y, Tanaka Y, Otsuka K (2018a) Impact of training on the intensification of rice farming: evidence from rainfed areas in Tanzania. Agric Econ 49:193–202

Nakano Y, Tsusaka TW, Aida T, Pede VO (2018b) Is farmer-to-farmer extension effective? The impact of training on technology adoption and rice farming productivity in Tanzania. World Dev 105:336–351

Niu C, Ragasa C (2018) Selective attention and information loss in the lab-to-farm knowledge chain: the case of Malawian agricultural extension programs. Agric Syst 165:147–163

Olounlade OA, Li GC, Kokoye SEH, Dossouhoui FV, Akpa KAA, Anshiso D, Biau G (2020) Impact of participation in contract farming on smallholder farmers' income and food security in rural Benin: PSM and LATE parameter combined. Sustainability 12:901

Otsuka K, Larson DF (2013) An African Green Revolution: finding ways to boost productivity on small farms. Springer, Dordrecht

Otsuka K, Larson DF (2016) In pursuit of an African Green Revolution: views from rice and maize farmers' fields. Springer, Dordrecht

Otsuka K, Nakano Y, Takahashi K (2016) Contract farming in developed and developing countries. Annu Rev Resour Econ 8:353–376

Ragasa C, Chapoto A (2017) Limits to Green Revolution in rice in Africa: the case of Ghana. Land Use Policy 66:304–321

Rogers EM (1995) Diffusion of innovations. Free Press, New York

Shikuku KM (2019) Information exchange links, knowledge exposure, and adoption of agricultural technologies in northern Uganda. World Dev 115:94–106

Suri T (2011) Selection and comparative advantage in technology adoption. Econometrica 79(1):159–209

Takahashi K, Barrett CB (2014) The system of rice intensification and its impacts on household income and child schooling: evidence from rural Indonesia. Am J Agric Econ 96(1):269–289

Takahashi K, Mano Y, Otsuka K (2019) Learning from experts and peer farmers about rice production: experimental evidence from Cote d'Ivoire. World Dev 122:157–169

Takahashi K, Muraoka R, Otsuka K (2020) Technology adoption, impact, and extension in developing country agriculture: a review of the recent literature. Agric Econ 51(1):31–45

Tjernstrom E (2017) Learning from others in heterogeneous environments. https://www.atai-res earch.org/wp-content/uploads/(2015)/11/Tjernstrom-(2017)-learning-from-others.pdf

van der Struijk LF (2013) Smallholders' strategies in response to contract farming programs in Chokwe, Mozambique, Master thesis, International Land and Water Management at Wageningen University, the Netherlands

Veldwisch GJ (2015) Contract farming and the reorganisation of agricultural production within the Chókwè Irrigation System. Mozambique. J Peasant Studies 45(5):1003–1028

Waddington H, Snilstveit B, Garcia Hombrados J, Vojtkova M, Phillips D, Davies P, White H (2014) Farmer field schools for improving farming practices and farmer outcomes: a systematic review. Campbell Systematic Revhttps://doi.org/10.4073/csr.2014.6

Weimann G (1994) The influentials. State University of New York Press, Albany

Kazushi Takahashi is a professor at the National Graduate Institute for Policy Studies (GRIPS) and is a director of the Global Governance Program at GRIPS, Japan. He received a Ph.D. in Development Economics from GRIPS. His scholarly interests include agricultural technology adoption, rural poverty dynamics, microfinance, human capital investment, and aid effectiveness in Asia and sub-Saharan African countries.

Keijiro Otsuka is a professor of development economics at the Graduate School of Economics, Kobe University and a chief senior researcher at the Institute of Developing Economies in Chiba. Japan since 2016. He received a Ph.D. in economics from the University of Chicago in 1979. He majors in Green Revolution, land tenure and land tenancy, natural resource management, poverty reduction, and industrial development in Asia and sub-Saharan Africa.

Chapter 3
The Case of Cote D'Ivoire: Learning from Experts of Rice Farming Management and Peer Farmers About Rice Production

Kazushi Takahashi, Yukichi Mano, and Keijiro Otsuka

Abstract Technological innovation is vital to economic growth and food security in sub-Saharan Africa where agricultural productivity has been stagnant for a long time. Extension services and learning from peer farmers are two common approaches to facilitate the diffusion of new technologies, but little is known about their relative effectiveness. Selection bias, whereby well-motivated training participants would perform better even without extension services, as well as knowledge spillovers, where non-participants can indirectly benefit from extension services, are among the major threats to causal inference. Using a unique sequential randomized experiment on agricultural training, this chapter attempts to meet the dual objectives of executing rigorous impact evaluation of extension services and subsequent spillovers on rice production in Cote d'Ivoire. Specifically, to reduce selection bias, we randomly assigned eligibility for training participation; and to satisfy the stable unit treatment value assumption, control-group farmers were initially restricted from exchanging information with treated-group farmers who had received rice management training. Once the positive impacts were confirmed one year after the training, information exchange between the treated and control farmers was encouraged. We found that the initial performance gaps created by the randomized assignment disappeared over time, due presumably to social learning from peer farmers. A detailed analysis concerning the information network and peer effects provided suggestive evidence

This chapter draws heavily on Takahashi et al. (2019b).

K. Takahashi (✉)
Graduate School of Policy Studies, National Graduate Institute for Policy Studies (GRIPS), 7-22-1, Roppongi, Room 1211, Minato-ku 106-8677, Tokyo, Japan
e-mail: kaz-takahashi@grips.ac.jp

Y. Mano
Graduate School of Economics, Hitotsubashi University, 2-1 Naka, Isono Building Room 324, Kunitachi-shi 186-8601, Tokyo, Japan
e-mail: yukichi.mano@r.hit-u.ac.jp

K. Otsuka
Graduate School of Economics, Kobe University, 2-1 Rokkodai-cho, Nada-ku, Fourth Academic Building, 5Th Floor, Room 504, Kobe 657-8501, Hyogo, Japan
e-mail: otsuka@econ.kobe-u.ac.jp

that there were information and technology spillovers from treated to control farmers after removing the information exchange restriction. Overall, our study demonstrates that information dissemination by farmers can be as effective in improving practices as the initial training provided by extension services.

3.1 Introduction

There has been increasing interest in replicating the Asian Green Revolution in sub-Saharan Africa (SSA), where stagnant agricultural productivity has long been an impediment to economic growth and food security (Evenson and Gollin 2003; Dawson et al. 2016; Bachewe et al. 2018). Emerging studies have identified lowland rice as the most promising staple crop among major cereals in the SSA, because of the high transferability of improved technology from Asia (Otsuka and Larson 2013, 2016). While the application of modern seeds and increased use of chemical fertilizer have been emphasized, Asian-type yield-enhancing rice-growing technologies also include improved management practices, such as bunding, leveling, and straight-row transplanting.[1] Since these management practices are knowledge-intensive and require deep understanding and careful execution to fully exploit yield potentials, the dissemination of those technologies may be difficult without credible and persuasive sources of information. How to reduce information friction in the diffusion process remains an important area of research.

Agricultural extension services are designed to play such a role by delivering advanced knowledge from lab and experimental fields to farmers. Yet, directly training millions of small farmers may be prohibitively costly. An alternative method is farmer-to-farmer training, in which selected progressive (contact) farmers participate in training organized by extension workers and then are encouraged to share skills learned with others in their network. While this contact farmer approach is common in many developing countries, evidence of its efficacy is mixed. For example, Feder et al. (2004), Tripp et al. (2005), and Kondylis et al. (2017) found that there is limited impact from the performance of contact farmers who have adopted new agricultural technologies on other farmers. On the other hand, Krishnan and Patnam (2014) and Nakano et al. (2018a, b) show that contact farmers increasingly adopt new technologies immediately after training, and demonstrated improvements spill over to other farmers later through farmer-to-farmer training.

An apparent empirical challenge to the examination of the effectiveness of the farmer-to-farmer extension system is selection bias, which might be one of the potential reasons for mixed empirical results: since progressive, contact farmers tend to be more motivated than ordinary farmers, they would perform better even without training. If this is the case, simply mimicking contact farmers' observed practices

---

[1] The existing studies show that with those technological packages rice yield can be more than 4 tons per hectare in SSA, which is comparable to or even higher than that in Asian countries like India (Otsuka and Larson 2013, 2016; Ragasa and Chapoto 2017; Nakano et al. 2018a, b).

may not necessarily improve fellow farmers' performance because there are differences in innate characteristics between the contact farmers and fellow farmers. Thus, it is of great importance to determine whether new technologies taught in the training period per se would have the intended positive impact. Rigorously implementing an impact evaluation of this kind, however, involves another empirical challenge in the presence of knowledge spillovers, because non-trained fellow farmers can indirectly benefit by imitation and learning from their trained peers, attenuating the true impact of training. This is a violation of the stable unit treatment value assumption (SUTVA) in social science, according to which the observation of one unit should not be affected by the treatment of other units.

To execute a rigorous impact evaluation of both extension services and the subsequent social learning in a unified framework, we implemented sequential field experiments related to rice production management training for this study, in collaboration with rice-growing farmers in Cote d'Ivoire. To mitigate selection bias, we conducted a randomized controlled trial (RCT), where a subset of randomly selected farmers from each sample site received training in the form of a short course in rice production management. To fulfill the SUTVA, farmers from the treated group were initially asked not to transmit information about their training to the control farmers and the latter were requested to refrain from asking treated farmers for agricultural advice. After one year of observation, we examined whether training had the intended positive impacts for trained farmers, such as the adoption of recommended agronomic practices, improved rice yields, and improved profit. Once the positive impacts were confirmed, we relaxed the restriction and started promoting spillovers by encouraging farmers to exchange information without any monetary incentive given to the trained farmers.

We conducted household surveys three times: at the baseline before the training (January 2015 to May 2015), at the midline one year after the start of training (March 2016 to May 2016), and at the endline two years after the end of the training (March 2017 to May 2017). Using these three-year panel data, we identified the evolution of both the intention-to-treat (ITT) effect and the treatment-on-the-treated (TOT) effect of the training. Furthermore, to explore the dynamics of information exchange between treated and control farmers in the first and second year of observation, we conducted a detailed analysis of information flows and the existence of peer effects using social network data collected in the mid- and endline surveys.

Our main findings are summarized as follows. We found that while the adoption rates of most improved rice management practices were unexpectedly high even at the baseline, the treated farmers were more likely to adopt improved practices by a year after training (i.e., the midline), such as transplanting in rows, canal/drainage construction, and field leveling after training. Higher adoption rates of those recommended agronomic practices lead to improved rice yield as well as increased income per hectare among treated farmers at the midline. Once all farmers were encouraged to exchange information later, the productivity gaps between treated and control farmers narrowed sharply by two years after training (i.e., the endline). It may seem reasonable to interpret this convergence as the sign of short-lived impacts of training where trained farmers dis-adopt new practices and return to traditional ones. However, we observed that trained farmers continued to adopt improved agronomic practices at the endline, and control farmers followed them. Our detailed network analysis based on a dyadic regression further revealed that information flows from treated to control farmers were less active than between control farmers (the reference group) at the midline but become more active at the endline. These results together suggest that farmers largely followed our guidance not to exchange information on rice production within the initial experimental phase, which mitigates estimation bias, if any, in our impact evaluation of training. Yet, once such a restriction is abolished and information exchange is encouraged, control farmers can successfully catch up with treated farmers through social learning. These results imply the importance of not only extension services to trigger the adoption of improved agricultural technologies, but also of social learning for their wider diffusion.

This study contributes to the literature on the role of social learning in the diffusion of agricultural technology (Foster and Rosenzweig 1995; Munshi 2004; Bandiera and Rasul 2006; Conley and Udry 2010). Most previous studies agree with the high potential of social learning, but there is little consensus as to the relative effectiveness of direct training by extension workers and learning from peer farmers. Moreover, a growing number of studies focus on determining who should be targeted to increase the initial adoption rate and to facilitate social learning (Beaman and Dillon 2018; BenYishay and Mobarak 2018; Macours 2019; Shikuku 2019). Our study demonstrates that information dissemination by farmers can be as effective at improving practices as the initial training provided by extension services, which is in line with the study of Nakano et al. (2018a, b). Unlike their study, however, our results show that the entry points to disseminate information are not necessarily the progressive, contact farmers, but can be the ordinary farmers who compose the majority of the rural community. Moreover, unlike BenYishay and Mobarak (2018), who revealed the importance of financial incentives for trained farmers to promote technology diffusion, our results also suggest that social learning can be facilitated by encouragement without any incentive.

The rest of this chapter is organized as follows. Section 3.2 explains the study setting, sampling framework, and experiment design, and examines the summary statistics of our sample. Section 3.3 explains our estimation strategy on the dynamic impact of training and discusses estimation results. Section 3.4 conducts a detailed analysis of the information network and explains the estimation results. Section 3.5 discusses the potentials and limitations of our study while referring to external validity and ethical concerns, and Sect. 3.6 concludes this chapter.

3.2 Survey and Experimental Design

3.2.1 The Study Setting

The study took place in the Bellier and Gbeke regions of Cote d'Ivoire, near the capital city of Yamoussoukro. Like other West African countries, rice is one of the major staple foods in this country and its consumption has exceeded domestic production (Fig. 3.1). The government has tried to increase rice yields to sustain food security and save foreign exchange reserves.

The adoption rate of modern inputs, such as improved seeds and chemical fertilizer, is higher in Cote d'Ivoire than in other rice-growing countries in the SSA. However, several recommended agronomic practices, including straight-row transplanting, that have been proven to boost rice yield in tropical Asia as well as other SSA countries, have not been adopted widely (David and Otsuka 1994; Otsuka and Larson 2013, 2016). There is thus room for management training to improve the performance of rice production.

The two regions, Bellier and Gbeke, were selected to improve domestic rice production and increase the quantity of marketed rice under a bilateral official development assistance (ODA) program between the Ivoirian and Japanese governments. Japanese technical experts were dispatched from 2014 to 2018 through an ODA scheme organized by the Japan International Cooperation Agency (JICA). There are a total of 107 production sites suitable for rice production within those two regions,

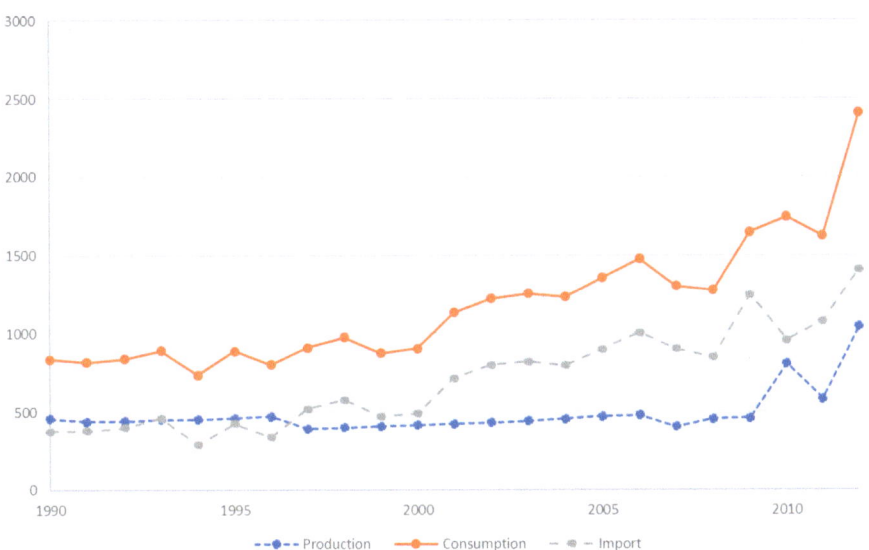

Fig. 3.1 Rice Production, Consumption, and Imports, Cote d'Ivoire (unit 1000 tons). *Note* The authors' calculation from FAOSTAT data. Rice consumption is measured by apparent consumption, that is production plus imports minus exports

which are all located in the lowlands. Some production sites have sufficient access to irrigation water and are able to cultivate rice twice in a good year. Others are in low-humidity zones (called bas-fonds) and are dependent on rainfall. The main rice cultivation season is roughly from July to December. If irrigated, a second cycle starts around January/February. When water is insufficient, farmers produce other crops, such as yams and peanuts, or leave the paddy field to fallow. Since these two regions are agro-climatically more favorable for rice production than other areas of the country, farmers have received various rice cultivation training packages provided by international donors, including JICA, World Bank, and AfricaRice (formerly known as WARDA [West Africa Rice Development Association], whose headquarters is located within Cote d'Ivoire), as well as local extension agencies, including the Agence Nationale d'Appui au Développement Rural (ANADER).

Out of 107 production sites, two were initially selected for the JICA project in 2014. Thereafter, the target area was expanded every year until 2018 to cover a total of 26 sites. This study relies on the data from the eight production sites selected in 2015. To choose our study sites, we closely collaborated with technical experts. Admittedly, the selection of study sites was not completely random because technical experts have a target to cover 1,500 hectares of land within the five-year project period. Thus, the study sites are relatively larger in operational size than the remaining sites in the Bellier and Gbeke regions. Since the impact of training may potentially vary by agro-ecological and institutional conditions, we classified all potential production sites in terms of access to irrigation and the existence of prior rice training. We selected two sites from each combination of with and without irrigation and prior training, generating a sample of eight production sites in total.[2]

3.2.2 Sampling Structure and Experimental Design

Prior to the experiment, we had meetings with farmers belonging to agricultural cooperatives at each selected site.[3] The objective of the meeting was to explain our implementation plan and obtain consent from farmers (see Fig. 3.2 for the timeline, data type, and sample size at each implementation period).

Based on an agreement with technical experts, we outlined our plan to the farmers as follows: (1) We would like to conduct a social experiment to assess the impact of training and ask farmers to cooperate with us; (2) We would group farmers randomly into two, with one eligible to receiving the training offered by JICA experts while the other was expected to apply the best management practice they had access to;

---

[2] Since the facility was generally weak and its capacity was small, heterogeneity was observed in accessibility of irrigation water within the same production site. Furthermore, even if some farmers at a production site received past training, this does not imply that all farmers in that community were eligible and received the same training.

[3] Virtually all rice-growing farmers belong to agricultural cooperatives, which manage the allocation of machinery and current inputs (i.e., seeds and chemical fertilizer) exclusive for cooperative members.

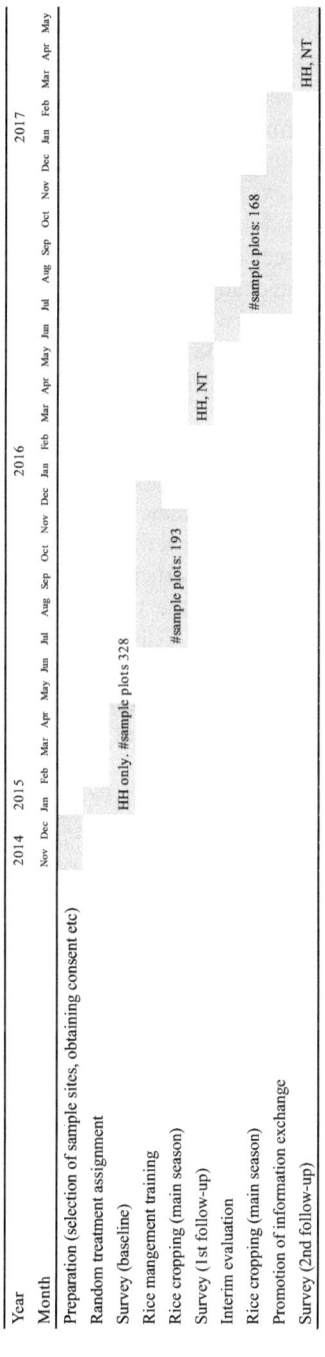

Fig. 3.2 Timeline of Implementation. *Note* HH refers to the household survey, while NT refers to the network survey

(3) All farmers including control farmers were to be provided with the necessary inputs, such as improved seeds, chemical fertilizer, and herbicide, on credit[4]; (4) The experimental phase was to last one year, during which farmers belonging to different groups would be expected not to exchange information about techniques and management practices taught in the training. Specifically, we requested treated farmers not to transmit such information and asked control farmers to refrain from asking treated farmers for agricultural advice; (5) Before and after the experiment, we would conduct household surveys for impact evaluation; (6) If farmers followed our guidance and treated farmers did not transmit information on rice production management, we would acquire precise and valuable knowledge regarding the effectiveness of the technological package taught in the training in their settings; (7) After the impact assessment, we would share which technology (i.e., conventional practice vs. one taught in training) was found to be superior; and (8) After the experimental phase, farmers would be encouraged to share information to facilitate knowledge diffusion.

While unequal treatment during the experimental phase could be a source of tension between treated and control farmers, we attempted to make them feel neither lucky nor unlucky in their treatment status. Rather, we emphasized that once we know which technology is better, everyone can benefit from such knowledge and that the success of this social experiment depends crucially on whether farmers exchange information or not within one year after the training, and also on whether they provide accurate information in the surveys. This sort of explanation seemed to ease tensions. Moreover, despite being uncommon, restrictions on information exchange were accepted by farmers without any revealed complaints once they understood the objectives of the research. Indeed, farmers showed a strong willingness to engage in this experiment.

After obtaining consent, we collected individual member lists from each agricultural cooperative. Out of 414 farmers on the shortlist, 275 households were found to be active rice producers who had cultivated rice at least once in the preceding year. These 275 households constitute the primary sample in this study to whom we assigned eligibility to participate in the training.[5] One half of the sample households were randomly selected as a treatment group and the other half selected as a control group at each site. Randomization was implemented at the farmer level within each site. Everyone in the sample knew not only own treatment status, but also that of their peers. We then conducted the baseline survey with those households from January 2015 to April 2015. The data obtained covered household demographic characteristics, details of rice production on all plots, other household income-generating activities, and household asset holdings.

---

[4] Interest of 3% per season was charged on these inputs, and farmers had to repay the loan to their agricultural cooperation after harvest.

[5] More precisely, 295 farmers were active rice producers in that they have at some time cultivated rice. 16 farmers were dropped from the analysis because of the lack of baseline data about their rice cultivation, although they were also candidates for random assignment and eligible to participate in training if they were in the treatment group. Out of the remaining 279, 4 households were dropped due to a lack of treatment information.

Technical experts provided short classroom training sessions to extension agents of ANADER and three key farmers who were selected from each site.[6] Those extension agents and key farmers in turn offered on-site training to eligible farmers under the supervision of technical experts. This training consisted of (a) land preparation, including bunding and land leveling, which is crucial to reduce the amount of water wasted and to promote the even growth of rice plants, (b) water control, including canal construction and maintenance, which is important in the management of water levels in rice fields during the growth period, (c) seed selection and incubation, (d) fertilizer and herbicide application, (e) transplanting in rows, which can be adopted to facilitate other complementary management practices such as hand or rotary weeding and even the application of fertilizers, herbicides, or insecticides, and (f) harvest and post-harvest management. To mitigate noncompliance, particularly in the participation of the control farmers, local counterparts visited every session of the training and recorded who participated in it. The on-site training proceeded gradually to meet the actual rice cultivation cycle, and, in total, it was held at least six times from June to November 2015 to cover the key practices.[7]

We conducted follow-up surveys twice, the first soon after the training period (March 2016 to May 2016) and, the second two years after training (March 2017 to May 2017), in which detailed data on information exchange across network members were also collected. Because there was a severe lack of rainfall during the 2015–2016 growing seasons, rice cultivation was difficult in those years. Most households could not cultivate rice in the sub-season. Even focusing on the main season, the number of sample plots cultivated for rice dropped sharply to 193 in the midline and further to 168 in the endline survey because of insufficient water. Due to the resulting lack of sufficient observations, we focused on the main season crop in the subsequent analysis and took attrition into consideration where relevant.

3.2.3 Descriptive Statistics and Balancing Test

Table 3.1, Columns (1)–(3) show the balance test on baseline characteristics for the sample household and plots in the main season. 275 households cultivated 328 rice fields in the baseline, of which 135 households and 160 plots belong to the treatment group.

About 72% of treated farmers had attended training at least once, while almost no (only two cases) control farmers had done so, indicating that control farmers largely adhered to our request and did not participate in the training. The attendance rate at

---

[6] These two-step approaches were proposed by JICA experts. Three key farmers selected for direct training by JICA experts were not included in our sample.

[7] A team of technical experts visited each production site frequently to organize the carefully designed on-site training in appropriate and timely agronomic techniques following the agricultural calendar.

Table 3.1 Baseline balance of sample plots by treatment and attrition status

	Treated (1)	Control (2)	Mean difference (3)	Attrition (4)	Non-attrition (5)	Mean difference (6)
Household Charactersitics						
Treatment (=1)	1.000 [0.000]	0.000 [0.000]	1.000	0.478 [0.047]	0.503 [0.040]	−0.025
Ever attend the training (=1)	0.719 [0.039]	0.014 [0.010]]	0.704***			
Attendance rate	0.383 [0.028]	0.002 [0.002]	0.380***			
Head's age (years)	44.014 [1.043]	44.341 [1.159]	0.326	43.611 [1.079]	44.543 [1.080]	−0.933
Head's education (years)	2.892 [0.340]	2.910 [0.323]	0.018	2.496 [0.345]	3.199 [0.314]	−0.703
Head is male (=1)	0.849 [0.030]	0.844 [0.031]	−0.004	0.788 [0.039]	0.889 [0.025]	−0.101**
HH size	8.712 [0.395]	9.000 [0.382]	0.288	2.892 2.892	8.383 [0.337]	1.157**
Number of plots	1.201 [0.034]	1.185 [0.034]	−0.016	1.096 [0.028]	1.262 [0.034]	−0.167***
Log asset value	4.532 [0.100]	4.470 [0.125]	−0.063	4.142 [0.100]	4.755 [0.111]	−0.613***
F-test of joint significance			0.090			5.829***
Number of household observations	135	140		113	162	
Plot Characteristics Plot size (ha)	0.571 [0.046]	0.472 [0.023]	0.099*	0.360 [0.026]	0.618 [0.036]	−0.258***
Owner (=1)	0.756 [0.034]	0.774 [0.032]	−0.018	0.847 [0.032]	0.712 [0.032]	0.135***
Leaseholder (=1)	0.188 [0.031]	0.161 [0.028]	0.027	0.089 [0.026]	0.229 [0.029]	−0.141***
Sharecropper (=1)	0.025 [0.012]	0.018 [0.010]	0.007	0.016 [0.011]	0.024 [0.011]	−0.008
Others (=1)	0.025 [0.012]	0.048 [0.016]	−0.023	0.040 [0.018]	0.034 [0.013]	0.006
F-test of joint significance			1.323			8.363***
Number of plot observations	160	168		124	204	

Standard deviations in brackets. *** p<0.01, ** p<0.05, * p<0.1

the training, a percentage representing how many times out of six key training opportunities each farmer attended, was about 38% among treated farmers. On average, households were large (about nine persons) and headed by a male in their mid-40 s with minimal or no formal education.

We conducted a joint significant test except for the treatment and attendance variables in Column (3), demonstrating that we failed to reject the zero-null hypothesis. This suggests that our randomization was mostly successful.

We present the remaining data using plots rather than households as the unit of observation because some characteristics vary at the plot level and because the main regression analysis is conducted at this level. The data include all rice plots cultivated by sample farmers. The average plot size was relatively small, approximately 0.5 hectare. Although the treatment status was randomized, the difference in the plot size between treatment and control farmers was found to be statistically significant. Most land was operated under owner cultivation. If rented, this was generally a fixed-rent contract. The joint significant test in Column (3) again shows that these plot-level variables are jointly statistically not different by the treatment status.

Columns (4)–(6) of Table 3.1 compare the baseline characteristics of attrition and non-attrition samples with a t-test of the equality of the mean between the two and the associated joint significance F-test. Out of 328 plots, 204 continued rice cultivation in either the midline or the endline year, or both.

While attrition is not sensitive to treatment status, many observable characteristics notably differ between attrition and non-attrition samples[8]: on average, attrition samples were more likely to be female-headed with less education, larger in household size, smaller in the number and size of cultivation plots, and were more likely to own rice plots. The joint significance test shows that the zero-null hypothesis is strictly rejected, implying that attrition is non-random. This non-random sample attrition is a potential threat to causal inference, and should be addressed in the econometric analysis.

Table 3.2 presents the changes in outcome variables of interest regarding the rice management practices and productivity of non-attrition samples over time. We again show the results of t- and F-tests for treated and control plots. In addition, columns (10) and (11) present an unconditional difference-in-differences (DID) regression estimate of the treatment effect (i.e., the difference in the time trend between treated and control plots).

The adoption of recommended management practices was generally quite high even in the baseline (Panel A). Because of its proximity to AfricaRice, adoption of the modern variety of rice was complete and uptake rates had reached 100%.[9] The use of chemical fertilizers was also remarkably high by SSA standards: on average, more than 200 kg/ha of fertilizer, such as NPK and UREA, were applied. In addition

---

[8] Farmers at the irrigation sites were more likely to continue rice cultivation, and therefore, in non-attritors. Attrition was caused partly because a household did not cultivate rice either at the midline or endline, and partly because a household used different plots over time.

[9] The vast majority of farmers use WITA-9, a high-yielding variety that is tolerant to rice yellow mottle virus and iron toxicity and has a maturity period of about 110 days.

Table 3.2 Changes in outcome variables by treatment status: baseline, midline, and endline

	Year 1			Year 2			Year 3			Unconditional DID	
	Treated (1)	Control (2)	Mean difference (3)	Treated (4)	Control (5)	Mean difference (6)	Treated (7)	Control (8)	Mean difference (9)	Year 2–Year 1	Year 3–Year 2
Panel A											
Fertilizer (kg/ha)	214.071 [19.979]	254.340 [32.869]	−40.269	248.822 [15.937]	261.288 [17.609]	−12.466	232.750 [21.745]	255.110 [17.994]	−22.360	27.803 [46.061]	−9.894 [36.572]
Seed selection (=1)	0.906 [0.029]	0.864 [0.034]	0.042	0.929 [0.026]	0.978 [0.015]	−0.050	0.976 [0.017]	0.976 [0.017]	−0.000	−0.092* [0.055]	0.050 [0.040]
Levelling (=1)	0.772 [0.040]	0.791 [0.039]	−0.019	0.857 [0.036]	0.677 [0.049]	0.180***	0.867 [0.037]	0.810 [0.043]	0.058	0.199** [0.081]	−0.122 [0.083]
Canal/drainage construction/repairing (=1)	0.906 [0.028]	0.879 [0.032]	0.027	0.867 [0.034]	0.731 [0.046]	0.136**	0.855 [0.039]	0.929 [0.028]	−0.073	0.109 [0.071]	−0.209*** [0.076]
Transplanting in row (=1)	0.054 [0.021]	0.019 [0.014]	0.035	0.378 [0.049]	0.108 [0.032]	0.270***	0.349 [0.053]	0.179 [0.042]	0.171**	0.235*** [0.063]	−0.099 [0.090]
Panel B											
Rice Yield (ton/ha)	3.440 [0.164]	3.940 [0.174]	−0.499**	4.052 [0.238]	3.671 [0.192]	0.382	3.416 [0.203]	3.724 [0.202]	−0.307	0.881** [0.387]	−0.689 [0.424]
Gross output value (000 CFAF/ha)	603.159 [32.452]	669.393 [31.737]	−66.233	645.433 [37.660]	582.479 [31.545]	62.954	536.090 [31.675]	597.471 [31.760]	−61.380	129.187* [66.960]	−124.334* [67.446]
Rice income (000 CFAF/ha)	405.308 [31.311]	405.091 [32.544]	0.217	413.726 [36.667]	353.458 [28.192]	60.268	232.174 [38.128]	292.198 [32.294]	−60.024	60.051 [64.909]	−120.292* [68.277]
Rice profits (000 CFAF/ha)	331.539 [29.863]	320.196 [34.177]	11.343	243.209 [51.905]	230.344 [32.074]	12.864	108.260 [40.102]	155.150 [31.341]	−46.890	1.522 [76.013]	−59.754 [81.382]

(continued)

Table 3.2 (continued)

	Year 1			Year 2			Year 3			Unconditional DID	
	Treated (1)	Control (2)	Mean difference (3)	Treated (4)	Control (5)	Mean difference (6)	Treated (7)	Control (8)	Mean difference (9)	Year 2–Year 1	Year 3–Year 2
F-test of joint significance			1.516			4.615***			1.777*		
Number of Household Observations	81	81		78	73		65	61		313	288
Number of Plot Observations	101	103		98	93		83	84		395	358

Standard deviations in brackets for mean values, and standard errors in brackets for unconditional difference-in-difference

*** $p<0.01$, ** $p<0.05$, * $p<0.1$

to these external inputs, the adoption rate of improved agronomic practices was mostly high. In our sample, about 90% of plots had selected better seeds by water or winnowing and constructed/repaired water canal/drainage systems in the baseline. It seems that the initial adoption rates were relatively low for leveling fields and greatly so for transplanting in a row, which might be technologies with some room for further improvement from training.

Panel B shows the rice productivity and profitability of sample plots. Gross production value per hectare is computed by multiplying the rice yield (1000 kg/ha) with the price received (CFAF/kg).[10] Rice income per hectare is equal to the gross production value minus paid-out costs, including land rent, irrigation fees, costs of purchased chemicals, and machinery rental, divided by the plot size. Profits per hectare are equal to rice income minus imputed family labor costs, divided by the plot size. To impute family labor costs, we used the typical prevailing hired wage rate for transplanting in each village.[11] The average yield exceeds 3.4 tons/ha which is significantly higher than the average of other countries in SSA of just above 2 tons/ha (Otsuka and Larson 2016). The average gross output value, rice income, and profits per hectare were about 600 thousand CFAF (or approximately 1,065 USD), 405 thousand CFAF (or 719 USD), and 320 thousand CFAF (or 568 USD), respectively.

The table also shows that while there were no statistically significant differences in the baseline adoption rate of recommended practices, treated farmers were more likely to adopt leveling, canal/drainage construction/repairs, and transplanting in rows at the time of the midline survey (Column (6)). Looking at each practice in detail, the increased adoption rate among treated farmers relative to control farmers stems partly from the fact that the control farmers did not continue several management practices, while the treated farmers did. This applies to leveling and canal/drainage construction/repairs. It is likely that insufficient rainfall, which tends to attenuate the impact of those improved practices on productivity and profitability, significantly reduced their adoption among control farmers, holding other things constant.

Meanwhile, control farmers increasingly adopted seed selection and transplanting in rows by the midline survey, and once information sharing was encouraged, the incremental adoption rate of most management practices between the mid- and endline was higher for control farmers (Column (11)). These observations suggest the existence of spillovers not only in the endline, but also the midline despite the restriction of information exchange during that period: indeed, there is room for seed selection and transplanting in rows to spill over because seed selection is relatively easy to imitate, and because transplanting depends largely on hired laborers who can assist both treatment and control farmers. Whether such technology dissemination, if any, significantly alters our view of the impact of training is one of the major issues addressed in later sections of this chapter.

---

[10] 1 USD is equivalent to 563 CFAF as of January 2015.

[11] One can use the different wage rate, such as for land preparation. However, since the land preparation contract is often made simultaneously with rental contract of a machine with an operator, it is difficult to extract only labor costs. On the other hand, our field observations revealed that wage rates for other activities, such as weeding, and harvesting are very close to those for transplanting.

On the other hand, all outcome variables except rice yield were not significantly different between the treated and control samples in the baseline, and no outcomes were significantly different in the midline and endline surveys. The unconditional DID estimates show that treatment plots increase rice yield and revenue between the baseline and midline more than control plots, while the reverse was true between the midline and endline surveys.

These results suggest that while the initial adoption rates of most management practices were already high, the treated farmers further improved or continued their rice management practices and performed better than control farmers in the first year after training. However, control farmers caught up with treated farmers in the following year. While these observations support the existence of positive training impacts in the first year and social learning in the following year, there are some reservations about the descriptive statistics. The following sections examine in more detail whether training has real impact and whether spillovers exist especially between the midline and endline surveys.

3.3 Dynamic Impacts of Training

3.3.1 Estimation Strategy

To identify the causal relationships between the provision of training and outcomes of interest, we estimate intention-to-treat (ITT) and treatment-on-the-treated (TOT) effects. We examine the average impacts of all production sites, allowing the impacts to vary across time. We are particularly interested in whether the training brings intended positive impacts in the first year after the training and whether the gap generated by the experiment decreases over time through spillovers in the next year. Following McKenzie (2012), we employ an analysis of covariance (ANCOVA) model in the form of:

$$Y_{ijt} = \beta_0 + \gamma Y_{ij0} + \beta_1 T_t + \beta_2 D_{ij} + \beta_3 (T_t \times D_{ij}) + X_{ij0}\delta + \mu_j + \varepsilon_{ijt} \quad (3.1)$$

where Y_{ijt} and Y_{ij0} are the post- and pre-treatment outcome variables of plot i in production site j at time t (i.e., either midline or endline) and time 0 (i.e., baseline), respectively; T_t is a dummy variable for the endline data; D_{ij} is a dummy variable equal to one if a household is eligible to participate in the training (ITT estimate) or a continuous variable for the attendance rate of training, instrumented by the treatment status (TOT estimate)[12]; X_{ij0} is a set of baseline control variables; μ_j is the time-invariant fixed effect at the production site; and ε_{ijt} is the unobserved error term. The parameters of interest are β_2 and β_3. The former is expected to capture the

---

[12] Strictly speaking, this is the local average treatment effect (LATE). However, because almost no control farmers attended on-site training, our estimate can be considered to be TOT (Angrist and Pische 2008).

short-term impacts of training under the imposition of the SUTVA, while the latter represents the mixture of the longer-term training impacts and spillover effects when the SUTVA is relaxed. We note that the pure training impact can be estimable only in the short term.

As outcome variables, we focus on the use of chemical fertilizer (kg/ha), the adoption of seed selection by water or winnowing (=1), leveling (=1), canal/drainage construction/repairing (=1), and transplanting in row (=1) as well as rice yield (ton/ha), gross output value ('000CFAF), rice income ('000CFAF), and rice profit ('000CFAF) per hectare. When the outcome is binary, we apply a linear probability model. As baseline control variables, we include household size, household head's characteristics (including age, gender, and years of education), plot characteristics (including parcel size and tenure status dummies), and the logged value of household assets at the baseline survey. We cluster all standard errors within production sites.[13]

While the random assignment of treatment status should make the treatment and control groups similar in all dimensions, the estimated parameters may be biased due to non-random sample attrition. To adjust for that, we use the inverse-probability weighting method suggested by Wooldridge (2010). Specifically, we ran a probit regression to compute the predicted probability of non-attrition at the plot level and used the inverse of this as weights in the main equation. This first-stage probit regression result is presented in Appendix 1.

3.3.2   Estimation Results

Table 3.3 shows the estimation results for the dynamic impacts of management training on rice productivity and profitability with coefficients on control variables suppressed for the sake of brevity. Note that the sample size here is 353, smaller by 8 from the sum of 193 (midline) and 168 (endline) observations due to missing explanatory variables in these few plots.

It is clear that training had positive and significant impacts on rice productivity by the midline, with the rice yield increasing by 0.75 ton/ha, the gross output value per hectare by 140 thousand CFAF, and rice income per hectare by 103 thousand CFAF. These improvements correspond to 20%, 24%, and 29% of the control means, respectively, suggesting that management training was effective in our context.[14] This improvement in productivity, however, did not lead to an increase in profits. As we will see, this might be because trained farmers test a larger number of improved management practices than control farmers, which require more family labor inputs. Qualitatively similar results were observed for TOT estimates. The fact that we

---

[13] Due to the small number of clusters, we also used unadjusted standard errors, assuming no serial correlations and heteroscedasticity. Statistical inference remains robust.

[14] According to experienced agricultural experts, impacts of recommended management practices on rice productivity are generally larger when there is sufficient water. Thus, our estimates could be considered the lower bound of the impacts that would be realized in a year with normal rainfall.

Table 3.3 Estimated results on the dynamic impacts of training: rice productivity

	Rice yield (ton/ha) (1)	Gross output value (000 CFAF/ha) (2)	Rice income (000 CFAF/ha) (3)	Rice profits (000 CFAF/ha) (4)
ITT				
Treatment (=1)	0.748* (0.335)	140.105** (51.348)	102.768* (45.640)	−8.362 (50.222)
×endline	−0.642* (0.279)	−126.704** (44.488)	−52.889 (61.507)	69.781 (70.156)
Endline (=1)	0.179 (0.231)	50.584 (47.283)	−48.217 (53.346)	−106.857** (42.896)
Wald test (Ho: total effect of treatment and its interaction is zero)	0.56	0.26	0.89	2.18
R-squared	0.465	0.418	0.678	0.709
TOT				
Attendance rate (instrumented)	1.453** (0.629)	270.438*** (94.168)	203.150** (81.749)	−15.249 (85.727)
×endline (instrumented)	−1.254** (0.527)	−245.705*** (80.283)	−114.245 (96.329)	123.358 (116.879)
Endline (=1)	0.187 (0.209)	51.485 (43.335)	−47.329 (48.608)	−107.506*** (39.799)
Wald test (Ho: total effect of treatment and its interaction is zero)	0.78	0.37	1.13	2.71*
R-squared	0.462	0.416	0.681	0.710

Sample size is 353. Clustered standard errors at the production site level in parentheses
*** $p < 0.01$, ** $p < 0.05$, * $p < 0.1$
Note Control variables included but not reported here were Year dummy, household size, head's characteristics (age and its square, years of education and the male dummy), plot characteristics (cultivation size, tenure dummies for owner, leaseholder, sharecroppers), log household asset value, and local fixed effect. Attendance rate was instrumented by the treatment dummy, while attendance × endline was instrumented by treatment × endline dummy

found quantitatively larger magnitudes of impacts in TOT than in the ITT estimates suggests that actual training participation rather than simple eligibility is important in the improvement of production performance.

Notably, the coefficient estimates on the interaction term were negative and significant for rice yield and gross output value per hectare. This indicates that the improvement of performance among treatment groups from the mid- to endline was lower than for the control groups. The Wald test shows that we cannot reject the null hypothesis that the total training effect is zero in most specifications, implying that treated farmers are no better than control farmers by the endline. We could interpret this negative interaction term, β_3, as reflecting either the short-lived training effects or the existence of spillover effects. If training impacts do not last long, however, we would have observed some signals, such as a declining adoption rate of improved management practices among treated farmers from the mid- to endline surveys. We did not observe clear dis-adoption patterns among the data presented in Table 3.2. Thus, this finding seems consistent with the operation of a mechanism wherein control farmers improve their performance by learning from treated farmers after the SUTVA is relaxed.

Table 3.4, which shows the estimated impacts of training on the adoption of improved agronomic practices, also provides supportive evidence for the existence of spillovers. When the information exchange between treated and control farmers was restricted during the year after the training, a positive training impact on the adoption of improved management practices, such as leveling, canal/drainage construction/repairing, and straight-row transplanting is observed among treated farmers (ITT estimate) and training participants (TOT estimate). These results are in line with the expectation drawn from Table 3.2 that leveling and straight-row transplanting have a relatively large opportunity to assist improvement.

On the other hand, we did not observe the same positive effects on fertilizer use and seed selection, presumably because there is little room for improvement due to the high initial adoption rates of those practices at the baseline. Also, given that the same amount of fertilizer was provided to both treated and control groups in the experimental phase, and that it is easy for farmers to imitate seed selection techniques, it seems reasonable to observe negligible effects on these practices.

Once the restriction was lifted two years after the training, control farmers caught up with treated farmers in the adoption of recommended practices, as reflected in the negative and significant coefficients on the interaction term, β_3. The Wald tests also revealed that in most outcomes we failed to reject the hypothesis of zero training impact in the longer term.

Note also that, if spillovers exist, the average performance of control groups would improve over time, which should be reflected in β_1 (the endline dummy) > 0. β_1 is positive for most outcome variables and statistically significant for the adoption of canal/drainage construction/repairing and straight-row planting for TOT estimation, further supporting our interpretation in favor of the existence of spillovers.[15]

Taken together, we confirm that, after removing selection bias using the randomized experiment, training has positive impacts in the short term not only on the adoption of improved rice management practices, but also on rice productivity. Spillovers do not exist or at least do not matter much to completely cancel out positive training impact in the initial phase of the experiment. Our further intervention encouraging farmers to spread information, however, improved control farmers'

---

[15] A similar explanation can be offered for outcomes in Table 3.3, although we could not find any statistically positive and significant effects there, presumably because of other time-fixed confounders.

Table 3.4 Estimated results on the dynamic impacts of training: agronomic practice

	Fertilzer (kg/ha) (1)	Seed Selection (2)	Levelling (3)	Canal/drainage (4)	Straight-row planting (5)
ITT					
Treatment (=1)	24.736 (17.358)	−0.033 (0.031)	0.178*** (0.050)	0.119* (0.062)	0.218** (0.067)
× endline	−27.566 (24.593)	0.041** (0.017)	−0.202 (0.134)	−0.236** (0.089)	−0.227* (0.099)
Endline (=1)	0.396 (11.579)	0.002 (0.016)	0.144 (0.104)	0.176 (0.113)	0.162 (0.096)
Wald test (Ho: total effect of treatment and its interaction is zero)	0.03	0.07	0.06	2.28	0.02
R-squared	0.416	0.133	0.313	0.427	0.627
TOT					
Attendance rate (instrumented)	47.437 (30.788)	−0.063 (0.056)	0.340*** (0.084)	0.227* (0.118)	0.417*** (0.095)
× endline (instrumented)	−52.111 (41.586)	0.076*** (0.030)	−0.380* (0.215)	−0.432*** (0.154)	−0.430*** (0.155)
Endline (=1)	0.273 (10.307)	0.002 (0.016)	0.143 (0.095)	0.177* (0.102)	0.161* (0.086)
Wald test (Ho: total effect of treatment and its interaction is zero)	0.04	0.08	0.06	2.81*	0.02
R-squared	0.415	0.129	0.295	0.418	0.628

Sample size is 353. Clustered standard errors at the production site level in parentheses
*** $p < 0.01$, ** $p < 0.05$, * $p < 0.1$
Note Control variables included but not reported here were Year dummy, household size, head's characteristics (age and its square, years of education, and the male dummy), plot characteristics (cultivation size, tenure dummies for owner, leaseholder, sharecroppers), log household asset value, and local fixed effect. Attendance rate was instrumented by the treatment dummy, while attendance endline was instrumented by treatment × endline dummy

practices and contributed to helping them catch up with treated farmers, presumably through spillovers.[16]

---

[16] Although the number of our outcome variables is not so large, one may wonder if we find false positives because we are testing multiple hypotheses. To address this concern, we computed false discovery rate sharpened q-values corrected multiple testing, following the Benjamini-Kreieger-Yekutieli method (Benjamini et al. 2006). All outcome variables that show statistically significant effects in Tables 3.6 and 3.7 remained significant at 10% or lower.

3.4 Spillover Effects

3.4.1 Information Network Analysis

3.4.1.1 Learning Link Data

Having outlined the treatment effects across time, we now examine whether social networks actually mediate information spillovers from treated to control farmers, using the detailed learning link data.

An empirical challenge on this topic is how to correctly specify social networks. Asking respondents about their social networks by arbitrarily setting a cap on the number of links may result in truncation bias, while asking an open-ended question tends to capture only the strong links, ignoring the weaker ones (see, for example, Maertens and Barrett (2013) for a thorough discussion of potential bias in empirically eliciting the true social network structure). To address this concern, we exploited a "random matching within sample" technique to elicit social networks, following, among others, Conley and Udry (2010), Maertens and Barrett (2013), and Mekonnen et al. (2018). More specifically, we matched each sample respondent with six other survey respondents randomly drawn from the sample at the same production sites and asked for details of the (non)existence of information exchange about agronomic practices between samples of farmers. We considered that a learning link between a respondent farmer i and a matched farmer j is established if i has *ever* asked j for advice at some time before interviews. This includes learning even before the training. We did not limit our interviews to only the post-training period, because respondents who violated the no-information-sharing rule, but were eager to satisfy the researchers' expectations, would likely manipulate their answers if they were asked about their behavior only after our intervention. If the reference period includes the non-intervention phase, they are more likely to freely provide the actual answer.[17]

To examine the differential roles played by treatment and control peers, we selected three matches from treated farmers and another three from control farmers. To capture changes in the network of interactions over time, we collected the learning link data in both the mid- and endline surveys.

Table 3.5 presents summary statistics of the learning link data for each year, separately for the probability of sample farmers knowing their match and the probability of sample farmers asking for agricultural information on the match, conditional on the

---

[17] One potential concern from this exercise is that respondents' self-reporting about information exchange patterns may be biased, reflecting their reluctance to tell the truth. Although we could not directly address such concerns, we reduced potential bias by asking about their past experiences rather than those of just the preceding year. We also observed that some cooperatives voluntarily created their own rules to keep control farmers from learning the management practices taught in the training program during the first year. Given that our experiments were executed by close collaboration with farmers to better understand suitable rice management practices in their contexts, we expected that such reporting errors may not be so serious in our study.

Table 3.5 Summary statistics of network data

	Year2 (mid-line)		Year3 (end-line)	
	Pr(Know family)	Pr(Ever ask agricultural advice I Know family)	Pr(Know family)	Pr(Ever ask agricultural advice I Know family)
Both control [Control, Control]	0.761 (0.427)	0.858 (0.349)	0.840 (0.367)	0.923 (0.268)
Both treat [Treat, Treat]	0.723 (0.448)	0.897 (0.304)	0.819 (0.385)	0.947 (0.224)
Own treat, pair control [Treat, Control]	0.769 (0.422)	0.874 (0.332)	0.821 (0.384)	0.975 (0.157)
Own control, pair treat [Control, Treat]	0.696 (0.460)	0.793 (0.406)	0.838 (0.369)	0.955 (0.208)
Number of observations	1664	1015	1607	1081

Standard deviations in parentheses

former knowing the latter. Because of the attrition of own and paired sample households, we had a total of 1,664 and 1,607 observations in the mid- and endline surveys, respectively. These samples included farmers who do not have rice production data in the baseline.

Conditional on sample farmers knowing their matches, about 80 to 90% of farmers asked their match for advice on agronomic practice, such as land preparation, transplanting, and fertilizer application in the midline. These results look high at first glance but seem to be reasonable because they reflect the probability that a sample farmer has *ever* asked matched farmers for agricultural advice. Thus, it may be more useful to focus on the differential probability of information exchange rather than the absolute level.

Compared with Control-Control pairs, Control-Treatment pairs are less likely to ask for advice in the midline, but more likely to do so in the endline. All other pairs have similar trends: they are more eager to ask advice by the endline surveys.

3.4.1.2 Dyadic Regression

We then ran a dyadic regression for those who know their matches to characterize the flow of information about management practices between farmers over time.[18] Formally, let L_{ijt} be equal to one if a respondent farmer i has ever asked a farmer j (conditional on i knows j) for advice by time t. We explore the correlates of learning links by including the attributes of a household i and j as:

---

[18] Using the full sample, including nonacquaintance pairs, did not alter our main findings.

$$L_{ijt} = \delta + \gamma T_t + \alpha_1 D^1_{ij} + \alpha_2 D^2_{ij} + \alpha_3 D^3_{ij}$$
$$+ \beta_1 (D^1_{ij} \times T_t) + \beta_2 (D^2_{ij} \times T_t) + \beta_3 (D^3_{ij} \times T_t)$$
$$+ (X_i + X_j)\rho + (X_i - X_j)\tau + W_{ij}\pi + \varphi + u_{ijt} \qquad (3.2)$$

where D^1_{ij}, D^2_{ij}, and D^3_{ij} are a combination of the treatment status of households i and j with [treated, treated], [treated, control], and [control, treated]. The remaining combination [control, control] is a reference group; T is a binary indicator for the endline survey; X_i and X_j denote a vector of baseline controls for farmers i and j characteristics, respectively[19]; W_{ij} describes a dummy equal to one if the gender of both farmers is the same; φ is the production site fixed effect; and u_{ijt} is a random disturbance. Following Attanasio et al. (2012) and Takahashi et al. (2019a, b), standard errors are clustered at the production site level to allow for possible correlations not only within dyadic pairs but also across all dyads in the same location.

Table 3.6 presents estimated results using a linear probability model. The coefficient estimate on the [treated, treated] dummy, α_1, is positive and statistically significant, but its interaction term with the endline data dummy, β_1, is statistically insignificant. This indicates that information exchange between treated farmers is more active than between control counterparts at the same production site, and this tendency does not systematically change over time. It might be that treated farmers are more likely to exchange information with each other to reinforce agricultural skills taught in the training, although we cannot completely deny the possibility that there was some baseline imbalance in the randomized matching process.

On the other hand, and consistent with our expectations, we observed a negative and significant coefficient on the [control, treated] dummy, α_3, and a positive and significant coefficient on the interaction term with the endline data dummy, β_3. The results suggest that controlled farmers refrained from asking agricultural advice from treated farmers or the latter refrained from disclosing management information to the former in the first year after training, but that they were eager and active in doing so in the second year after training. This indicates that impact evaluation in the initial phase was less likely to be undermined by spillovers, supporting our claim that the recommended practices are more productive. It also supports our main finding that there were information spillovers after the relaxation of the SUTVA in the two years after training, which would facilitate control farmers to improve their rice management practices and performance through social learning.

The insignificant effects of the [treated, control] dummy, α_2, and its interaction with the endline dummy, β_2, are also broadly consistent with our expectations because treatment farmers may have no more incentive to ask control farmers for advice than control farmers do. Considering that our network data were intended to capture the one directional flow of information, non-symmetric results of α_2 and α_3 seem reasonable.

---

[19] If L_{ij} is bidirectional (i.e., $L_{ij} = L_{ji}$), $\beta X_{ij} = \beta X_{ji}$ should be imposed: In such a case, $|X_i - X_j|$ instead of $(X_i - X_j)$ is more relevant as regressors (Fafchamps and Gubert 2007).

Table 3.6 Estimated results on the dyadic regression

	Ask agriculture advice = 1
Both treat [Treat, Treat]	0.045* (0.020)
×endline	−0.018 (0.031)
Own treat, pair control [Treat, Control]	0.019 (0.030)
×endline	0.036 (0.040)
Own control, pair treat [Control, Treat]	−0.063** (0.027)
×endline	0.091** (0.033)
N	2096
R-squared	0.063

Clustered standard errors at the production site level are in parentheses

*** $p < 0.01$, ** $p < 0.05$, * $p < 0.1$

Note Control variables included but not reported here were: The sum and differences of household size, heads' age, head's years of education, cultivation land size, asset values, a dummy equal to one if the household heads are same gender, and local fixed effects

3.4.2 Extension to the Linear-In-Mean Model

While our analysis so far supports the existence of social learning, one may wonder if social learning actually plays a significant role. If this were so, we might observe the influences of peer behavior and performance on one's own behavior. As a final robustness check to verify this possibility, we employed an extended linear-in-mean model.

We restricted the observation of this analysis to the endline year as this reflects the normal condition without the prohibition of information exchange in which spillovers are more likely to take place. We also restricted the outcome variables to rice yield, gross output value per hectare, the adoption of field leveling, canal/drainage construction/repairing, and straight-row planting, for which strong information spillovers from treated to control farmers seem to exist as observed in Tables 3.3 and 3.4.

To disentangle social effects from other confounders, we modify Eq. (1) as follows:

$$Y_i = \gamma_0 + X_{i0}\delta_1 + \overline{X}^N_{-i0}\delta_2 + \gamma_1 Y_{i0} + \gamma_2 \overline{Y}^N_{-i,t-1}$$
$$+ \text{Networ} k_i + \frac{\#\text{Treatment}^N_i}{\#\text{Network}_i} + \mu + \epsilon_i \qquad (3.3)$$

where $\overline{X}^N_{-i0}$ denotes the average values of baseline observable characteristics in i's information network excluding i's own value, regardless of whether network peers are treated or control farmers. For simplicity, we omit the subscript j to denote a production site. As in the previous sub-section, we define network peers as those persons i had asked for agricultural advice by the time of the end-survey. $\overline{X}^N_{-i0}$ are then computed using the baseline data for each network peer's characteristics. This, along with the production site fixed effects, μ, serves to control for environmental and institutional factors that lead farmers to behave in a similar fashion[20]; $\overline{Y}^N_{-i,t-1}$ is the average productivity or technology adoption in i's network at the midline, regardless of their treatment status. This allows us to explore whether their peers' average behavior and performance directly affect farmer i's performance. Following Mekonnen et al. (2018), we use lagged rather than contemporaneous values of mean group performance or behavior in the recognition that information on agricultural technology cannot be diffused quickly; and $Network_i$ is the network size (i.e., max six), while $\frac{\#Treatment^N_i}{\#Network_i}$ is the share of treated farmers in i's information network. We expect the latter to capture peer effects, especially those mediated by treated farmers. This is akin to the methodology used by Kremer and Miguel (2007) and Oster and Thornton (2012). The original intuition behind this method is that once we control for network size (which could be potentially endogenous),[21] the share of network peers in the treatment group is random because of the randomized experiment. This exogenous variation can be then used to identify peer effects.

Note that the average peer performance and the share of treated farmers in i's network are expected to reflect different channels of peer effects; the former may partly capture learning by direct observation even without mouth-to-mouth communication, while the latter may partly capture knowledge transmission from treated farmers even when treated farmers do not actually adopt new technologies.

The estimated results in Table 3.7 show that the share of treated farmers significantly and positively affected their own behavior and performance for gross output value per hectare, leveling, and straight-row planting. We also observed a positive effect on the average performance of network members in most specifications, although this was statistically insignificant except for leveling, perhaps due partly to low statistical power and partly to difficulties in mimicking new technologies without learning through deep communication. While these results should be interpreted with caution to avoid strong causal inference, it seems to be no exaggeration to argue that the results provide further suggestive evidence on the existence of spillovers, especially mediated through treated farmers.[22]

---

[20] As discussed by Manski (1993), impact of social network is generally difficult to identify due to reflection problems. Our estimation method attempts to overcome these problems.

[21] In the random matching within sample method, network size should not be interpreted literally, but rather as a proxy for one's social connectedness where the more random matches a household has, the larger will be their true social network (Murendo et al. 2018).

[22] Although we attempted to minimize concerns about spurious correlation, we are aware of a potential endogeneity issue in this exercise. For example, control farmers who are more motivated, if all else is held constant, may be more willing to establish information links with treated farmers

Table 3.7 Estimated results on the linear-in-mean model

	Rice yield (ton/ha) (1)	Gross output value (000 CFAF/ha) (2)	Levelling (4)	Canal/drainage (5)	Sraight-row planting (3)
The average outcome value (lagged) in network	0.279 (0.197)	0.206 (0.155)	0.273* (0.133)	−0.085 (0.116)	0.079 (0.070)
Network size	−0.107 (0.129)	−12.155 (20.075)	−0.024 (0.017)	0.027* (0.012)	0.025 (0.021)
Share of treatment in network	2.426 (1.312)	494.419* (207.006)	0.537** (0.202)	0.256 (0.224)	0.586** (0.217)
R-squared	0.448	0.386	0.388	0.503	0.737

Sample size is 144. Clustered standard errors at the production site level are in parentheses
*** $p < 0.01$, ** $p < 0.05$, * $p < 0.1$
Note Control variables included but not reported here were: baseline respondent's and average values in respondent's network of household size, head's characteristics (age and its square, years of education and the male dummy), plot characteristics (cultivation size, tenure dummies for owner, leaseholder, sharecroppers), log household asset value, and local fixed effects, as well as the treatment dummy for the respondent

3.5 Discussion

Before concluding this chapter, we must note the several caveats in our study. First, the performance of the rice-growing farmers in our study was better than we expected, due presumably to past training provided by local governmental and international organizations such as AfricaRice. Therefore, many recommended practices were known and practiced by sample farmers even before the training program, except for straight-row transplanting and, to a lesser extent, leveling. This limited our scope, since we could not explore the variations and magnitudes of the spillover effects of different cultivation practices when they are introduced to "virgin land."

Second, while rice production is sensitive to weather conditions, especially rainfall, there was a significant lack of rainfall during the growing seasons in the midline and endline surveys, which resulted in many farmers halting their rice production during our observation periods. Weather conditions cannot be controlled, so we made attempts to mitigate potential estimation bias. Nevertheless, the conclusions may have been more solid and credible if the experiments had been conducted in more ideal settings.

Third, we proposed a new experimental design to implement rigorous impact evaluation and the promotion of spillovers in a unified framework. To achieve the

---

who know the new technique or with peers who actually adopt it. Given the possibility that such interventions can alter the underlying network structure (Advani and Malde 2018), we admit that our constructed variables to capture social learning effects may not be free from endogeneity concerns. Most likely, if anything, our results would underestimate the true effects.

same dual objectives, many existing studies use two-step randomization, in which they randomly select treated and control villages first, and then treat and control individuals within treated villages, allowing spillovers within the treated village. However, this type of cluster-level randomization is often costly, since it requires a larger sample size than individual-level randomization to have sufficient statistical power. We added to the literature by showing an alternative, less costly approach in collaboration with farmers.

One may argue that our approach, especially the restriction of information exchange between farmers, may pose an ethical concern if the training impacts are a priori known to be positive. Although our expectation of training impact was positive as recommended rice practices were mostly established in experimental stations and several countries in SSA have successfully improved rice productivity (Otsuka and Larson 2013, 2016), we were not sure whether that was the case at our study sites. As a result, we felt it was important to evaluate the training impact through an RCT because it is common to observe differences between on-farm and on-station results as well as across countries.

We also wondered whether management practices taught in our training were ineffective for those who had already received similar training in the past or those whose productivity was already close to the production possibility frontier. Thus, we believe that our approach did not conflict with the "do no harm" principle. Rather, once positive impacts were confirmed, we encouraged information exchange between treated and control farmers. By doing so, we were able to successfully reduce any inequalities between treated and control farmers generated by our experiment. It might be valuable to build in this kind of mechanism in other models to allow control groups to catch up with treated ones, which is often overlooked in existing RCTs.

Finally, while we carefully executed an RCT to establish internal validity, we are not fully confident of the external validity of our method. First, our sample farmers were extremely collaborative, which may not always be the case. Second, we provided control farmers with improved seeds, fertilizer, and herbicide on credit. We took this approach because we wanted to isolate the impact of management training from the use of current inputs and also enhance the cooperation of control farmers. But if the provision of credit induces farmers to be more keen to learn technologies, or there is complementarity between current inputs and management practices (e.g., Ragasa and Mazunda 2018), our findings may not be reproduceable in other experiments or scaled-up implementation without this input provision.

3.6 Conclusions

This chapter executed an RCT to examine whether rice production management training has positive impacts on the adoption of recommended practices and productivity in the short term as well as whether social learning can be effective for the wider diffusion of recommended practices by facilitating information spillovers from treated to control farmers in the subsequent period. By using a random assignment of farmer's eligibility for training participation we attempted to reduce selection bias in impact evaluation. Also, by asking farmers not to exchange information in the initial phase of impact evaluation, we attempted to maintain the SUTVA.

We found positive and significant short-term effects from this training, which widen the gap in yield by 20%, the gross output value per hectare by 24%, and the adoption rates of selected rice management practices between treated and control farmers. However, after the restriction on information exchange was removed, control farmers improved their performance significantly, and, as a result, the gap between treatment and control-group farmers becomes virtually zero in the longer term. This suggests that information dissemination by farmers can be as effective in improving practices as the initial training provided by extension services. Although the generalizability of our findings may be questioned, Nakano et al. (2018a, b) found similar results in Tanzania. Our detailed analysis of learning link data and peer effects provides further supportive evidence for the existence of information and technology spillovers.

Our experiment relies on random assignments of management training without any monetary incentive scheme. For future research, it seems to be vital to inquire in different contexts how to best select treated nodes and whether monetary or other incentives should be given to them to maximize social benefit as well as to examine the external validity of our research findings (Kondylis et al. 2017; Maertens 2017; Barrett et al. 2018; Beaman and Dillon 2018; BenYishay and Mobarak 2018; Shikuku 2019).

Appendix 1: Estimation Results for the Non-attrition Probit Model

	Year 2 (1)	Year 3 (2)
Head's age (years)	−0.015	0.024
	(0.043)	(0.035)
Head's age squared (years)	0.000	−0.000
	(0.000)	(0.000)
Head's educatio n (years)	0.037	0.021
	(0.023)	(0.022)
Head is male (=1)	0.570**	−0.177
	(0.283)	(0.271)
HH size	−0.011	−0.011
	(0.021)	(0.021)
Plot size (ha)	0.116	0.641**
	(0.217)	(0.270)

(continued)

(continued)

	Year 2 (1)	Year 3 (2)
Owner (=1)	−0.078	0.191
	(0.376)	(0.381)
Leaseholder (=1)	−0.283	0.314
	(0.412)	(0.410)
Log asset value	0.194***	0.081
	(0.068)	(0.067)
Constant	−0.633	−2.175**
	(1.142)	(1.042)
Production site fixed effects	Yes	Yes

Sample size is 328. Standard errors in parentheses
*** $p < 0.01$, ** $p < 0.05$, * $p < 0.1$

References

Advani A, Malde B (2018) Credibly identifying social effects: accounting for network formation and measurement error. J Econ Survey 32(4):1016–1044

Angrist JD, Pischke J (2008) Mostly harmless econometrics: an empiricist's companion. Princeton University Press, New York

Attanasio O, Barr A, Cardenas JC, Genicot G, Meghir C (2012) Risk pooling, risk preferences, and social networks. Am Econ J Appl Econ 4(2):134–167

Bachewe FN, Berhane G, Minten B, Taffesse AS (2018) Agricultural transformation in Africa? Assessing the evidence in Ethiopia. World Dev 105:286–298

Bandiera O, Rasul I (2006) Social networks and technology adoption in northern Mozambique. Econ J 116(514):869–902

Barrett CB, Christian P, Shimeles A (2018) The processes of structural transformation of African agriculture and rural spaces. World Dev 105:283–285

Beaman L, Dillon A (2018) Diffusion of agricultural information within social networks: evidence on gender inequalities from Mali. J Dev Econ 133:147–161

Benjamini Y, Krieger AM, Yekutieli D (2006) Adaptive linear step-up procedures that control the false discovery rate. Biometrika 93(3):491–507

BenYishay A, Mobarak AM (2018) Social learning and incentive for experimentation and communication. forthcoming in Rev Econ Studies

Conley TG, Udry CR (2010) Learning about a new technology: pineapples in Ghana. Am Econ Rev 100(1):35–69

David CC, Otsuka K (1994) Modern rice technology and income distribution in Asia. Lynne Rienner Publishers, Boulder

Dawson N, Martin A, Sikor T (2016) Green revolution in sub-saharan Africa: implications of imposed innovation for the wellbeing of rural smallholders. World Dev 78:204–218

Evenson RE, Gollin D (2003) Assessing the impact of the green relution, 1960 to 2000. Science 300(5620):758–762

Fafchamps M, Gubert F (2007) The formation of risk sharing networks. J Dev Econ 83(2):326–350

Feder G, Murgai R, Quizon JB (2004) The acquisition and diffusion of knowledge: the case of pest management training in farmer field schools. Indonesia. J Agric Econ 55(2):221–243

Foster AD, Rosenzweig MR (1995) Learning by doing and learning from others: human capital and technical change in agriculture. J Political Econ 103(6):1176–1209

Kondylis F, Mueller V, Zhu J (2017) Seeing is believing? Evidence from an extension network experiment. J Dev Econ 125:1–20

Kremer M, Miguel E (2007) The illusion of sustainability. Quarterly J Econ 132(4):1007–1065

Krishnan P, Patnam M (2014) Neighbors and extension agents in Ethiopia: who matters more for technology adoption? Am J Agric Econ 96(1):308–327

Macours K (2019) Farmers' demand and the traits and diffusion of agricultural innovations in developing countries. forthcoming in Ann Rev Resource Econ

Maertens A (2017) Who cares what others think (or do)? Social learning and social pressures in cotton farming in India. Am J Agric Econ 99(4):988–1007

Maertens A, Barrett CB (2013) Measuring social networks' effects on agricultural technology adoption. Am J Agric Econ 95(2):353–359

Mansk CF (1993) Identification of endogenous social effects: the reflection problem. Rev Econ Studies 60(3):531–542

McKenzie D (2012) Beyond baseline and follow-up: the case for more T in experiments. J Dev Econ 99(2):210–221

Mekonnen DA, Gerber N, Matz JA (2018) Gendered social networks, agricultural innovations, and farm productivity in Ethiopia. World Dev 105:321–335

Munshi K (2004) Social learning in a heterogeneous population: technology diffusion in the Indian green revolution. J Dev Econ 73(1):185–213

Murendo C, Wollni M, Brauw AD, Mugabi N (2018) Social network effects on mobile money adoption in Uganda. J Dev Studies 54(2):327–342

Nakano Y, Tanaka Y, Otsuka K (2018a) Impact of training on the intensification of rice farming: evidence from rain-fed areas in Tanzania. Agric Econ 49(2):193–202

Nakano Y, Tsusaka TW, Aida T, Pede VO (2018b) Is farmer-to-farmer extension effective? The impact of training on technology adoption and rice farming productivity in Tanzania. World Dev 105:336–351

Otsuka K, Larson DF (eds) (2013) An African green revolution: finding ways to boost productivity on small farms. Springer, Dordrecht

Otsuka K, Larson DF (eds) (2016) In pursuit of an African green revolution: views from rice and maize farmers' fields. Springer, Dordrecht

Oster E, Thornton R (2012) Determinants of technology adoption: private value and peer effects in menstrual cup take-up. J Europ Econ Assoc 10(6):1263–1293

Ragasa C, Chapoto A (2017) Limits to green revolution in rice in Africa: the case of Ghana. Land Use Policy 66:304–321

Ragasa C, Mazunda J (2018) The impact of agricultural extension services in the context of a heavily subsidized input system: the case of Malawi. World Dev 105:25–47

Shikuku KM (2019) Information exchange links, knowledge exposure, and adoption of agricultural technologies in northern Uganda. World Dev 115:94–106

Takahashi K, Barrett CB, Ikegami M (2019a) Does index insurance crowd in or crowd out informal risk sharing? Evidence from rural Ethiopia. Am J Agric Econ 101(3):672–691

Takahashi K, Mano Y, Otsuka K (2019b) Learning from experts and peer farmers about rice production: experimental evidence from Cote d'Ivoire. World Dev 122:157–169

Tripp R, Wijeratne M, Piyadasa VH (2005) What should we expect from farmer field schools? A Sri Lanka case study. World Dev 33(10):1705–1720

Wooldridge JM (2010) Econometric analysis of cross section and panel data, 2nd edn. The MIT Press, Cambridge

Kazushi Takahashi is a professor at the National Graduate Institute for Policy Studies (GRIPS) and is a director of the Global Governance Program at GRIPS, Japan. He received a Ph.D. in Development Economics from GRIPS. His scholarly interests include agricultural technology adoption, rural poverty dynamics, microfinance, human capital investment, and aid effectiveness in Asia and sub-Saharan African countries.

Yukichi Mano is a professor at Hitotsubashi University, Japan, and is a fellow at Tokyo Center for Economic Research (TCER). He received a Ph.D. in Economics from the University of Chicago in 2007. His scholarly interests include agricultural technology adoption, horticulture, and high-value crop production, business, and management training (KAIZEN), human capital investment, migration and remittance, and universal health coverage in Asia and sub-Saharan Africa.

Keijiro Otsuka is a professor of development economics at the Graduate School of Economics, Kobe University and a chief senior researcher at the Institute of Developing Economies in Chiba. Japan since 2016. He received a Ph.D. in economics from the University of Chicago in 1979. He majors in Green Revolution, land tenure and land tenancy, natural resource management, poverty reduction, and industrial development in Asia and sub-Saharan Africa.

Chapter 4
The Case of Tanzania: Effectiveness of Management Training on Rice Framing and Farmer-to-Farmer Extension

Yuko Nakano

Abstract This chapter discusses the effectiveness of agricultural training and farmer-to-farmer extension (F2FE) on rice cultivation technologies and productivity in irrigated and rain-fed rice-growing areas. For this purpose, the chapter describes the empirical results of two studies conducted in Tanzania. The first study examines management training for the Modified System of Rice Intensification (MSRI) by assessing its impacts on technology adoption and rice farming productivity in a rain-fed area. The second study focuses on the effectiveness of F2FE training in an irrigated area. For both studies, we consistently find positive impacts of training on technology adoption and productivity of participating farmers. We also observe spillover effects from participants on non-participants in irrigated areas.

4.1 Introduction

Agricultural training has been considered a promising way to diffuse new technologies (Anderson and Feder 2004; Otsuka and Larson 2013; Takahashi et al. 2020). As discussed in Chap. 1, the training is particularly important for rice cultivation, which requires the adoption of improved management practices. Several studies have already found that agricultural training effectively increases technology adoption and the productivity of rice cultivation (deGraft-Johnson et al. 2014; Kijima et al. 2012). However, given that the impact of agricultural training may depend on agro-ecological conditions (Mgendi et al. 2021), it is important to accumulate

This chapter draws heavily on Nakano et al. (2018a, b) with the permission from John Wiley and Sons and Elsevier. We highly appreciate the kind cooperation from TANRICE project team members, including Mr. Motonori Tomitaka, Nobuhito Sekiya, Nobuaki Oizumi of JICA, and the staff of the JICA Tanzania and JICA Ogata Research Institute.

Y. Nakano (✉)
Faculty of Humanities and Social Sciences, University of Tsukuba, 1-1-1, Tennodai, Jinsya Buliding A309, Tsukuba 305-8577, Ibaraki, Japan
e-mail: nakano.yuko.fn@u.tsukuba.ac.jp

© JICA Ogata Sadako Research Institute for Peace and Development 2023
K. Otsuka et al. (eds.), *Rice Green Revolution in Sub-Saharan Africa*, Natural Resource Management and Policy 56, https://doi.org/10.1007/978-981-19-8046-6_4

evidence in different areas to ensure the external validity of the results of previous studies.

As discussed in Chap. 2, since it would be prohibitively expensive for extension workers to train all farmers, it is important to find appropriate methods for diffusing technologies taught to a small number of farmers who then pass them on to non-trained farmers. Considering this, growing attention has recently been paid to the effectiveness of farmer-to-farmer extension (F2FE). The empirical results on the effectiveness of F2FE, however, are still inconclusive. Several studies have shown that the knowledge taught to the lead farmers is adopted and disseminated to other farmers through social learning (Emerick and Dar 2021; Fafchamp et al. 2020; Lee et al. 2019; Morgan et al. 2020; Nakano et al. 2018b Takahashi et al. 2019; Yamada et al. 2015). On the other hand, Kondylis et al. (2017) showed that direct training to lead farmers enhanced the adoption of technologies by lead farmers, while it did not affect the adoption of the surrounding farmers.

The purpose of this chapter is to discuss the effectiveness of agricultural training and F2FE on technology adoption and productivity of rice cultivation. For that purpose, we show the results of two case studies conducted by the author in irrigated and rain- fed areas in Tanzania (Nakano et al. 2018a, b). The first study is on the impact of the Modified System of Rice Intensification (MSRI), provided by a private company called Kilombero Plantation Limited (KPL) in a rain-fed area (Nakano et al. 2018a). The SRI is a set of low-input rice cultivation technologies developed in the 1980s in Madagascar. It is said to produce higher paddy yields by adopting several agronomic practices without additional external inputs.[1] In our study site, the major recommended practices include the use of modern varieties (MVs) called SARO5 and chemical fertilizer, as well as improved agronomic practices. Since these recommended practices differ from the original SRI, which prescribes no MVs or chemical fertilizers, we call this set of recommended technologies the MSRI. This study mainly examined the impact of training on participants and found that training successfully increased the adoption of technologies and paddy yield in rain-fed areas.

The second study is on the effectiveness of TANRICE training, which is a regular F2F training conducted in an irrigated area by the Japan International Coopera-tion Agency (JICA) and the Ministry of Agriculture Training Institute (MATI) of Tanzania (Nakano et al. 2018b).[2] In TANRICE training, intensively trained farmers (designated "key farmers") were responsible for inviting five additional farmers to training sessions at a demonstration plot in the village. The invited farmers were referred to as "intermediate farmers" and were expected to train other non-trained "ordinary farmers" later. We found that there are direct positive impacts of training

---

[1] Recommended practices of SRI include (1) raising seedlings in a carefully managed, garden-like nursery; (2) early transplanting of eight to 15-day-old seedlings; (3) adopting single, widely spaced transplanting; (4) early and regular weeding; (5) carefully controlled water management; and (6) using compost as much as possible, without adopting new varieties or other purchased chemical inputs (Stoop et al. 2002).

[2] The formal name for the TANRICE training program is Technical Cooperation in Supporting Service Delivery Systems of Irrigated Agriculture (TC-SDIA).

on key farmers, and ordinary farmers caught up with key farmers in a few years in terms of technology adoption and paddy yield because of this F2F diffusion system.

In sum, we observed the positive training impact on technology adoption and productivity among participants in both irrigated and rain-fed areas. Furthermore, we observed spillover effects from key to intermediate and ordinary farmers in TANRICE training, which implies the effectiveness of the F2FE program in the irrigated area.

The remaining part of this chapter is organized as follows. Section 4.2 explains the first study on the training of MSRI in a rain-fed area; Sect. 4.3 discusses the effectiveness of TANRICE training in the irrigated area; Sect. 4.4 summarizes our main conclusions and offers some policy implications as well as suggestions for further research.

4.2   The Impact of Direct Training in Favorable Rain-Fed Areas[3]

4.2.1   Study Sites

The first study was conducted in Kilombero district, Morogoro region. While there is no irrigation infrastructure, the study site is in the Kilombero valley, where farmers enjoy plenty of rainfall and thus, can be classified as a favorable rain-fed area. A private company called Kilombero Plantation Limited (KPL) provided training on rice cultivation technologies to surrounding farmers. KPL operated a large-scale plantation of approximately 5,000 ha and a rice miller as their primary business. KPL offers extension services at the request of the Tanzanian government, which was responding to the complaints of neighboring farmers that a single large company cultivates such a huge area.

The recommended practices in our study site include (1) use of MV called SARO5, (2) chemical fertilizer use, (3) seed selection in salty water, (4) straight-row dibbling[4] or transplanting, and (5) spacing of 25 cm by 25 cm or more. In addition, the use of a rotary weeder and dibbling or transplanting one to two seeds or seedlings per hole or hill was suggested. However, the adoption rates of the latter two technologies are generally low. As a result, only one key component of SRI technologies is adopted in our study site: wide spacing of 25 cm by 25 cm. Furthermore, the recommended practices are substantially different from the original SRI, requiring no MVs or chemical fertilizers. Thus, we call the recommended technologies in our study site MSRI technologies.

The SRI office, established as a section of KPL, is in charge of extension services to the local farmers. They trained 25 farmers in a village in 2010 and expanded

---

[3] This section mostly relies on Nakano et al. (2018a).

[4] Dibbling is a method of sewing. Farmers make small holes in the ground and sew seeds (not seedlings).

their extension service to an additional 1,350 farmers in 2011, 2,850 farmers in 2012, and 2,250 farmers in 2013. The extension services provided by the SRI office are financially supported by the United States Agency for International Development (USAID) and operate in 10 surrounding villages. When they start the training program, officers call for a village meeting and ask those interested in training to form a group of 25 farmers. The criteria for the participants are that they must be residents of the villages, must be farmers, and must not have been trained by the SRI office before. A group of participants has to provide a quarter-acre piece of land called a demo plot. The extension officers, qualified agronomists hired by KPL and USAID, provide training on the demo plot during the cultivation season. During the training, each participant is provided with 26 kg of chemical fertilizer and 4 kg of seeds of SARO5, which are recommended amounts for a quarter acre. Each farmer is recommended to cultivate a quarter acre of his or her own land following the technology and management practices taught by the training program.

One year after receiving training, instead of free modern inputs, trainees are eligible to receive in-kind credit of chemical fertilizer and seed from NGOs associated with the SRI office. Farmers are obliged to repay part of the loan every two weeks during the cultivation season for five months, resulting in 10 installments. In addition, farmers need to sell six bags (approximately 600 kg) of paddy at the agreed price to KPL at the time of harvest so that KPL can repay its remaining balance to the lending NGOs. However, this credit service has not been popular among farmers. First, it is difficult for the farmers to repay the loan every two weeks during the cultivating season, as they generate most of their cash income at harvest time. Furthermore, there is sometimes disagreement over the selling price of the rice between farmers and KPL due to a fluctuation of the market price of paddy in the harvesting season. Only 11 households out of 25 eligible farmers received the loan from NGOs associated with KPL in 2013.

4.2.2 Data Collection

Data collection was carried out from February to March 2014 and covered the cultivation season from October 2012 to May 2013. To examine the impact of the training program, we selected three villages where training was held (henceforth referred to as training villages). We also covered two nearby villages where no training was held (we refer to these as non-training villages). Training villages and non-training villages were adjacent and in a similar agro-ecological condition. In each training village, we interviewed on average 37 training participants and 35 non-participants. In addition, we interviewed on average 35 farmers per village in non-training villages, generating a total sample size of 283 households.

We asked farmers to list all of their farming plots during the interviews. Among those listed, we selected two paddy plots for plot-level analysis. In our study sites, trainees differentiated between plots where they adopted MSRI technologies (called

MSRI plots) and plots where they did not adopt these technologies (called non-MSRI plots). Presumably, because the technologies are newly introduced, farmers cultivate at most one plot using MSRI technologies. Thus, for sample farmers who have attended MSRI training, we automatically selected the MSRI plot and selected at most one more plot randomly where rice is grown using traditional cultivation methods.[5] For sample farmers who have not attended MSRI training, we randomly selected up to two plots where rice is grown. Since farmers do not necessarily adopt all the MSRI technologies even on MSRI plots, we investigate the adoption rate of each component of MSRI technologies.

The number of sample plots for the cultivation season of 2013 was 406 cultivated by 283 households. After dropping households and plots with missing values in crucial variables, the total sample size became 398 plots of 281 households. Note that a significant number of farmers cultivate only one plot either by MSRI or traditional methods. We also collected recall data on paddy yield and the adoption of critical technologies from 2010 to 2013 to construct a panel data set before and after the training. Our panel data sample size was 398 plots for four years, generating a total sample size of 1351 plots. Note that our sample is unbalanced because some farmers do not grow rice in some years.[6]

Out of 110 training participants in our sample, no farmers were trained before 2011, 25 farmers were trained in the main season of 2012, and 85 farmers were trained in 2013. This implies that 25 farmers trained in 2012 and 85 trainees in 2013 received the free inputs for a quarter acre from KPL in 2012 and 2013, respectively, while 25 trainees in 2012 were eligible for the KPL credit program in 2013. Since trainees in 2012 and 2013 received different support (i.e., credit and free inputs) from KPL in 2013, we name the trainees from 2012 early trainees and those from 2013 late trainees. In the following analyses, we differentiate the training impacts on each group. Note that the impact of the MSRI training in 2013 for early trainees partially includes the impact of the credit service. Regardless of this, it is difficult to distinguish the effects of training and credit statistically.

4.2.3   Descriptive Analyses

Table 4.1 compares the adoption of modern inputs and improved practices—including straight-row dibbling or transplanting, wide spacing, and seed selection in salty water—separately by early trainees, late trainees, non-trainees in training

---

[5] There are seven MSRI trainees who did not declare any MSRI plots. In these cases, we randomly chose one plot as the MSRI plot in order not to overestimate the impact of training on technology adoption and productivity. Even if we count them as non-MSRI plots, however, the main results do not change.

[6] We confirmed that the main results of our analyses remain unchanged even when we use the balanced panel data by omitting those households who did not cultivate rice in any single year within the 4 years under study.

villages, and farmers in non-training villages in 2013. The most important observation is that trainees, regardless of their training year, achieved an average yield as high as 4.7 tons/ha on their MSRI plots in 2012 and 2013. This yield is remarkably high compared to the average yield of 2.9 tons/ha on the trainees' non-MSRI plots and 2.6 tons/ha for the non-trainees in the training village.

Importantly, we do not observe a significant difference between yields before training (2010–2011) on the MSRI plots (2.6 tons per ha) and the non-MSRI plots of trainees (2.7 tons per ha). This suggests that farmers do not necessarily select plots of good quality to apply MSRI technologies. The high yield of trainees on MSRI plots may be attributed to the high adoption rate of new technologies on these plots. Of the five MSRI technologies,[7] SRI trainees adopt 3.7 of them on average on their MSRI plots but only 0.3 on their non-MSRI plots. However, there is some variation in the adoption rate of each technology: 90.9% for MV, 78.2% for straight-row dibbling, 56.4% for wide spacing, and 71.8% for seed selection in salty water. Trainees apply much more chemical fertilizer (52.4 kg/ha) on their MSRI plots than on their non-MSRI plots (6.1 kg/ha). Note also that there is no significant difference in the performance between early and late trainees in 2012 and 2013 (Table 4.1, columns b-c). This suggests that the high yield and high rate of technology adoption are not due to the free inputs distributed only for the 2013 trainees in our survey year. In fact, 2012 trainees achieve as high a yield and technology adoption rate as the 2013 trainees without receiving the free inputs.

Lastly, the yield of the non-MSRI plots of trainees (2.9 tons/ha) is slightly higher than that of non-trainees (2.6 tons/ha). However, the adoption rate of technologies on the non-MSRI plots of trainees is not significantly higher than those of non-trainees in the training villages (columns d-g), except for slightly higher chemical fertilizer use and seed selection in salty water. Furthermore, non-trainees in the training village do not show higher yield or technology adoption rates than farmers in the non-training villages (columns g-h). These observations suggest that spillover effects from the MSRI plots to the non-MSRI plots of trainees and from trainees to non-trainees are limited, at least during our observation periods. Whether MSRI technologies further diffuse to non-MSRI plots of the trainees and all of the plots of non-trainees over a longer period is an important remaining issue.

4.2.4 Estimation Methods

To investigate the impact of MSRI training on the adoption of rice cultivation technologies and paddy yield, we estimate the difference-in-differences (DID) model by using recall panel data (Imbens and Wooldridge 2009). The dependent variables are paddy yield (tons/ha) and technology adoption using a dummy variable that takes 1

---

[7] Among the five technologies listed on page 6, farmers adopt either straight-row transplanting or straight-row dibbling. Since it is important to differentiate between dibbling and transplanting, we show six technologies here.

Table 4.1 Yield and technology adoption in the sampled rice plots in 2013 by Modified System of Rice Intensification (MSRI) training participation

	Training village							Non-training village					
	Trainee						Non-trainee						
	MSRI plot in 2013			Non-MSRI plot in 2013									
	Average	2012 trainee	2013 trainee	Average	2012 trainee	2013 trainee							
	(a)	(b)	(c)	(d)	(e)	(f)	(g)	(h)	a–d	b–c	d–g	g–h	a–h
Paddy yield (tons/ha)	4.7	4.7	4.7	2.9	3.1	2.8	2.6	2.9	1.8****	0.0	0.3**	−0.4**	1.8***
Average paddy yield before training (2010–2011) (tons/ha)	2.6	2.7	2.5	2.7	2.8	2.6	2.3	2.3	−0.1	0.1	0.4**	0.0	0.3*
MSRI adoption in 2013													
Share of modern variety (%)	90.9	88.0	91.8	10.1	10.0	10.2	5.6	2.4	80.7***	3.8	4.6	3.1	88.5***
Chemical fertilizer use (kg/ha)	52.4	57.9	50.8	6.1	9.1	5.1	2.5	2.5	46.3***	7.1	3.6*	0.0	49.9***
Share of straight-row dibbling (%)	78.2	80.0	77.6	0.0	0.0	0.0	0.8	2.4	78.2***	−2.3	−0.8	−1.6	75.8***
Share of straight-row transplanting (%)	7.3	8.0	7.1	0.0	0.0	0.0	0.8	1.2	7.3***	−0.9	−0.8	−0.4	6.1**

(continued)

Table 4.1 (continued)

	Training village							Non-training village					
	Trainee						Non-trainee						
	MSRI plot in 2013			Non-MSRI plot in 2013									
	Average	2012 trainee	2013 trainee	Average	2012 trainee	2013 trainee							
	(a)	(b)	(c)	(d)	(e)	(f)	(g)	(h)	a–d	b–c	d–g	g–h	a–h
Share of plots adopting spacing of 25 cm x 25 cm or more (%)	56.4	60.0	55.3	0.0	0.0	0.0	1.6	2.4	56.36***	4.7	−1.6	−0.8	54.0***
Seed selection by salty water (%)	71.8	68.0	72.9	3.8	10.0	1.7	0.0	1.2	68.0***	4.9	3.8**	−1.2	70.6***
Number of technologies adopted	3.7	3.7	3.7	0.3	0.4	0.2	0.2	0.1	3.5***	0.0	0.1**	0.0	3.6***
Size of cultivated area (ha)	0.4	0.5	0.4	1.1	1.4	1.0	1.0	1.2	−0.7***	−0.1*	1.4	−0.2**	−0.8***
Observations	110	25	85	79	20	59	126	83					

Source Nakano et al. (2018a) *** denotes significant at 1%, ** significant at 5%, and * significant at 10% in t-test comparison between the labeled categories

if a farmer adopts MVs, dibbling or transplanting in rows, a recommended spacing of 25 cm by 25 cm, or seed selection in salty water, and chemical fertilizer use (kg/ha). The base model is:

$$Y_{ijt} = \tau \left(MSRI_{ij} * trainee_i * post\text{-}training_{it} \right) + \delta \left(trainee_i * post\text{-}training_{it} \right)$$
$$+ \eta trainee_i + \lambda MSRI_{ij} + \theta year_t + p_{ij} + u_{ijt}, \tag{4.1}$$

where Y_{ijt} is the outcome variable of individual i's plot j at time t; $MSRI_{ij}$ is the time-invariant dummy variable that takes 1 if the plot is an MSRI plot in any single year; $trainee_i$ is a dummy variable that takes 1 if the cultivator of the plot is either an early or late trainee; $post\text{-}training_{it}$ is a $year_t$ 2012 dummy for early trainees and year 2013 dummy for early and late trainees; year is a year dummy; p_{ij} is a plot- and household-specific time-invariant characteristics, and u_{ijt} is an error term.

We estimate the DID model controlling for plot-level fixed effects. By doing so, we attempt to control for plot- and household-specific, time-invariant characteristics (p_{ij}) that may affect a farmer's endogenous selection of the MSRI plot and innate household characteristics. Note that in the basic DID model, the terms of the time-invariant training status dummy for $trainee_i$ and $MSRI_{ij}$ are included. In our case, however, these terms are absorbed by p_{ij}, as we control for the plot fixed effects.

The most important independent variable in the model is the interaction term of the MSRI plot, trainee, and post-training dummies. This interaction term is intended to measure changes in the outcome values induced by the MSRI training. Note that the MSRI plot dummy is a time-invariant variable that takes 1 if the plot is an MSRI plot in any single year. We include post-training dummies (i.e., the year 2012 dummy for early trainees and the year 2013 dummy for both early and late trainees) since trainees have attended the training by the indicated years. As discussed earlier, farmers receive free input during the training period and are eligible for the credit program in subsequent seasons. Thus, to examine the differential impact of the training in 2012 and 2013, we constructed three interaction terms representing, respectively, (1) the interaction of the MSRI plot, the early trainees, and the year 2012 dummies; (2) the interaction of the MSRI plot, early trainee, and the year 2013 dummies; and (3) the interaction of the MSRI plot, the late trainee, and the year 2013 dummies. Since only three households discontinued their adoption of MSRI technologies after their initial adoption, we consider the coefficient τ to estimate the impact of the training on the MSRI plots relative to non-MSRI plots of the trainees.

We also include interaction terms for the trainee and the post-training dummies. The coefficient δ was expected to capture the impact of the training on productivity and technology adoption in the non-MSRI plots of trainees due to their labor real-location from non-MSRI plots to MSRI plots or the positive spillover effect of the training on the non-MSRI plots of the trainees. Again, to estimate the impact of training on the trainees in 2012 and 2013 separately, we include the interaction term of the 2012 trainee and year 2012 dummies, that of the 2012 trainee and year 2013 dummies, and that of the 2013 trainee and year 2013 dummies. We also include year dummies to capture the effects of general trends, including non-trainees in both training and non-training villages.

4.2.5 Estimation Results

Table 4.2 presents DID estimation results for the training impact on paddy yield and technology adoption with plot fixed effects. All three interaction terms of the MSRI plot, trainee, and post-training dummies have positive and significant coefficients in all the regressions except for the chemical fertilizer use of the 2012 trainees in 2012. This suggests significant effects of MSRI training on paddy yield and improved technology adoption. Compared to the non-MSRI plots of trainees, adoption rates are higher on the MSRI plots to the order of 50 to 90 percentage point, even after we control for plot fixed effects. Trainees also increased their chemical fertilizer application by 20–27 kg/ha on their MSRI plots in 2013.

As a result of the high adoption rate of the new technologies, trainees' paddy yield increases by 1.3–1.4 tons/ha on their MSRI plots. This result strongly indicates that rice cultivation training has a significant yield-enhancing effect even under rain-fed conditions. Note that trainees in 2012 received free inputs in 2012 and trainees in 2013 received the same in 2013, while trainees in 2012 were eligible for the credit program instead of receiving free inputs in 2013. However, for paddy yield, the difference between the estimated coefficient of the interaction term for the MSRI plot, the year 2012 trainee, and year 2012 dummies (indicated as a) and that for the MSRI plot, the year 2012 trainee, and year 2013 dummies (indicated as b) is not statistically significant, as the F-test statistics shown in Table 4.2 indicate. This result implies that the training was effective even after KPL stopped providing free inputs to the trainees.

The interaction terms of the trainee and the post-training dummies, which capture the impact of the training on productivity and technology adoption on trainees' non-MSRI plots compared to non-trainee plots, have negative coefficients for some technologies such as recommended spacing and positive coefficients for chemical fertilizer use. This suggests that the adoption rates of labor-intensive technologies such as wide spacing may be lower due to the increased labor requirement on MSRI plots, while there is some positive spillover effect on the adoption of chemical fertilizer. However, the coefficients are insignificant for paddy yield, suggesting no significant training impact on the productivity of non-MSRI plots of the trainees relative to non-trainees' plots.

Table 4.2 Difference-in-differences estimates of impact of Modified System of Rice Intensification (MSRI) training on yield and technology adoption in 2010–2013 (Plot fixed effect—Unbalanced panel)

	Paddy yield (tons/ha)	Modern variety (=1)	Chemical fertilizer (kg/ha)	Dibbling/transplanting in rows (=1)	Recommended spacing (=1)	Seed selection in salty water (=1)
MSRI plot x 2012Trainee x year 2012 dummy (a)	1.386**	0.630***	8.289	0.313*	0.559***	0.579***
	(0.593)	(0.113)	(17.057)	(0.173)	(0.122)	(0.118)
MSRI plot x 2012Trainee x year 2013 dummy (b)	1.379***	0.821***	19.952*	0.861***	0.682***	0.625***
	(0.526)	(0.085)	(10.327)	(0.078)	(0.106)	(0.125)
MSRI plot x 2013Trainee x year 2013 dummy (c)	1.279***	0.766***	27.077***	0.862***	0.536***	0.634***
	(0.354)	(0.060)	(6.243)	(0.046)	(0.072)	(0.077)
Effect on non-MSRI plot of trainees						
2012Trainee x year 2012 dummy	−0.154	−0.032**	23.208**	0.247*	−0.013	−0.022**
	(0.397)	(0.016)	(9.629)	(0.126)	(0.009)	(0.011)
2012Trainee x year 2013 dummy	−0.299	0.031	4.961	−0.013	−0.018**	0.079
	(0.354)	(0.042)	(3.982)	(0.011)	(0.008)	(0.062)
2013Trainee x year 2013 dummy	−0.349	0.077*	0.582	−0.018**	−0.016**	0.016
	(0.249)	(0.041)	(1.209)	(0.008)	(0.008)	(0.021)
Year dummy	Included	Included	Included	Included	Included	Included
Constant	2.482***	0.007	−0.002	−0.000	0.001	0.001
	(0.040)	(0.007)	(0.417)	(0.005)	(0.006)	(0.005)

(continued)

Table 4.2 (continued)

	Paddy yield (tons/ha)	Modern variety (=1)	Chemical fertilizer (kg/ha)	Dibbling/transplanting in rows (=1)	Recommended spacing (=1)	Seed selection in salty water (=1)
Observations	1,326	1,351	1,305	1,351	1,351	1,351
R-squared	0.218	0.656	0.306	0.723	0.502	0.548
Equality of coefficients (F-statistics)						
(a) = (b)	0.00	2.99*	0.57	11.11***	1.71	0.09
(a) & (b) = (c)	0.04	0.13	0.77	5.57**	0.46	0.06

Source Nakano et al. (2018a) Base year is 2010. ***p < 0.01, **p < 0.05, *p < 0.1. Clustered standard errors at plot level in parentheses

4.3   The Impact of Farmer-to-Farmer Training in Irrigated Areas[8]

4.3.1   Study Site and Data

Having established the positive impact of agricultural training on rice productivity, this section moves on to the question of whether F2FE can complement direct training by extension workers. It specifically relies on the case of agricultural training on rice production technologies conducted by JICA in the Ilonga irrigation scheme in the Kilosa district, Morogoro region of Tanzania before and during the main crop season from November 2008 to May 2009 (hereafter, this particular crop season will be referred to as the 2009 crop season). The irrigation scheme is located nearly 15 km from the nearest town of Kilosa. The program, called the TANRICE training, covered several technologies: the use of MVs of rice, the application of chemical fertilizer, improved bund construction, plot leveling, and transplanting in rows. Improved bund construction entails piling soil solidly around the plots, while plot leveling involves flattening the ground for better storage and uniform water distribution on paddy fields. Transplanting seedlings in rows allows rice growers to control plant density precisely and remove weeds easily.

Intensive training was offered to 20 farmers, called key farmers, at the nearby training institute (MATI Ilonga) over 12 days in November 2008, prior to the 2009 crop season. Subsequently, during the 2009 main crop season, each key farmer was requested to invite five intermediate farmers to training sessions held at a demonstration plot within the irrigation scheme. The key farmers and MATI jointly provided three-day training sessions to the intermediate farmers at three different stages of farming—nursery preparation, transplanting, and harvesting. Following these "in-field training" sessions, key and intermediate farmers were expected to disseminate technologies to the remaining farmers (i.e., the ordinary farmers). One day of in-field training was open to all the farmers in the scheme, including the ordinary farmers. The key farmers were competent and leading farmers selected by MATI based on such criteria as age, literacy, gender balance, residence within the irrigation scheme, and the practice of rice farming and were confirmed at an all-villagers meeting. The intermediate farmers were selected personally by the key farmers with no formal involvement by MATI. Thus, the selection of the key and intermediate farmers was rather purposive. Neither the key nor the intermediate farmers were paid for attending the training.

---

[8] This section largely depends on Nakano et al. (2018b).

4.3.2   Data

Three rounds of the annual survey were implemented in 2010, 2011, and 2012. In the first survey, we interviewed 208 randomly selected farmers from the farmer roster in the irrigation scheme. We asked the respondents to identify the most important rice plot for their livelihood, hereafter referred to as the farmer's "sample plot." Farmers were asked in detail about rice cultivation on their sample plot, including their use of labor, capital, and other inputs in 2010. Similar information on the sample plot was also collected for the 2011 and 2012 crop seasons. During the first survey in 2010, we collected recall data on rice cultivation on the sample plot for the 2008 and 2009 main crop seasons, before and during the TANRICE training, respectively.

In the main analysis, we dropped households that took erroneous values in important variables and those that did not grow rice on the sample plot. The attrited households from the survey interviewed in the first round but not found in the second and third rounds were also omitted. This resulted in 171 observations for 2008, 182 for 2009, 202 for 2010, 168 for 2011, and 167 for 2012. We estimated an attrition probit model[9] and confirmed that the attrition had occurred randomly concerning the observed set of variables. This result implies that analysis using the available observations (i.e., balanced and unbalanced panel data) will not suffer serious attrition bias. Thus, we use unbalanced panel data with more information due to the larger sample size.

4.3.3   Descriptive Analyses

Table 4.3 presents the changes in the average paddy yields and the technology adoption by the key, intermediate, and ordinary farmers from 2008 to 2012. The t-tests and χ^2-tests comparing the key and intermediate farmers to the ordinary farmers are also shown. Note that the TANRICE training was conducted immediately before and during the 2009 main crop season. The table shows that even before the training (i.e., in 2008), the key farmers attained a slightly higher yield than the ordinary farmers, presumably due to the higher adoption rates of improved technologies and some innate abilities for rice production. Due to their increased technology adoption rates, the key farmers' yield increased immediately after the training, from 3.1 tons/ha in the pre-training year 2008 to 4.4 tons/ha in 2009. They continued to achieve higher yields than the ordinary farmers, reaching 5.3 tons/ha in 2011 and 4.7 tons/ha in 2012. The key farmers' adoption rates for modern varieties, improved bund construction, transplanting in rows, and chemical fertilizer use also rapidly increased in 2009 and

---

[9] When households drop out of the sample over time for some non-random reasons, it can cause a bias in the estimates, known as attrition bias. In order to examine if sample attrition occurs randomly, we estimated an attrition probit model. The dependent variable is a dummy variable which takes one if the household remains in the sample and the independent variables are the set of observed variables used in the main equation.

remained significantly higher than the ordinary farmers until 2012, contributing to a high yield each year.

In contrast, the change in yield from the 2008 base year for the intermediate farmers is not as rapid as that of the key farmers. Soon after receiving the training during the 2009 season, the intermediate farmers' technology adoption rates, including MVs, improved bund, and transplanting in rows, began increasing, eventually boosting the yield to a significantly higher level than the ordinary farmers in 2011. These results indicate that, although the effect of the training—both in terms of magnitude and speed—is more significant for the key farmers than for the intermediate farmers, the intermediate farmers also caught up with the key farmers in the years following the training.

It is remarkable to observe that the paddy yield of the ordinary farmers also rose from 2.6 tons/ha in 2008 to 3.7 tons/ha in 2012, even though the change was neither rapid nor drastic compared with the key and intermediate farmers. This increment can be attributed to the increased use of chemical fertilizers and improved agronomic practices. The belated yet significant technological changes seen in the behaviors of the ordinary farmers indicate that technologies taught in the TANRICE training spilled over from the key and intermediate farmers to the ordinary farmers over the years. The yield gap between the key and ordinary farmers ranged from 1.7 to 2.3 tons/ha between 2009 and 2011, while it diminished to one ton per hectare in 2012. These results suggest that the key farmers' performance improved rapidly after the training. In contrast, the ordinary and intermediate farmers' performance improved by learning from the key farmers, resulting in a smaller gap in yield and technology adoption in later years.

4.3.4 Estimation Methods

To evaluate the effects of TANRICE training on the adoption of rice cultivation technologies and paddy yield, we estimate fixed effect difference-in-differences (FE-DID) models with multiple periods and multiple treatment groups (Imbens and Wooldridge 2009; Meyer 1995) using our five-year panel data. In FE-DID, we use the panel structure of our data set to control for unobservable time-invariant household-specific characteristics that may influence training participation and the trends in the outcomes. Namely, the following econometric model is estimated:

$$Y_{it} = \alpha + \beta T_t + \gamma T_t S_i + C_i + u_{it} \qquad (4.2)$$

The dependent variables are paddy yield (tons per hectare) and the following set of technology adoption variables: a dummy variable for MV adoption, the amount of chemical fertilizer use (kg per hectare), and dummy variables for the adoption of improved bund construction, plot leveling, and transplanting in rows, respectively, in the sample plot; T_t is a vector of four-year dummies in year t, with the base year of 2008; S_i is a vector of two training status dummies (i.e., key farmer and intermediate

Table 4.3 Changes in paddy yield and technology adoption by training status for TANRICE training in irrigated area (key and intermediary farmers)

	2008	2009	2010	2011	2012
	Pre-training	During training	Post-training		
Key farmer					
Paddy yield (tons/ha)	3.07* [1.37]	4.40*** [1.32]	4.81*** [1.43]	5.34*** [2.36]	4.67** [2.43]
Adoption rate of MVs (%)	46.15 [51.89]	69.23*** [48.04]	75.00*** [44.72]	54.44*** [46.92]	66.67*** [47.14]
Chemical fertilizer use (kg/ha)	63.42 [71.81]	115.82*** [86.07]	137.73*** [74.45]	178.26*** [89.52]	131.28*** [67.07]
Adoption rate of improved bund (%)	15.38** [37.55]	23.08** [43.29]	31.25*** [47.87]	40.00** [50.71]	15.38 [37.55]
Adoption rate of plot leveling (%)	46.15 [51.89]	76.92 [43.85]	81.25 [40.31]	86.67 [35.19]	76.92 [43.85]
Adoption rate of transplanting in rows (%)	23.08 [43.85]	76.92*** [43.85]	93.75*** [25.00]	93.33*** [25.82]	92.31*** [27.74]
Observations	13	13	16	15	13
Intermediary farmers					
Paddy yield (tons/ha)	2.47 [1.13]	2.57 [1.39]	2.84 [1.39]	4.63*** [2.40]	3.93 [2.15]
Adoption rate of MVs (%)	30.43 [47.05]	44.44* [50.64]	54.84** [50.59]	34.38 [46.52]	49.48** [48.97]
Chemical fertilizer use (kg/ha)	22.20** [34.86]	49.00 [41.31]	79.05 [50.44]	103.85** [63.94]	95.23 [58.63]
Adoption rate of improved bund (%)	13.04** [34.44]	18.52** [39.58]	22.58** [42.50]	33.33** [48.15]	33.33*** [48.15]
Adoption rate of plot leveling (%)	43.48 [50.69]	70.37 [46.53]	74.19 [44.48]	79.17 [41.49]	62.50 [49.45]
Adoption rate of transplanting in rows (%)	13.04 [34.44]	44.44*** [50.64]	64.52*** [48.64]	45.83** [50.90]	58.33** [50.36]
Observations	23	27	31	24	31
Ordinary farmers					
Paddy yield (tons/ha)	2.57 [1.34]	2.67 [1.41]	2.53 [1.36]	3.58 [1.70]	3.67 [2.00]
Adoption rate of MVs (%)	26.67 [44.39]	26.76 [44.43]	32.26 [46.90]	23.62 [39.33]	32.85 [44.04]
Chemical fertilizer use (kg/ha)	46.52 [54.63]	58.31 [62.95]	69.72 [67.59]	85.79 [59.49]	83.16 [61.57]
Adoption rate of improved bund (%)	2.96 [17.02]	4.93 [21.73]	7.74 [26.81]	16.15 [36.95]	11.54 [32.07]

(continued)

Table 4.3 (continued)

	2008	2009	2010	2011	2012
	Pre-training	During training	Post-training		
Adoption rate of plot leveling (%)	54.81 [49.95]	64.08 [48.15]	69.03 [46.39]	76.15 [42.78]	66.92 [47.23]
Adoption rate of transplanting in rows (%)	11.11 [31.54]	19.01 [39.38]	25.81 [43.90]	26.92 [44.53]	36.92 [48.45]
Observations	135	142	155	130	130
Annual rainfall (mm)	1027.4	869.2	917.3	1546.9	651.1
Rainfall during the main season (mm)	980.9	925.7	966.6	1326.0	783.6

Source Nakano et al. (2018b)
Standard deviations in brackets. *** Statistically significant at 1%, ** 5%, and * 10% in *t*-test comparisons of key and intermediary farmers to ordinary farmers (i.e., paddy yield, and chemical fertilizer use) or in a *chi-square* test in the case of dummy variables (i.e., adoption rates of MVs, improved bund, plot leveling, and transplanting in rows, respectively)

farmer dummies, with their base group being ordinary farmers); $T_t S_i$ is a vector of all pairwise interactions between T_t and S_i; C_i is the time-invariant household-specific effect for household i and u_{it} is the stochastic error term. Since we sample one plot for each household, we use subscript i for the outcome variable while it is measured at the plot level. Note that in basic DID models, the terms of time-invariant training status dummies S_i are included. In our case, however, this term is absorbed by C_i, as we control for the household fixed effects.

Years 2009 to 2012 are all post-treatment years, while 2008 is pre-treatment. Thus, coefficients γ associated with the interaction between the year dummies and training status dummies are the DID estimates of interest to capture the gap in the training effects between the trained (key and intermediate) farmers and the ordinary farmers. The strength of this model is that the term C_i absorbs the unobservable time-invariant household characteristics, which are likely to affect training participation. In other words, a potential selection bias is largely addressed. The year-specific effects represented by β capture the changes in the outcome variables for the ordinary farmers. These year dummies are assumed to capture the indirect effects of training on ordinary farmers through knowledge spillover from the trained farmers and other year-specific characteristics such as weather.

It is important to note that, in our case, γ the vector of DID estimators should not be interpreted as the "pure" training impact, i.e., the difference in growth between the factual and counterfactual situations for the key and intermediate farmers with and without the training intervention, respectively. Instead, γ is designed to capture the differences between the effect on the key and intermediate farmers and the effect on the ordinary farmers, since β captures the changes in the performance of the ordinary farmers, which incorporates the spillover from the key and intermediate farmers. Thus, as the ordinary farmers catch up with the key and intermediate farmers, γ is expected to become smaller. In Nakano et al. (2018b), we also estimate a similar model by using the Propensity Score Matching-DID method. The results are largely the same for both cases.

4.3.5 Results

Table 4.4 presents the results of the FE-DID estimation on paddy yield and technology adoption. The year fixed effects are positive and significant in 2009–2012 for the use of chemical fertilizer and the adoption of plot leveling and transplanting in rows; in 2010–12, for improved bund construction; and in 2011–2012, for paddy yield. The adoption of MVs also increased in 2012. This indicates a steady increase in the technology adoption and paddy yield for the ordinary farmers, suggesting positive spillover effects of training over time.[10]

For paddy yield, the DID estimates are significant for the key farmers in 2009 and 2010, indicating that training for the key farmers took immediate effect. Our results suggest that the training impact on key farmers' yield is larger by 1.2–1.7 tons/ha than on ordinary farmers' yield in 2009 and 2010. This rapid increase in paddy yield can be predominantly attributed to the fast technology adoption by the key farmers. The increase in chemical fertilizer use by key farmers relative to ordinary farmers is 41.8 kg in 2009, 56.3 kg in 2010, and 78.0 kg in 2011. The increase in the adoption rate of transplanting in rows is also steadily higher for key farmers from 2009 to 2012.

A more striking finding is the absence of significant yield effects captured by the interaction terms between the key farmer dummy and the 2011 and 2012 dummies. This suggests that the key farmers' "yield premium" disappeared by 2011 and 2012. Given that the performance of the ordinary farmers was steadily improving from 2010 to 2012, this suggests that the ordinary farmers caught up with the key farmers presumably because of knowledge spillover from the key and intermediate farmers to ordinary farmers.

We do not observe significant coefficients for the interaction terms of intermediate and year dummies for paddy yield. On the other hand, these coefficients are positive and significant for chemical fertilizer use in 2010 and 2011 and transplanting in rows in 2009 and 2010. The increase in chemical fertilizer use by intermediate farmers is larger than that of ordinary farmers by 22.4 kg in 2009 and 28.5 kg in 2010. The training impact on the adoption rate of transplanting in rows is also larger for intermediate farmers, by 21% in 2009 and 34% in 2010 (vs. ordinary farmers). These results imply that intermediate farmers adopted new technologies more rapidly than ordinary farmers, although their productivity increase is no faster than those of ordinary farmers. Furthermore, consistent with the previous observation, we find

---

[10] It is also important to note that this steady increase in yield and technology adoption was not solely attributed to rainfall conditions, since the annual rainfall in 2008 was not lower than in the following years.

Table 4.4 Estimation results of difference-in-differences with household fixed effects models for paddy yield (tons/ha) and technology adoption for TANRICE training in irrigated areas. (2008–12)

	(1)	(2)	(3)	(4)	(5)	(6)
	Paddy yield (tons/ha)	MVs	Chemical fertilizer use (kg/ha)	Improved bund	Plot leveling	Transplanting in rows
Key * 2009	1.214*** [0.355]	0.233* [0.121]	41.706** [18.514]	0.057 [0.076]	0.204 [0.133]	0.480*** [0.179]
Key * 2010	1.662*** [0.459]	0.173 [0.161]	56.250*** [20.969]	0.057 [0.080]	0.158 [0.134]	0.594*** [0.129]
Key * 2011	1.135 [0.743]	0.138 [0.190]	78.019** [33.401]	0.047 [0.188]	0.099 [0.204]	0.596*** [0.137]
Key * 2012	0.280 [0.651]	0.028 [0.175]	36.415 [26.750]	−0.143 [0.147]	0.147 [0.181]	0.460*** [0.142]
Intermediary * 2009	−0.226 [0.243]	0.079 [0.088]	3.640 [9.963]	0.030 [0.050]	0.145 [0.097]	0.213** [0.094]
Intermediary * 2010	0.141 [0.329]	0.095 [0.116]	22.391* [12.319]	0.014 [0.054]	0.111 [0.102]	0.340*** [0.105]
Intermediary * 2011	0.843 [0.525]	−0.026 [0.117]	28.446* [14.794]	−0.006 [0.147]	0.100 [0.133]	0.139 [0.117]
Intermediary * 2012	−0.108 [0.574]	0.037 [0.148]	17.926 [15.020]	0.093 [0.149]	0.023 [0.134]	0.173 [0.119]
Year 2009	0.119 [0.084]	−0.002 [0.027]	10.691*** [3.801]	0.020 [0.016]	0.104*** [0.032]	0.059** [0.024]
Year 2010	−0.085 [0.107]	0.049 [0.036]	19.169*** [4.874]	0.030** [0.015]	0.150*** [0.034]	0.122*** [0.033]
Year 2011	0.973*** [0.153]	0.022 [0.039]	36.652*** [5.617]	0.111*** [0.032]	0.232*** [0.052]	0.119*** [0.039]
Year 2012	1.101*** [0.178]	0.130*** [0.049]	34.333*** [6.152]	0.056* [0.031]	0.133** [0.061]	0.233*** [0.046]
Constant	2.653*** [0.073]	0.305*** [0.023]	47.778*** [2.943]	0.075*** [0.013]	0.524*** [0.025]	0.146*** [0.021]
Observations	891	891	891	891	891	891
R-squared	0.194	0.034	0.164	0.031	0.059	0.144
Number of households	202	202	202	202	202	202

Source Nakano et al. (2018b)

2008–2012 unbalanced panel. Standard error clustered at household level in brackets. ***$p < 0.01$, **$p < 0.05$, *$p < 0.1$. Base year is 2008

that ordinary farmers gradually catch up with intermediate farmers in terms of the technology adoption, and significant differences between them tend to disappear in 2012 for chemical fertilizer use and in 2011 and thereafter for transplanting in rows. Although results are not shown, we obtained consistent results even when we controlled for the participation in other programs, including fertilizer subsidy and credit program, suggesting the robustness of our results.

4.4 Conclusion

This chapter shows two case studies on the impact of management training on the participants and that of F2FE on non-participants in Tanzania. The first notable finding is that in both cases, trained farmers achieved as high as 4.7 to 5.3 tons/ha, which is high even compared to the Asian standard. This suggests the high potential of the African rice Green Revolution, or we can even say that Green Revolution has taken place in limited areas. In the first study, we examined the impact of MSRI training provided by a private company in a rain-fed area of Kilombero district in Tanzania. We found that the training effectively enhances the technology adoption and increases the paddy yield of trainees by 1.4–2.5 tons/ha. As a result, the farmers who apply recommended MSRI technologies achieve as high a yield as 4.7 tons/ha on average. Given that the average yield of non-trainees in the study sites is 2.6 tons/ha, this is a remarkably high yield.

In the second study, we investigated the impact of F2FE in an irrigated area. We found that participants of the training increased the adoption of recommended technologies, and their paddy yields increased significantly. In addition, the performance of non-participants also improves later, and these farmers catch up with the participants. This suggests that F2FE effectively induced spillover effects from participants to non-participants in the irrigated area. Nakano et al. (2018b) further revealed that, by using spatial econometric techniques, the social relationships between key and ordinary and between intermediate and ordinary farmers play a significant role in the adoption of technologies by ordinary farmers.

In sum, we observed the positive impact of training on technology adoption and productivity of training participants in both cases. Our results are consistent with previous studies that found a positive training impact on rice productivity in both irrigated and rain-fed areas in other African countries (deGraft-Johnson et al. 2014; Kijima et al. 2012; Takahashi et al. 2019). We also found that F2FE successfully improves the performance of non-participants, possibly due to spillover effects from participants of the training, especially in irrigated areas, which is also consistent with Takahashi et al. (2019; and Chap. 3 in this volume), who found positive spillover effects of rice cultivation training in irrigated and rain-fed areas in Côte d'Ivoire.

By contrast, we did not find strong evidence of spillover effects in rain-fed areas. The limitation of the study in the rain-fed area is that the survey was conducted soon after the training. Since it may take some time for the spillover effects to be observed, it is still not conclusive that spillovers can occur in rain-fed areas. Our study in an irrigated area also observed that the spillover from key farmers to ordinary farmers gradually took place. Further investigation is needed on the differential impact of agricultural training and F2F extensions in different agro-ecological conditions in the mid- and long-term.

References

Anderson JR, Feder G (2004) Agricultural extension: good intentions and hard realities. World Bank Res Obs 19:41–60

deGraft-Johnson M, Suzuki A, Sakurai T, Otsuka K (2014) On the transferability of the Asian rice green revolution to rainfed areas in sub-Saharan Africa: an assessment of technology intervention in Northern Ghana. Agric Econ 45(5):555–570

Emerick K, Dar MH (2021) Farmer field days and demonstrator selection for increasing technology adoption. Rev Econ Stat 103(4):680–693

Fafchamps M, Islam A, Malek MA, Pakrashi D (2020) Can referral improve targeting? Evidence from an agricultural training experiment. J Dev Econ 144:102436

Imbens GW, Wooldridge JM (2009) Recent developments in the econometrics of program evaluation. J Econ Lit 47(1):5–86

Kijima Y, Ito Y, Otsuka K (2012) Assessing the impact of training on lowland rice productivity in an African setting: evidence from Uganda. World Dev 40(8):1610–1618

Kondylis F, Mueller V, Zhu J (2017) Seeing is believing? Evidence from an extension network experiment. J Dev Econ 125:1–20

Lee G, Suzuki A, Nam VH (2019) Effect of network-based targeting on the diffusion of good aquaculture practices among shrimp producers in Vietnam. World Dev 124:104641

Meyer BD (1995) Natural and quasi-experiments in economics. J Bus Econ Stat 13(2):151–161

Mgendi G, Mao S, Qiao F (2021) Is a training program sufficient to improve the smallholder farmers' productivity in Africa? Empirical evidence from a Chinese agricultural technology demonstration center in Tanzania. Sustain 13(3):1527

Morgan SN, Mason NM, Maredia MK (2020) Lead-farmer extension and smallholder valuation of new agricultural technologies in Tanzania. Food Policy 97:101955

Nakano Y, Tanaka Y, Otsuka K (2018a) Impact of training on the intensification of rice farming: evidence from rainfed areas in Tanzania. Agric Econ 49(2):193–202

Nakano Y, Tsusaka TW, Aida T, Pede VO (2018b) Is farmer-to-farmer extension effective? The impact of training on technology adoption and rice farming productivity in Tanzania. World Dev 105:336–351

Otsuka K, Larson DF (2013) An African green revolution: finding ways to boost productivity on small farms. Springer, Dordrecht

Stoop WA, Uphoff N, Kassam A (2002) A review of agricultural research issues raised by the system of rice intensification (SRI) from Madagascar: opportunities for improving farming systems for resource-poor farmers. Agric Syst 71:249–274

Takahashi K, Mano Y, Otsuka K (2019) Learning from experts and peer farmers about rice production: experimental evidence from Côte d'Ivoire. World Dev 122:157–169

Takahashi K, Muraoka R, Otsuka K (2020) Technology adoption, impact, and extension in developing countries' agriculture: a review of the recent literature. Agric Econ 51(1):31–45

Yamada H, Shimamoto D, Wakano A (2015) Importance of informal training for the spread of agricultural technologies: farmers as in-residence extension workers and their motivation for sustainable development. Sustain Dev 23(2):124–134

Yuko Nakano is associate professor at Facluty of Humanities ans Social Science, University of Tsukuba, Japan. She received Ph.D. degree in Development Economics in 2009 from National Graduate Institute for Policy Studies, Japan. Her specialization is development and agricultural economics.

Chapter 5
The Case of Uganda: Long-Term and Spillover Effects of Rice Production Training

Yoko Kijima

Abstract Using the case of rice production training in the rainfed lowlands of Eastern Uganda, this chapter examines the extent to which training continues to enhance participants' technology adoption and productivity five years after the provision of training and the extent to which the training effect spills over to non-training participants. Rice production data was collected from training participants and non-participants in program villages and rice farmers in non-program villages one year before and one year and five years after the training. According to descriptive statistics, the gap in the average rice yield between the training participants and non-participants within a program village opens up right after the training, but it disappears in the long term. To identify program and spillover effects, propensity score matching and difference-in-differences method were used (PSM–DID). This study finds that training enhanced adoption rates for improved cultivation practices not only in the short term but also long term, while rice yield increased only in the long term. Although the adoption rate of improved cultivation practices did not increase among non-participants in training villages relative to their counterparts in non-program villages, rice yield increased after five years, which suggests signs of spillover within training villages in the long term.

5.1 Introduction

This chapter examines the long-term and spillover effects of rice cultivation training on technology adoption and rice productivity in rainfed-lowland production areas of Eastern Uganda. In Uganda, rainfed-lowland areas underpin the main rice production system, accounting for 52% of the production area and 58% of total rice production in 2018 (estimated) (MAAIF 2009). As shown in Chaps. 3 and 4, the information spillover from training participants to non-participants is likely to happen within

Y. Kijima (✉)
Graduate School of Policy Studies, National Graduate Institute for Policy Studies (GRIPS), 7-22-1 Roppongi, Room 1221, Minato-ku, Tokyo 106-8677, Japan
e-mail: kijima@grips.ac.jp

© JICA Ogata Sadako Research Institute for Peace and Development 2023
K. Otsuka et al. (eds.), *Rice Green Revolution in Sub-Saharan Africa*, Natural Resource Management and Policy 56, https://doi.org/10.1007/978-981-19-8046-6_5

irrigation schemes where rice farmers regularly interact for canal cleaning and maintenance organized by a water user association. However, it is common for farmers from different villages to rent plots in the same rainfed-lowland production area and plant rice without any coordination.[1] Given such differences in farmers' interactions between rainfed-lowland and irrigated areas, it is not obvious to what extent the information imparted by the training is shared among training participants and non-participants in rainfed-lowland areas. As discussed in Chap. 2, no study has identified the spillover effects of rice cultivation training in the rainfed-lowland rice production system in sub-Saharan Africa (SSA).

In a rainfed-lowland production system, farmers cannot control water availability (quantity and timing) well. Insufficient water during the critical growth period of rice plants is likely to depreciate the returns to recommended cultivation practices, potentially leading to the disadoption of such practices. Thus, there is no guarantee that the training effect on participants' technology adoption and productivity persists in the long term, particularly in rainfed areas. While project sustainability is crucial, long-term impacts are rarely assessed, primarily because of data limitations. In addition, the short-term impact may not capture the spillover effects in rainfed areas where the learning speed is expected to be slow if it occurs at all. In the context of upland non-rice crops, Kondylis et al. (2017) analyzed the effect of training for sustainable land management in the Zambezi Valley of Mozambique and found that only training participants adopted the practices, and there was no evidence suggesting that the training changed practices among non-participants. Since their endline survey was conducted two years after the training, the process of spillover might not have occurred sufficiently before the endline survey was conducted.

As discussed in Chap. 2, program evaluation without random assignment presents a methodological challenge, as program participants and non-participants are not usually comparable. I adopt the propensity score matching and difference-in-differences method (PSM–DID) to address potential selection bias. Furthermore, the difference between training participants and non-participants within program villages includes both the direct training effect and spillover effect, and each effect cannot be identified separately. In other words, when non-participants seem to catch up with training participants, this can be explained by decreasing direct training effects on participants and positive spillover effects on non-participants. To address this identification problem, I measure the direct training effect and spillover effect separately by using a comparison group outside program villages, arguably not affected by the training program or spillovers. Training participants are compared to similar rice farmers residing in non-program villages to assess the direct impact of the training. Training non-participants in program villages are compared with similar rice farmers in non-program villages to measure the spillover effects.

Section 5.2 describes the data collected, the study areas, and the rice training program implemented in Uganda. Section 5.3 explains the empirical framework, and the results are discussed in Sect. 5.4. Section 5.5 summarizes this chapter.

---

[1] In some cases, the plot size in rainfed-lowland areas is larger than the average plot size in the irrigation scheme areas.

5.2 Data, Study Area, and Rice Training

This chapter takes the case of a lowland rice farming training project implemented by the Japan International Cooperation Agency (JICA) and Uganda's Ministry of Agriculture under a sustainable irrigated agricultural development (SIAD) project in Eastern Uganda.[2] This project provides training in lowland rice cultivation practices based on experiences in Asia designed to enhance rice production and productivity by introducing sustainable rice cultivation practices. Such practices have been widely adopted in Asia but are not commonly employed in Uganda. The study area covered the Eastern region of Uganda in the second cropping season of 2009: two districts where the training was provided and five districts where the training was not offered.[3] JICA experts selected one lowland area as a project site in each district.

As the training aims at improving rice cultivation practices in rainfed lowlands, site selection was not random but targeted lowland areas with seasonal or year-round streams. Once such ecological conditions were met, JICA experts approached the rice growers and asked about their interest in participating in a training project. Once their interest was confirmed, JICA experts asked the farmers to form a group to ease communication. Lowland areas are approximately 20–30 ha and are cultivated by 90–150 rice growers in 7–11 different villages. In these villages, not all households grow rice. In the uplands, maize, cassava, and beans are mainly cultivated. Rice cultivation started around the early 2000s.

Field training was offered in a demonstration plot of each site following a cropping calendar on a learning-by-doing basis with simple explanations using flip charts to ensure that participants understood the contents. Primary trainers in the field training are local extension workers who took three-day training sessions on rice cultivation provided by JICA at the National Crops Resources Research Institute, a national agricultural research organization in Uganda. Firstly, in the 1–2 months before the planting season, training participants learned how to establish a demonstration plot followed by the trainer's instruction and prepared the demo plot for field training. Secondly, 2–3 weeks before the planting time, training participants prepared nursery beds and grew seedlings, constructed bunds around the demo plot, and leveled the demo plot. Thirdly, the improved transplanting method (straight-line planting) and weeding (timings and method) were taught and practiced in the field. Finally, trainees harvested rice in the demo plot and learned improved threshing techniques with a simple threshing device. The first training took 2–3 days while the other training took half to one day. Since the application of chemical fertilizer was not a part of JICA training, chemical fertilizer was not provided during the training. At the time of the project, there was no lowland rice variety in Uganda that was officially

---

[2] Please see Kijima (2022) for more details of the project.

[3] In these five districts, there are comparable rainfed-lowland areas. In selecting these districts, we considered different rice cultivation experiences.

recommended by the government. Thus, improved variety seeds were not given to participants during the training.[4]

The baseline survey was conducted in August 2009, before the training started in September 2009, and it collected information about farming activities from August 2008 to July 2009. The first follow-up survey was conducted in September 2011, gathering information about rice production from September 2010 to August 2011. Because the training period lasted until March 2010, this first follow-up survey captures the program's immediate impact. The second follow-up survey was conducted in September 2015, likely capturing the long-term training effect.

The sampling scheme differed between program villages and non-program villages. In each program site, sample households were randomly selected every 25 m based on the distance from the demonstration plot to their own rice plots. The total number of sample households in the program villages is 150. The share of training participants in each project site is different. For sampling households in non-program villages, the first five districts where the training had not yet been provided by 2009 were selected. In each district, two sub-counties with rice production and access to rainfed-lowland areas were selected. In each subcounty, six villages were randomly selected. In each village, ten households were randomly selected.

As shown in Table 5.1, the pre-program characteristics of training participants and non-participants within the program villages are similar. The means of all these variables are not statistically different between participants and non-participants in the program villages and between participants and control households in the non-program villages. The only difference found is that participants' rice plots are closer to the demonstration plot than non-participants. Although the training program was not assigned randomly, this table suggests that training participants are neither more educated/experienced nor more connected with other community members. Other than these observed characteristics, time and risk preferences are not statistically different, at least on average.[5] This finding demonstrates that the project did not select training participants based on their characteristics. Furthermore, households in non-project villages are comparable in terms of these pre-program characteristics.

Figure 5.1 presents the adoption rate of selected rice cultivation practices and rice yield per hectare separately for training participants (participant), non-participants in the program villages (non-participant), and households in the non-program villages

---

[4] Lowland rice seeds tend to be self-produced by farmers and locally traded among farmers. There are two popular lowland varieties: the first is comprised of modern rice varieties crossed with local varieties and a popular variety called "K5," "K85," or "Kaiso," which was developed initially for the Kibimba Rice Scheme. The other is a local variety called "Supar" (meaning rice), which has been widely adopted in the lowland areas of Eastern Uganda, as well as in Tanzania. While K5 had its origins as one of the early Asian modern varieties, the origin of Supar is less clear.

[5] Time and risk preference measures are obtained from hypothetical questions. The household takes 1 for patience if it prefers to wait for 30 days to receive 10,000 shillings rather than a lower amount today. The hypothetical lottery (coin toss) offers five choices with different expected values: (a) Sh. 50,000 (heads) and Sh. 50,000 (tails), (b) Sh. 40,000 (heads) and Sh. 100,000 (tails), (c) Sh. 30,000 (heads) and Sh. 130,000 (tails), (d) Sh. 20,000 (heads) and Sh. 160,000 (tails), (e) Sh. 10,000 (heads) and Sh. 190,000 (tails). Risk averse is defined as a household selecting choice (a), while a household is considered to be risk loving if taking choice (e).

Table 5.1 Selected household characteristics in 2009 by training participation status

	Participants in JICA training	Non-participants in training	Non-JICA training villages
Rice experience in years	11.24 (8.10)	11.00 (9.78)	10.58 (9.56)
HH head's age	39.60 (12.38)	40.27 (13.82)	42.82 (11.87)
Head's years of education	6.069 (3.999)	5.854 (3.513)	6.083 (3.164)
Num. of HH members	8.517 (3.516)	8.089 (3.735)	8.569 (3.944)
Share of males aged 15–64	0.223 (0.109)	0.244 (0.167)	0.252 (0.136)
Share of females aged 15–64	0.235 (0.108)	0.266 (0.169)	0.231 (0.103)
Size of land owned (ha)	2.139 (1.725)	1.987 (1.703)	2.059 (5.227)
Share of lowland size owned	0.204 (0.319)	0.231 (0.282)	0.200 (0.315)
Local group member (dummy)	0.724 (0.451)	0.646 (0.481)	0.511 (0.501)
Own a bull (dummy)	0.466 (0.503)	0.519 (0.503)	0.330 (0.471)
Has patience (dummy)	0.741 (0.442)	0.737 (0.443)	0.666 (0.473)
Risk averse (dummy)	0.466 (0.503)	0.329 (0.473)	0.359 (0.481)
Distance to demo plot (km)	0.651 (0.552)	1.313* (0.482)	–
Population density per squared km (village level)	0.604 (0.350)	0.577 (0.332)	0.539 (0.367)
Number of observations	58	79	327

Source Authors' calculations. Figures are means, and numbers in parentheses are standard deviations
* Indicates mean differences between training participants and non-participants at a 5% significance level

(control) over the survey years. As Panel A shows, the transplanting method, rather than direct seeding, was relatively common in the training villages (about 65% of the rice growers) even before the training project. However, straight-row transplanting was not adopted before the program (Panel B). After the training program in 2011, the participants' adoption rates of transplanting and transplanting in rows jumped to 80 and 20%, respectively. In contrast, adoption rates did not change in the short term among the non-participants in the training villages and farmers in the control villages. The non-participants' adoption of transplanting increased in 2015 to reach

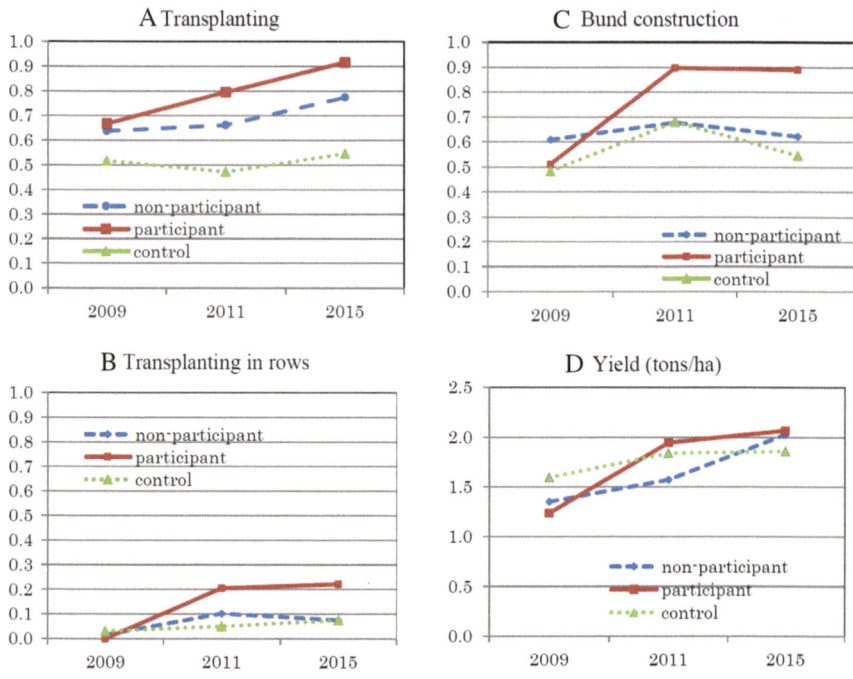

Fig. 5.1 Adoption of rice cultivation practices and productivity over time and training category

nearly 80%, while that of control households remained low. Training participants increased the adoption of transplanting to 90% by 2015, whereas the adoption of transplanting in rows stagnated after 2011. Similarly, training participants increased bund construction from 50 to 90% in the short term and maintained the same level until 2015. Although about 60% of non-participants constructed bunds in 2009, no further increase was observed.

Regarding rice yield, Panel D shows an interesting pattern. For both participants and non-participants, rice yield was about 1.3 tons per ha in 2009. After the training, it increased significantly to 2 tons per ha for participants, while non-participants experienced a moderate increase to 1.5 tons per ha in 2011. However, the yield gap disappeared in 2015. Notice that the speed of yield enhancement among participants slowed down after 2011, consistent with the stable adoption of the transplanting in rows method and bunds after 2011. In the case of non-participants, yield enhancement in 2015 may be due to the shift from broadcast planting methods to transplanting, among other factors. The emergence of a substantial yield gap between participants and non-participants in 2011 and its disappearance in 2015 suggests that non-participants learned new production methods from participants with a time lag. In other words, information spillover is likely to take place relatively slowly in rainfed areas.

5.3 Empirical Framework

This chapter first estimates the average training impact on the adoption of cultivation practices and rice yield (the average treatment effect on the treated, ATT). To estimate ATT, we need to obtain the counterfactual outcome of the training participants had they not participated in the training.

There are two empirical issues for estimating the training impact on adopting cultivation practices and rice yield in the current setting. The first empirical issue is the non-random assignment of training. Although the descriptive statistics show that participants and non-participants had similar observed characteristics before the training, unobserved characteristics may be different, and hence, they may have affected the training participation and outcomes. The second empirical issue is the knowledge spillover from participants to non-participants within the program village, potentially violating the stable unit treatment values assumption (SUTVA) necessary for appropriate program evaluation (Imbens and Rubin 2015).

To address the non-random program assignment, I use the propensity score matching (PSM) method to assure that the training participants are compared to similar rice farmers in non-program villages regarding the observed characteristics in the pre-training period. When the strong ignorability assumption holds, outcomes are independent of treatment once conditioning on the probability of participating in the training is included (Rosenbaum and Rubin 1983). ATT is identified, assuming that outcome variables are independent of treatment assignment once a set of observable characteristics before the training are controlled for:

$$\text{ATT}_t = E(Y_{it}(1)|T_i = 1, P(X_{i0})) - E(Y_{it}(0)|T_i = 0, P(X_{i0})) \qquad (5.1)$$

where E() is an expectation operator, Y(1) is an outcome of household i with participating in the training, Y(0) is an outcome of the household i without participating in the training, T is an indicator variable taking unity if household i participated in the training, and P(X) is the propensity score or probability of training participation given observed pre-training characteristics X. The propensity scores are estimated by a probit model using pre-training observable characteristics as explanatory variables. Since farmers in non-program villages cannot participate in training due to the program design, I will use only samples in the program village to estimate the probit model and apply the estimated coefficients to non-program villages to predict their propensity scores.[6] This paper uses a kernel matching method and a common support condition for constructing a comparison group.

PSM estimator of ATT is denoted as

---

[6] In eight non-program villages, there have been training and/or programs related to rice cultivation such as NAADS. Although actual participants in such trainings and programs comprised just eight households in our sample, there might be spillover effects from the program to non-participants in such communities. To avoid this possibility, we dropped these eight communities to construct control groups as a way of estimating the direct and indirect effects of the training. The results are qualitatively similar to the main results. We believe there is no problem with the contamination from other rice programs.

$$\text{ATT}_t^{\text{PSM}} = \frac{1}{N_1} \sum_{i \in N_1} \left(Y_{it}(1) - \sum_{j \in N_0} W_{ij} Y_{jt}(0) \right) \quad (5.2)$$

where N_1 and N_0 are the numbers of matched treatment and control households, and W is the weights calculated from PSM. The validity of PSM is based on conditional independence and overlap in propensity scores across the participants and non-participants. To assure conditional independence, we use preference measures (hypothetically asked about risk aversion and time discount) to calculate the propensity scores that are usually unobserved to researchers but likely to affect participation.

Because unobserved characteristics may have affected the training participation and we have panel data before and after the training, I employ a PSM–difference-in-differences (DID) estimator of the ATT to mitigate the selection problem due to time-invariant unobservables (Smith and Todd 2005). We examine the training effect on the change in the outcomes from baseline (before the training), $\Delta Y_{it} \equiv Y_{it} - Y_{i0}$. The PSM–DID estimator is denoted as

$$\text{ATT}_t^{\text{PSMDID}} = \frac{1}{N} \sum_{i \in N_1} \left(\Delta Y_{it}(1) - \sum_{j \in N_0} W_{ij} \Delta Y_{jt}(0) \right) \quad (5.3)$$

To measure the direct impact of training, we exclude non-participants in the training villages from estimation and compare the training participants with non-program village farmers who have similar propensity scores. In contrast, to estimate the spillover effects, we compare the non-participants in the program villages with their counterparts in non-program villages who have similar propensity scores.

5.4 Results

5.4.1 Determinants of Program Participation

By using the data of training participants and non-participants in the program villages, the probability of participating in the training is estimated by the probit model. The estimation results are given in Table 5.2. Significant factors affecting the training participation are rice cultivation experience, risk aversion, and population density. Households with longer years of rice cultivation experience are more likely to participate in the training. This may be because they are more interested in rice cultivation or have previous experience with the challenges. Those with higher risk aversion are more likely to participate in the training, probably because the training is expected to reduce the risk of low production. The negative sign of a coefficient of population density may be because villages with higher population density in lowland rice areas have less room to expand or because higher population density is associated with the closeness of the town and the opportunity cost of attending the training is high. The coefficients estimated by the probit model are used to estimate the propensity scores of the households in control villages. The distribution of the propensity scores for training participants and households in control villages and for non-participants and households in control villages are shown in Panel A and Panel B of Fig. 5.2, respectively. After the matching based on propensity scores, 8% of the treatment households were not matched and dropped from the analyses. Table 5.3 shows balancing test results, indicating that matching was successful, although one of the variables shows unbalance between the participant and control groups.

Table 5.2 Determinants of participation in the training program in 2009 (probit model)		Training participants versus non-participants
	ln (rice experience)	0.103+ (1.83)
	ln (head age)	0.055 (0.31)
	Head years of education	−0.037 (1.10)
	Head education squared	0.004 (1.49)
	ln(number of HH members)	0.058 (0.49)
	Share of males aged 15–64	−0.241 (0.68)
	Share of females aged 15–64	−0.192 (0.46)
	Size of land owned (ha)	−0.033 (1.00)
	Share of lowland owned	−0.083 (0.50)
	Local group membership	0.096 (0.97)
	Owns a bull	−0.006 (0.06)
	Risk averse	0.164+ (1.77)
	Has patience	−0.092 (0.92)
	Population density per squared km (village level)	−0.363** (2.90)
	N	372

**, *, and + represent statistical significance at the 1, 5, and 10% levels, respectively. Marginal effects are shown. The numbers in brackets are z-statistics

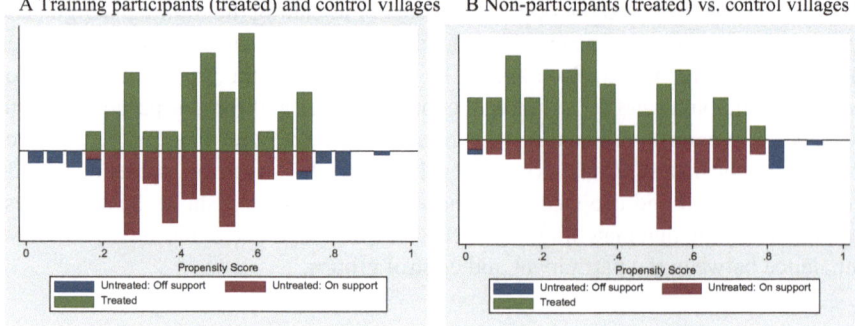

Fig. 5.2 Distributions of propensity scores

5.4.2 Training Impact on the Adoption of Cultivation Practices and Rice Yield

The estimated ATT by PSM–DID on cultivation practices and rice yield are presented in Table 5.4. Column 3 of Panel A shows the average direct effect of the training, while Column 3 of Panel B indicates the average spillover effect of the training. Regarding the adoption of transplanting in rows, the direct short-term and long-term average effects are 13 and 20 percentage points, respectively. The corresponding spillover effects are 5 and 2 percentage points, but they are not significantly different from zero. There was a long-term direct effect on bund construction and maintenance, whereas the short-term direct effect and the spillover effect in both short and long terms were not found.

Regarding rice productivity, both in the direct and spillover effects, the short-term impact is not statistically significant, while the long-term direct and spillover effects of training on rice yield are 0.84 and 0.47 tons per ha, respectively. Although the short-term direct impact on rice yield was not significant, the direct impacts on rice yield have gradually turned to positive. The improved productivity would likely attract non-participants to adopt the transplanting method, leading to a positive indirect effect on rice yield in the long term. As the training participants increased the adoption of transplanting in rows over time, it is probable that the direct effect on rice yield in the long term was brought about by the adoption of better transplanting methods.[7]

Although the training program resulted in enhancing productivity among the non-participants in the program villages in the long term, we did not find evidence that the adoption rate of the better cultivation practices increased among them. This seems puzzling but it can be explained by the fact that non-participants shifted from

---

[7] According to Kijima (2022), training participants increased the application of chemical fertilizer. Therefore, the increased yield might not be induced solely by the adoption of the transplanting in rows method.

Table 5.3 Balancing test results

	Participants (T)	Control C0	t-stats	Non-participants C1	Control C0	t-stats
Log (rice experience years)	2.203	1.915	1.42+	1.907	1.778	1.54
Log (HH hea"s age)	3.612	3.709	1.36	3.667	3.680	0.82
Head's years of education	6.061	6.185	0.14	5.527	5.792	0.63
Education squared	51.52	48.12	0.28	39.08	40.95	0.27
Log (num. of HH members)	2.042	2.157	0.98	1.976	2.082	1.12
Share of males aged 15–64	0.242	0.236	0.20	0.251	0.230	0.70+
Share of females aged 15–64	0.231	0.227	0.16	0.264	0.228	1.42+
Land ownership (ha)	1.708	1.857	0.37	2.251	1.931	0.90
Share of lowland size	0.182	0.189	0.10	0.270	0.236	0.58
Local group member	0.697	0.505	1.60	0.630	0.479	1.58
Owns a bull	0.485	0.347	1.13	0.519	0.321	2.11
Risk averse	0.485	0.379	0.86	0.370	0.238	1.50
Has patience	0.667	0.634	0.27	0.741	0.695	0.52
Population density	0.660	0.470	2.29*	0.556	0.631	1.48
# obs. on support	43	137		54	155	

t-stats for the mean difference between the 2 groups (* indicates the means between treatment and control groups are significantly different at the 5% level). + indicates that the variance ratios between the 2 groups (V(T)/V(C)) are outside of [0.47; 2.13]

broadcast planting to transplanting in the long term. It is also possible that non-participants learned a multiplicity of improved cultivation practices from participants through conversations and observations of participants' fields, which are not captured by selected cultivation practices in this study.

Table 5.4 Average impact of training (PSM–DID)

Panel A. Direct training effect		Participants	Control	ATT Direct	t-stat
		(1)	(2)	(3)	(4)
Transplanting	2011–2009	0.061	−0.026	0.086	0.65
	2015–2009	0.114	−0.068	0.182*	2.43
Transplanting in rows	2011–2009	0.152	0.022	0.129*	1.92
	2015–2009	0.229	0.028	0.200*	2.63
Bunds	2011–2009	0.333	0.202	0.131	1.32
	2015–2009	0.286	0.098	0.187+	1.83
Yield (ton/ha)	2011–2009	0.417	0.135	0.283	1.32
	2015–2009	0.861	0.024	0.837*	3.01
Panel B. Spillover effect		Non-Participants	Control	ATT Indirect	t-stat
		(1)	(2)	(3)	(4)
Transplanting	2011–2009	0.056	0.027	0.028	0.42
	2015–2009	0.061	−0.041	0.103+	1.81
Transplanting in rows	2011–2009	0.093	0.039	0.054	1.16
	2015–2009	0.082	0.060	0.021	0.45
Bunds	2011–2009	−0.074	0.162	−0.088	1.02
	2015–2009	−0.082	0.032	−0.114	1.27
Yield (ton/ha)	2011–2009	0.171	0.269	−0.098	0.57
	2015–2009	0.589	0.120	0.470*	2.17

* and + represents statistical significance at the 5% and 10% levels, respectively

5.5 Conclusion

This study examined the short-term and long-term impact of agricultural training. Furthermore, we assessed the program's direct and indirect impacts on cultivation practices and rice yield using the PSM–DID method. The results show that training enhanced the adoption of cultivation practices taught in the training session among training participants but not among non-participants in the program villages. We did not observe the disadoption of such cultivation practices even after five years among the training participants. The average direct impact of training on rice yield was 0.84 tons per ha in the long term. Since the pre-program average rice yield was about 1.5 tons per ha, the direct impact accounts for more than 50% of the increase. Given that the program did not provide chemical fertilizer or high-yielding varieties, this impact is surprisingly high. Thus, the training that imparts improved rice cultivation practices with training participants is considered to be effective and sustainable, even in the long term.

In terms of spillover effects, the average rice yield increased in the long term by 0.47 tons per ha among non-participants in the program villages. This finding suggests that there are spillover effects in the long term. Although non-participants adopted transplanting (not in line) in the long term, the adoption rate of the recommended cultivation practices (transplanting in rows and construction of bunds) was not enhanced even in the long term.

Can we conclude that there is a spillover effect? Kijima (2022) estimated the ATT of the same project by applying the difference-in-differences inverse probability weighting approach (Imbens and Wooldridge 2009) and found similar results to this chapter. The likely reason explained in the paper was that non-participants did not learn key concepts but mimicked the transplanting method and other numerous cultivation methods by observation, such as the appropriate timing of an appropriately shallow transplanting. This explanation was based on further analyses showing that knowledge of the transplanting method increased among training participants but was not enhanced among non-participants. Combined with these findings, what seemed to happen was that the yield enhancement among non-participants occurred as a result of the shift from the broadcast planting method to transplanting, among other factors. Such a shift can be induced by a higher adoption rate of transplanting in rows in the program villages, since non-participants can observe such changes in fields. Since non-participants did not adopt the entire set of better cultivation practices taught in the program, it may not be reasonable to expect that they will wholly catch up with the participants unless they can acquire the entire set of improved practices from participants. For such learning to be effective, non-participants need to know who took part in the training and be able to raise appropriate questions with former participants regarding their knowledge of improved cultivation practices. Developing ways to further enhance learning between training participants and non-participants, especially in the rainfed lowlands, is an important area for future research.

References

Imbens G, Rubin D (2015) Causal inference for statistics, social, and biomedical sciences: an introduction. Cambridge University Press, Cambridge

Imbens G, Wooldridge J (2009) Recent developments in the econometrics of program evaluation. J Econ Lit 47(1):5–86

Kijima Y (2022) Long-term and spillover effects of rice production training in Uganda. J Dev Eff 14(4):395–415. https://doi.org/10.1080/19439342.2022.2047763

Kondylis F, Mueller V, Zhu J (2017) Seeing is believing? Evidence from an extension network experiment. J Dev Econ 125:1–20

MAAIF (2009). Uganda National Rice Development Strategy (UNRDS). Ministry of Agriculture Animal, Industry, and Fisheries, Government of the Republic of Uganda. https://www.jica.go.jp/english/our_work/thematic_issues/agricultural/pdf/uganda_en.pdf

Rosenbaum PR, Rubin DB (1983) The central role of the propensity score in observational studies for causal effects. Biometrika 70(1):41–55

Smith JA, Todd PE (2005) Does matching overcome laLonde's critique of nonexperimental estimators? J Econometrics 30(9):1621–1638

Yoko Kijima is a professor and vice president of National Graduate Institute for Policy Studies (GRIPS), Tokyo. She obtained dual Ph.D. in Agricultural Economics and Economics from Michigan State University in 2003. Her area of research is Development Economics.

Chapter 6
The Case of Mozambique: The Importance of Management Training for Rice Farming in Rainfed Areas

Kei Kajisa and Trang Thu Vu

Abstract This chapter assesses the results of a randomized controlled trial (RCT) of management training for rice farming in remote rainfed lowland areas of Mozambique. The training taught basic practices but did not require the use of modern purchased inputs such as inorganic fertilizers or modern varieties, which are not easily available to poor farmers in remote areas. The intention-to-treat (ITT) effect on paddy yield was 447–546 kg/ha (29–36% of the control group average yield) with statistical significance at 7–8%. Our analysis also demonstrates that this increase was achieved when key improved management practices were adopted as a package because of the complementarity of the improved practices. These results indicate that the adoption of the practice package alone can improve rice yield substantially even without modern inputs.

6.1 Introduction

Rice yield in Mozambique has remained low at 1 to 1.5 tons/ha of paddy for several decades. Meanwhile, rice consumption has continued to grow rapidly (USDA 2021), with rice imports increasing at a rate of 9.0% annually from 44 thousand tons in 1990 to 650 thousand tons in 2020. This has increased foreign exchange expenditures that could otherwise be used to finance local development projects. Therefore, finding ways to increase the country's rice productivity can provide an important component of its food security strategies (Kajisa 2015; Kajisa and Payongayong 2011; Otsuka and Larson 2013, 2016). The strategy should be designed for a rainfed area, at least in the short or medium term, because the proportion of areas equipped with irrigation facilities remains marginal at about 2% of the country (FAO 2021).

K. Kajisa
School of International Politics, Economics and Communication, Aoyama Gakuin University, 4-4-25 Shibuya, W403, Shibuya-ku, Tokyo 150-8366, Japan

T. T. Vu (✉)
Research Department, Asian Development Bank Institute (ADBI), Kasumigaseki Building 8F, 3-2-5, Kasumigaseki, Chiyoda-ku, Tokyo 100-6008, Japan
e-mail: trangvumpi@gmail.com; tvu1@adbi.org

© JICA Ogata Sadako Research Institute for Peace and Development 2023
K. Otsuka et al. (eds.), *Rice Green Revolution in Sub-Saharan Africa*, Natural Resource Management and Policy 56, https://doi.org/10.1007/978-981-19-8046-6_6

It has been argued that the dissemination of basic management practices is a key element in increasing rice productivity in rainfed lowland areas in sub-Saharan Africa (SSA), including Mozambique (Kijima et al. 2012; Balasubramanian et al. 2007; Barker and Hardt 1985). The basic management practices include seed selection and nursery bed setup (for quality seedlings), field leveling and bund construction (for even water distribution), straight-row transplanting (for easier crop management and weeding), timely weeding, and water management. The rationale for this strategy is twofold. First, even these basic practices—already common in Asia during its Green Revolution—are not commonly observed or standardized in SSA, as rice was not a dominant staple crop there. Second, this strategy can improve productivity without relying on modern, purchased inputs such as inorganic fertilizers and modern high-yielding varieties. Hence, even cash and market-access-constrained remote farmers could increase rice production if they adopted this strategy. A standard approach to realizing this strategy is the provision of training in basic practices, and thus, we need a better understanding of training effectiveness.

However, empirical evidence on the impact of training in basic rice farm management practices is scarce.[1] The aim of this chapter is to assess the impact of such training provided by the Japan International Cooperation Agency (JICA) in remote rainfed lowland areas in Mozambique using a randomized controlled trial (RCT). The training has three features. First, the training comprised the combination of a conventional approach (farmer field schools (FFS) at demonstration plots) and a contemporary approach (farmer-to-farmer extension (F2FE) through social learning). Second, the training did not provide any performance-based monetary incentives to accelerate technology diffusion. Third, the training did not rely on modern inputs, such as the newly developed improved varieties or inorganic fertilizers. Our study contributes to the literature on agricultural training by assessing the effectiveness of the JICA project with the above characteristics for farmers experiencing cash- and market-access constraints in remote rainfed areas in SSA.

6.2   Rice in Mozambique

Among the major cereals, maize has been the dominant staple in Mozambique, but rice has also been growing in importance. As a result of increased urbanization and the convenience of preparing rice meals, Mozambique, like other African countries, has seen a shift in consumer preference for rice (Hossain 2006). Therefore, rice consumption in Mozambique has rapidly increased by 8.9% annually from 1990 to 2020, faster than the growth in maize consumption at 4.5% or wheat at 6.1% (USDA 2021). In response to this increase, production initially grew at 12.1% annually from 1993 to 1998, but growth has largely stagnated since then (Fig. 6.1). As shown in Fig. 6.2, the modest growth in production can be attributed to the expansion of the

---

[1] Exceptions include studies on rainfed rice by Nakano et al. (2018) in Tanzania, deGraft-Johnson et al. (2014) in Ghana, and Kijima et al. (2012) in Uganda.

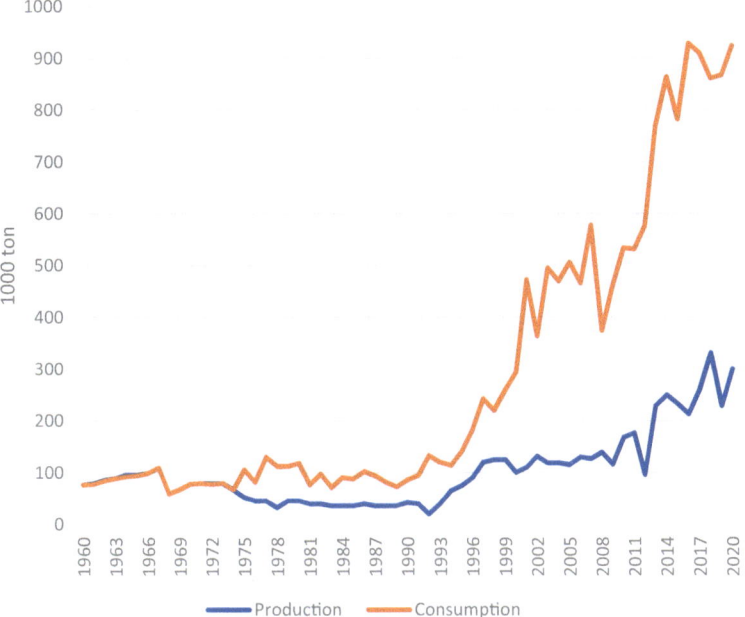

Fig. 6.1 Production and consumption of rice (milled bases) in Mozambique, 1960–2020. *Data Sources* USDA: PS&D Online April 2021; USBC: International Data Base, August 2006

harvested area rather than yield improvements. Paddy yield has stagnated at a level of around 1 to 1.5 tons per hectare, which is below the average yield of 2.2 tons per hectare in SSA (see Fig. 1.5). As mentioned in the introduction, this stagnation has led to a rapid increase in rice imports, as indicated by the widening gap between consumption and production (Fig. 6.1).

Rice in Mozambique is produced mostly in the rainfed lowland ecological regions, where farmers follow traditional cultivation practices. The area equipped with irrigation facilities accounts for only 2% of the arable land in the country. Among the rainfed lowland areas, Zambézia Province, including the Zambézi River basin, is the dominant rice producing province (48% of the total rice area), followed by Nampula (14%), Sofala (12%), and Cebo Delgado (10%) (Ministério da Agricultura e Segurança Alimentar 2015) (Fig. 6.3).

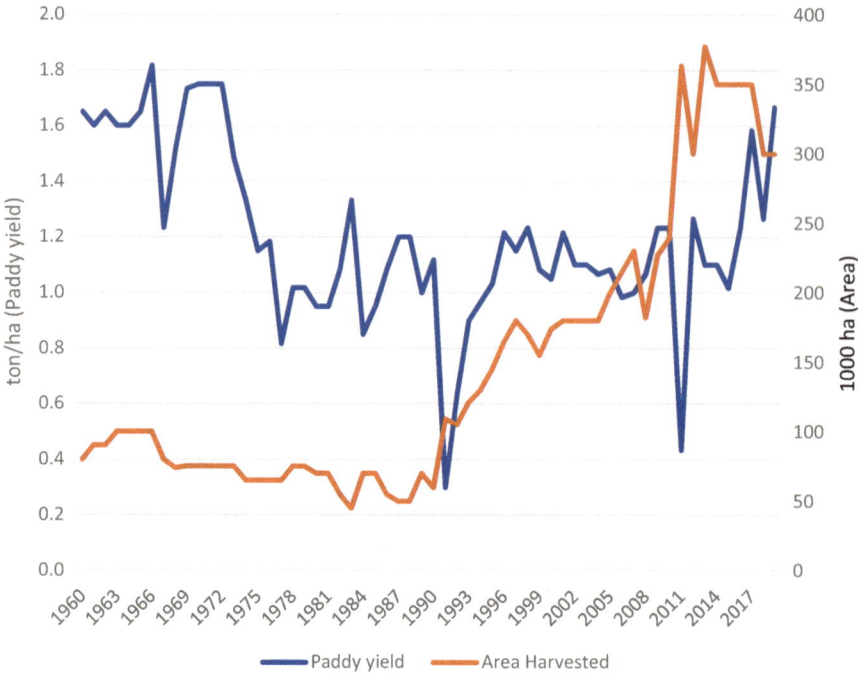

Fig. 6.2 Area harvested and paddy yield in Mozambique, 1960–2020. *Note* Milled rice yields in the original data set were converted to paddy yields at 60% milling recovery rate. *Data Sources* USDA: PS&D Online April 2021; USBC: International Data Base, August 2006

6.3 Experimental Design

6.3.1 JICA Rice Training

The project to provide training on rice farm management practice in Zambézia Province started in 2016 with financial support from JICA. The unit of intervention was the farmer's association. The JICA rice training project, in consultation with the Provincial Directorate of Agriculture and Fisheries (Direcção Provincial de Agricultura e Pescas, DPAP), selected 17 farmer's associations in six local units (*localidade*) in the rainfed area and five associations in five local units in the irrigated area. In this impact assessment study, we focused on the 17 rainfed associations, given the purpose of the study and delays in the rehabilitation projects for irrigation facilities in the selected area.

The project established demonstration plots in each association, using the association's common plots, usually located at an accessible and observable location in the

Fig. 6.3 Study site. *Source* d-map.com, (https://d-maps.com/carte.php?num_car=35360&lan g=en), Accessed June 28th, 2022

association's rice area.[2] In collaboration with the staff of the National Directorate of Assistance to Family Farming (Direcção Nacional de Assistência a Agricultura Familiar, DNAAF), the project provided four training sessions in the demonstration plots. The training sessions provided training in (1) the use of recommended

---

[2] If the associations did not have common plots, the project leased private plots suitable for demonstration.

varieties, (2) the seed selection method, (3) the nursery bed setup for seedlings, (4) land leveling, (5) bund construction, (6) straight-row transplanting or straight-row direct sowing, (7) weeding at the proper time, and (8) harvesting at the bottom of the plant, rather than the panicles. The recommended seed selection method was to remove empty seeds floating in the water. All the recommended rice varieties are *local* varieties, rather than modern varieties, which have been developed recently and are usually sold at markets in towns. This is because modern varieties are not easily accessible to the cash and market-access-constrained farmers in remote areas. For the same reason, the use of inorganic fertilizers was not included in the training in the rainfed areas.

To disseminate the improved management practices at the demo plot, the project selected demo farmers who were invited to the demo plot for training and expected to pass on the new practices and technologies to the other member farmers. In this study, we refer to these invitees as "lead farmers" (LF) to ensure that the terms are comparable to the existing literature. Later, due to strong requests from the other member farmers, any other members who wanted to participate in the training were invited to join the project. These participant farmers were called "replica farmers" by the project. However, as they were supposed to be less capable than LF in terms of farming skills and network formulation but eventually participated in the training, we refer to this group as "participant ordinary farmers" (POF). The remaining farmers in the group are called "ordinary farmers" (OF). The ordinary farmers could still observe and learn new practices voluntarily at the demo plot. Moreover, farmers from any group (LF, POF, and OF) could learn the new practices from others at any time. In this regard, the training can be summarized as a hybrid of two approaches: implementing farmer field schools (FFS) in demonstration plots and disseminating learned practices through farmer-to-farmer extension (F2FE).

6.3.2 Experimental Design and Sample

There are three to four target associations in each local unit, and we randomized the order of association-level training within each local unit (cluster RCT). This means that one association was randomly selected from each of six local units in the first project year, generating six treated associations. They are labeled Demo 1. The other six associations from each local unit were selected in the second year, and they are labeled Demo 2. This leaves five associations as the control group. Note that Demo 1, Demo 2, and the control group associations are not concentrated in a particular local unit because we randomized the order of training within each local unit. We conducted a pre-training baseline survey in 2017 based on the 2016–17 rice season, and after completing the training in the Demo 1 and Demo 2 groups, a follow-up survey in 2019 on the 2018–19 rice season. Since the associations are far apart and little spillover effect exists between them, we believe that the stable unit treatment value assumption (SUTVA) is not violated. The weather in the baseline rice season was normal, but the follow-up season had irregular rainfall. Hence, on average, rice yield decreased at the time of the follow-up survey.

Given the number of associations (clusters) in each experimental arm, we conducted a power calculation to obtain an appropriate sample size in each cluster.[3] We collected a random sample of 13–25 farmers proportionate to the size of each association, generating 311 observations in the baseline survey. In the follow-up survey, we collected data from 257 farmers in the baseline survey, with the attrition of 54 farmers. Our statistical analysis relies on a balanced panel of these 257 farmers in two periods (514 observations) while statistically controlling for attrition bias.

6.4   Impact of the Training

6.4.1   Balance Test and Outcome

Columns (1)–(5) in Table 6.1 show the baseline balance of sample households by treatment. Of the 257 farmers, 78 farmers were under the treatment of the demonstration plot in the first year (Demo 1), and 101 farmers were added in the second year (Demo 2), while the 78 farmers in the control group were not receiving any treatment. The household characteristics consist of household size (heads), household head's schooling years (years), the log of household total asset value (000 MT), total plot area (ha) including non-survey plots, the proportion of known members (%), weather shock in the rice season of the survey year (dummy), and weather shock in the non-rice season immediately before the rice season of the survey year (dummy). The variable "proportion of known members" measures what percentage of sample farmers in the association is known by an interviewed sample farmer, indicating individual network size within the association. The dummy variable "weather shock" takes the value 1 if farmers self-reported that their rice crop suffered from flood, drought, or irregular rainfall.

The table shows that all the household characteristics—either in Demo 1 or Demo 2, except for the proportion of known members—are not statistically different from those of the control group. A joint significance test between Demo 2 and the control

---

[3] A project consultant conducted a pilot study in the study site before our baseline survey, providing useful summary statistics for a power calculation. Using these, we set the mean yield at 1 t/ha, the standard deviation at 1 t/ha, the number of clusters in one experimental arm at 6, intra-cluster correlation (ICC) at 0.15 and, being conventional, the proportion of the yield explained by baseline covariates at 0. We set the significance level at 0.05 and the power of test at 0.8. Under these settings, the sample size of 15 in each cluster generates the statistically detectable change of yield by 0.81 t/ha. Moreover, since we took the baseline data in this project, if the proportion explained by the baseline covariates improves from 0 to 0.4, we can detect the change by 0.74 t/ha. Since the target of the project was to increase yield by 1 t/ha, we decided to set our target sample size in each cluster (association) at 15.

Table 6.1 Baseline balance of sample households by treatment and attrition status

Variable	(1) Demo 1 Mean/SE	(2) Demo 2 Mean/SE	(3) Control Mean/SE	(4) Difference (1)–(3)	(5) Difference (2)–(3)	(6) Non-attrition Mean/SE	(7) Attrition Mean/SE	(8) Difference (6)–(7)
Treated (= 1)						0.696	0.796	−0.100
						[0.029]	[0.055]	
Household size	3.718	4.050	4.282	−0.564	−0.233	3.634	4.352	−0.718
	[0.226]	[0.211]	[0.299]			[0.220]	[0.542]	
Head's education (years)	3.846	3.574	3.500	0.346	0.074	3.634	4.352	−0.718
	[0.427]	[0.334]	[0.398]			[0.220]	[0.542]	
Log of asset values	7.563	7.677	7.477	0.085	0.199	7.581	7.208	0.374
	[0.184]	[0.139]	[0.247]			[0.108]	[0.259]	
Total plot area (ha)	0.813	0.621	0.703	0.110	−0.082	0.704	0.413	0.291**
	[0.122]	[0.078]	[0.107]			[0.058]	[0.044]	
Proportion of known members (%)	32.869	55.789	41.660	−8.79*	14.13**	44.545	72.627	−28.082***
	[2.933]	[3.713]	[4.028]			[2.179]	[4.679]	
Weather shock in the last rice season (= 1)	0.115	0.149	0.154	−0.038	−0.005	0.140	0.315	−0.175***
	[0.036]	[0.036]	[0.041]			[0.022]	[0.064]	
Weather shock in the last non-rice season (= 1)	0.795	0.772	0.833	−0.038	−0.061	0.798	0.907	−0.110*
	[0.046]	[0.042]	[0.042]			[0.025]	[0.040]	

(continued)

Table 6.1 (continued)

Variable	(1) Demo 1 Mean/SE	(2) Demo 2 Mean/SE	(3) Control Mean/SE	(4) Difference (1)–(3)	(5) Difference (2)–(3)	(6) Non-attrition Mean/SE	(7) Attrition Mean/SE	(8) Difference (6)–(7)
N	78	101	78			257	54	
F-test of joint significance (F-stat)				1.641	2.577**			4.216***
F-test, number of observations				156	179			311

The values displayed for *t*-tests are the differences in the means across the groups. The values displayed for *F*-tests are the F-statistics. ***, **, and * indicate significance at the 1, 5, and 10% critical levels
Source Authors

(shown at the bottom of the table) was statistically significant, but it became insignificant if we removed the variable of the proportion of known members (the result is not shown in the table).

Columns (6)–(8) in Table 6.1 compare the household characteristics by attrition status, in which we additionally compare the dummy of treatment. The table shows that, although attrition had little to do with treatment, it occurred non-randomly because non-attrition households operated larger areas of farmland, knew fewer farmers in the same association, and were less likely to have experienced weather shocks in both the rice and non-rice seasons. These differences might constitute a source of bias in the impact assessment, which needs to be managed with an appropriate econometric technique.

Table 6.2 shows differences in outcome variables by treatment status at the baseline season (columns (1)–(5)) and the follow-up season (columns (6)–(10)). The outcome variables we examine are the adoption of the practices demonstrated by the training, namely, the adoption of seed selection by water ($= 1$), setup of the nursery bed ($= 1$), bund construction ($= 1$), leveling ($= 1$), straight-row transplanting ($= 1$), conducting weeding at least once ($= 1$), harvesting at the bottom of the plant ($= 1$), use of sickle for harvesting ($= 1$), and use of a recommended rice variety of either Chupa ($= 1$), Mocuba ($= 1$), or Mamina ($= 1$). These varieties are local varieties that possess the characteristics of late maturity and high yield, unlike the other popular local variety Nene, which has the features of early maturity and low yield. The adoption of these three varieties is used as our outcome variable because these are the varieties preferred by farmers and recommended by the project. We also compare paddy yield (kg/ha) as the outcome of the project. Note that the weeding variable is empty in the baseline because we failed to collect this information correctly.

The table shows that, at the time of the baseline survey, the adoption of improved practices was quite low (at most about 30%), and the differences by treatment status were statistically insignificant, except for two variables related to harvesting (harvesting at the bottom of the plant and the use of sickle) in the Demo 2 group. Nevertheless, the adoption of these two practices was lower in Demo 2 group than in the control group at the pre-training time. Thus, a possible higher adoption rate at post-treatment does not mean that it was higher from the beginning. Meanwhile, we observe significant differences in rice variety choices.

The paddy yields were low at 1,940 kg/ha in Demo 1, 1,527 kg/ha in Demo 2, and 1,975 kg/ha in the control group, which was understandable under rainfed conditions even for a normal weather season. The low yield of Demo 2 was statistically different from that of the control group at the 10% significance level. We can still use this result to claim that, even if the yield became higher after the training in the Demo 2 group, it was not higher from the beginning.

In the follow-up survey, the adoption rate of recommended practices increased sharply among the treated groups, resulting in statistically significant differences compared to the control group in most cases (about 30–50 percentage points higher

Table 6.2 Changes in outcome variables by treatment status: baseline and follow-up

Variable	Baseline					Follow-up				
	(1)	(2)	(3)	(4)	(5)	(6)	(7)	(8)	(9)	(10)
	Demo 1	Demo 2	Control	Difference	Difference	Demo 1	Demo 2	Control	Difference	Difference
	Mean/SE	Mean/SE	Mean/SE	(1)–(3)	(2)–(3)	Mean/SE	Mean/SE	Mean/SE	(6)–(8)	(7)–(8)
Seedling preparation practices										
Seed test, water (= 1)	0.256	0.337	0.231	0.026	0.106	0.769	0.644	0.141	0.628***	0.503***
	[0.050]	[0.047]	[0.048]			[0.048]	[0.048]	[0.040]		
Nursery bed set up (= 1)	0.269	0.386	0.333	−0.064	0.053	0.872	0.812	0.333	0.538***	0.479***
	[0.051]	[0.049]	[0.054]			[0.038]	[0.039]	[0.054]		
Land preparation practices										
Plot bunding (= 1)	0.192	0.267	0.218	−0.026	0.049	0.474	0.495	0.192	0.282***	0.303***
	[0.045]	[0.044]	[0.047]			[0.057]	[0.050]	[0.045]		
Plot leveling (= 1)	0.141	0.188	0.244	−0.103	−0.055	0.667	0.455	0.038	0.628***	0.417***
	[0.040]	[0.039]	[0.049]			[0.054]	[0.050]	[0.022]		
Crop care practices										
Straight-row transplanting (= 1)	0.013	0.000	0.000	0.013	N/A	0.462	0.356	0.000	0.462***	0.356***
	[0.013]	[0.000]	[0.000]			[0.057]	[0.048]	[0.000]		
Weeding at least once (= 1)	N/A	N/A	N/A	N/A	N/A	0.628	0.455	0.359	0.269***	0.096
						[0.055]	[0.050]	[0.055]		

(continued)

Table 6.2 (continued)

	Baseline					Follow-up				
	(1)	(2)	(3)	(4)	(5)	(6)	(7)	(8)	(9)	(10)
	Demo 1	Demo 2	Control	Difference	Difference	Demo 1	Demo 2	Control	Difference	Difference
Variable	Mean/SE	Mean/SE	Mean/SE	(1)–(3)	(2)–(3)	Mean/SE	Mean/SE	Mean/SE	(6)–(8)	(7)–(8)
Harvesting practices										
Harvesting at the bottom of plant (= 1)	0.038	0.010	0.051	−0.013	−0.041*	0.526	0.465	0.192	0.333***	0.273***
	[0.022]	[0.010]	[0.025]			[0.057]	[0.050]	[0.045]		
Using sickle to harvest	0.295	0.277	0.410	−0.115	−0.133*	0.615	0.426	0.321	0.295***	0.105
	[0.052]	[0.045]	[0.056]			[0.055]	[0.049]	[0.053]		
Rice varieties										
Using Chupa variety (= 1)	0.128	0.050	0.026	0.103**	0.024	0.231	0.337	0.064	0.167***	0.273***
	[0.038]	[0.022]	[0.018]			[0.048]	[0.047]	[0.028]		
Using Mocuba variety (= 1)	0.179	0.168	0.295	−0.115*	−0.127**	0.359	0.168	0.231	0.128*	−0.062
	[0.044]	[0.037]	[0.052]			[0.055]	[0.037]	[0.048]		
Using Mamima variety (= 1)	0.179	0.139	0.269	−0.090	−0.131**	0.179	0.119	0.167	0.013	−0.048
	[0.044]	[0.035]	[0.051]			[0.044]	[0.032]	[0.042]		
Output										
Paddy yield (ha)	1939.7	1527.1	1974.9	−35.191	−447.794*	1782.5	1751.5	1535.8	246.659	215.661
	[172.671]	[139.327]	[197.380]			[126.150]	[109.771]	[131.771]		
N	78	101	78			78	101	78		
F-test of joint significance (F-stat)				2.096**	2.294***				18.186***	11.436***
F-test, number of observations				156	179				156	179

N/A: No data available or no statistical comparison possible. The values displayed for *t*-tests are the differences in the means across the groups; The values displayed for *F*-tests are the F-statistics; ***, **, and * indicate significance at the 1, 5, and 10% critical levels
Source Authors

than the control group's adoption levels). When comparing yield, we must note that the follow-up season suffered from irregular rainfall, and thus the *overall* average at the study site decreased slightly from approximately 1,800 kg/ha at the baseline to about 1,700 kg/ha at follow-up. However, we can still observe differential outcomes by treatment status: the reduction for Demo 1 was marginal and Demo 1 achieved 1,783 kg/ha. Furthermore, Demo 2 improved its yield to 1,752 kg/ha, while the yield of the control group decreased to 1,536 kg/ha. This implies that Demo 1 and 2 associations were able to mitigate the weather shock. As a result, the yields of Demo 1 and Demo 2 were approximately 200 kg/ha higher than those of the control group, although the differences were not statistically significant at any conventional level. We will examine these impacts in a more statistically rigorous manner in the next sub-section.

6.4.2 Econometric Analysis

To assess the causal influence of the provision of training on the outcomes of our interest, we estimate intention-to-treat (ITT) effects by employing an analysis of covariance (ANCOVA) model specified below (McKenzie 2012).

$$Y_{ijk1} = \beta_0 + \gamma Y_{ijk0} + \beta_1 D^1_{jk} + \beta_2 D^2_{jk} + X_{ijk0}\delta + \eta_k + \varepsilon_{ijk1} \qquad (6.1)$$

where Y_{ijk1} and Y_{ijk0} are the follow-up and baseline outcome variables of the most important rice plot of household i in association j in local unit (*localidade*) k; D^1_{jk} and D^2_{jk} are the treatment dummy variables, equal to 1 if association j in local unit k sets up the demonstration plot in the first round (Demo 1) or the second round (Demo 2), respectively; X_{ijk0} is a set of baseline control variables; η_k is the local unit fixed effect; ε_{ijk1} is the unobserved error term. Our primary outcome variable Y_{ijkt} is the paddy yield (kg/ha). Our Y_{ijkt} also includes individual management practices and variety adoption. For management practices, we focus on five essential ones: seed test by water (S), nursery bed setup (N), bund construction (B), field leveling (L), and straight-row transplanting (TP). We cannot include weeding in the set of crop care practices due to the lack of baseline data.[4] In addition, we do not include the two recommended harvesting practices because they are not yield improving practices. Meanwhile, we include the dummy of adoption of five practices as a package in order to identify the complementarity effects among them. When the outcome is binary, the employed model is a linear probability model. Our baseline control variables (X_{ijk0}) are the variables used in the balance test in Table 6.1, and the squared terms for household size and total plot area.

---

[4] It is possible to show the status of weeding adoption and its impact at follow-up. The trend of this practice is similar to those of the other practices: The yield of weeding adopters is lower than the non-adopters in the follow-up. This is partly due to self-selection: farmers who suffered weed problems did weeding more frequently.

A possible attrition bias was adjusted using the inverse-probability weighting method suggested by Wooldridge (2010). We run a probit regression model that estimates the probability of non-attrition, while using the inverse of the probability as weights in Eq. (6.1).[5] The probit regression results are presented in Appendix Table 6.7.

Table 6.3 shows the estimation results of the treatment effects (β_1 and β_2) in Eq. (6.1). Hereafter, all the results present wild bootstrap cluster robust p-values because the number of clusters in our data is less than 42, the threshold for the use of cluster robust standard errors suggested by Angrist and Pischke (2009).[6] The t-test of an equal impact between Demo 1 and Demo 2 (i.e., $\beta_1=\beta_2$, ,) is shown in the lower part of the table. The full regression results with the other control variables are listed in Table 6.7 in the Appendix.

The results on the yield in column (1) in Table 6.3 indicate that the project increased the yield of the Demo 1 group by 545.5 kg/ha at a p-value of 7.95% and that of the Demo 2 group by 447.5 kg/ha at a p-value of 6.50%, which corresponds to a 35.5% or 29.1% increase from the control group yield, respectively (see the control group mean of 1,535 kg/ha at the lower part of the table).[7] The t-test of equal impact does not reject the null hypothesis, indicating that a one-year lag in training implementation did not create a significant disadvantage. However, the magnitude is higher in Demo 1 by 98 kg/ha.

As the high adoption rates of the improved management practices in Demos 1 and 2 in Table 6.2 suggest, the impact of the training on those outcomes is positive and statistically highly significant (columns (2)–(6)), with no statistical difference between β_1 and β_2. The impact of training for the full adoption of five practices (Column (7)) shows a significant result in the Demo 1 group at a p-value of 6.3%, while Demo 2 gives a positive coefficient at 20% of the p-value, suggesting that a sequential adoption of all five practices requires time. The results for variety adoption (columns (8)–(10)) are ambiguous.

In summary, the training enhanced the adoption of recommended basic practices and increased the yield by 0.4 or 0.5 t/ha among the farmers in the treated associations. A remaining question is: How did the farmers in the treated associations increase yield?

---

[5] The explanatory variables consist of the same variables in Xs and the squared term of the head's education.

[6] For wild bootstrap, see Roodman et al. (2019) and Wooldridge (2010).

[7] As a robustness check, we combine Demo 1 and Demo 2 dummies and estimate the impact of the training as a whole. The estimate is 481.9 kg/ha at a p-value of 3.7%.

Table 6.3 Estimated results of ANCOVA model on the impact of training: rice productivity and technology adoption

Variables	(1) Paddy yield	(2) (S) Seed test by water	(3) (N) Nursery bed set up	(4) (B) Bund	(5) (L) Leveling	(6) (S) Straight-row TP	(7) Use all 5	(8) Use Mamima	(9) Use Mocuba	(10) Use Chupa
Demo 1 (treatment)	545.5*	0.570***	0.592***	0.376**	0.609**	0.508*	0.367*	0.0903*	0.0895	0.0899
	[0.0795]	[0.00085]	[0.0005]	[0.0440]	[0.0390]	[0.0750]	[0.0635]	[0.0985]	[0.3710]	[0.6010]
Demo 2 (treatment)	447.5*	0.479*	0.461***	0.326**	0.416**	0.449**	0.200	−0.00583	−0.0568	0.289
	[0.0650]	[0.0710]	[0.0000]	[0.0265]	[0.0400]	[0.0100]	[0.2015]	[0.8485]	[0.7730]	[0.1380]
Control variables	Yes	Yes	Yes	Yes	Yes	Yes	Yes	Yes	Yes	Yes
Local unit FE	Yes	Yes	Yes	Yes	Yes	Yes	Yes	Yes	Yes	Yes
t-test (Demo 1 = Demo 2)	0.8372	0.7911	1.3768	0.5446	1.3559	0.4146	1.7405	2.0472*	1.8529	−1.7335
	[0.5005]	[0.5795]	[0.3330]	[0.7695]	[0.4005]	[0.7605]	[0.1900]	[0.0570]	[0.1910]	[0.2235]
Control mean value	1535	0.141	0.333	0.192	0.038	0.00	NA	0.167	0.231	0.064
Observations	257	257	257	257	257	257	257	257	257	257
R-squared	0.363	0.300	0.413	0.510	0.395	0.380	0.418	0.403	0.525	0.302

Wild bootstrap cluster robust p-values in brackets; Inverse probability weights are used to control for attrition bias (see Appendix Table 6.7 for the probit analysis of non-attrition)

*** $p < 0.01$, ** $p < 0.05$, * $p < 0.1$

See Appendix Table 6.8 for full regression results

Source Authors

6.5 Practice Adoption and Diffusion

6.5.1 Adoption and Yield Increase

To answer the above question, we examine what practices and rice varieties increased the yield. Panel A in Table 6.4 shows the percentage of adopters of individual practices or their packages and corresponding yields among the entire sample ($n = 257$) at the baseline and follow-up seasons. The asterisks on the yield values indicate the significant mean difference from the yield under no adoption based on the t-test.

Regarding the impact of adoption, one of the key questions is whether yield increases resulted from farmers adopting all five practices as a package or whether single or partial adoption still increases yield. The answer to this question is practically important because it determines the specific recommendations given to farmers in the training. Consequently, while Table 6.4 shows the yield under the solo or partial adoption from the five practices, we do not include the farmers who adopted all five practices in this data. For example, in the case of the adoption of the Seed test by water ((S) in the table), the results do not include the farmers who adopted all five practices—only the farmers who adopted the seed test alone or the seed test plus some other practices but not all the other practices. If the adoption of (S) alone still has an impact, yield under (S) is expected to be higher than in the case of no adoption. The table also shows the case of combining any single or partial adoptions of five practices in one row above the case of full adoption. Hence, the sum of "No adoption," "Any single or partial adoption," and "All 5 practices" is 100%. Henceforth, we refer to the farmers who adopted all five practices as full adopters.

Panel A shows these three features. First, unexpectedly, at baseline, the case of no adoption shows the highest yield. This may be because farmers experiencing very favorable agro-ecological conditions were able to achieve high productivity with conventional practices. Second, at the baseline, there was no full adopter at all, while there were some single and a few partial adopters. Third, at the follow-up survey, the proportion of full adopters increased to 12% and they achieved the highest yield (2,206 kg/ha), although the difference was not statistically significant due to the small sample size.

Panel B in Table 6.4 shows the impact of variety adoption. We did not find significant differences in yield except for the use of the Mocuba variety at baseline. Mocuba again shows the highest yield at the follow-up with almost the same proportion of users. This may be because each farmer was already using a variety suitable for their local conditions before the training. Our data strongly suggest, at least in our study site, that rice variety adoption was not a major driving force of yield improvement. From this point, we focus on the exploration of improved practice adoption only.

Table 6.4 Improved management practices, variety adoption and paddy yield in the follow-up survey

Panel A: Key practices

Adoption status	----------Baseline----------		----------Follow-up----------	
	Percentage of farmers (%)	Paddy yield (kg/ha)	Percentage of farmers (%)	Paddy yield (kg/ha)
No adoption	37	2098	20	1805
Partial Adoption[a]				
Seedling preparation practices				
(S) Seed test by water	28	1295***	41	1536
(N) Nursery bed set up	33	1611*	56	1596
Land preparation practices				
(B) Bund construction	23	1262***	28	1614
(L) Leveling	19	1740	27	1507
Crop care practice				
(TP) Straight-row planting	0.4	2442	16	1326**
Combinations				
(S) + (N)	11	657***	35	1552
(B) + (L)	8	1924	11	1596
(S) + (N) + (B) + (L)	2	1276	5	1384
(S) + (N) + (TP)	0	Na	14	1227
(B) + (L) + (TP)	0	Na	1	2158
Any single or partial adoption	63	1609**	67	1571
Full Adoption				
All 5 practices (S) + (N) + (B) + (L) + (TP)	0	Na	12	2206

Panel B: Key varieties

Adoption status	----------Baseline----------		----------Follow-up----------	
	Percentage of farmers (%)	Paddy yield (kg/ha)	Percentage of farmers (%)	Paddy yield (kg/ha)
Neither Chupa, Mamima, nor Mocuba	53	1678	38	1698

(continued)

Table 6.4 (continued)

Panel B: Key varieties

| | ----------Baseline---------- | | ----------Follow-up---------- | |
Adoption status	Percentage of farmers (%)	Paddy yield (kg/ha)	Percentage of farmers (%)	Paddy yield (kg/ha)
Variety Chupa	7	1792	22	1493
Variety Mamima	19	1486	15	1572
Variety Mocuba	21	2316**	25	1949

[a] Individual or partial adoption does not include the case of all 5 adoptions; *** $p < 0.01$, ** $p <$ 0.05, the mean difference from the case of "No adoption"(0); Sample size = 257
** $p < 0.05$, the mean difference from the case of "Neither Chupa, Mamima, nor Mocuba"; Sample size = 257
Source Authors

6.5.2 Characteristics of the Full Adopters

The fact that full adopters achieved the highest yield warrants special attention. Table 6.5 compares the full adopters with the non or incomplete adopters of the five key practices by three types of farmers, namely LF, POF, and OF in the treated associations ($n = 179$). Since the number of non-adopters among each farmer type is very small, the qualitative results are the same even if we separate non- and incomplete adopters. Seven features can be identified from the table. First, the proportions of full adopters shown at the bottom of the table indicate that LF achieved the highest adoption (23%), followed by similar proportions by POF (15%) and OF (16%). Given the intensity of the training, it is naturally expected to observe the highest proportion for LF, followed by that of POF. The 16% total for full adoption among OF indicates the existence of farmer-to-farmer diffusion mechanisms or OFs' voluntary training participation.

Second, the full adopters achieved the highest yield for any type. Interestingly, OF shows the largest improvement, and this increase was the only one to achieve statistical significance among the three types of farmers.

Third, we do not find advantages among the full adopters in terms of their socio-economic and agro-ecological conditions, such as household size, education, asset holdings, plot size, or weather conditions. Some variables show statistically significant differences between the full adopters and the non or incomplete adopters, but the differences are not consistent across the three types of farmers.

Fourth, the size of the baseline social networks was measured by the proportion of known LF, POF, or OF among the sample members at the baseline. The results indicate that the full adopters' networks were generally smaller than those of the non or incomplete adopters. This is contrary to our presumption of a social learning mechanism.

Table 6.5 Characteristics of full adopters by farmer's training status

	LF			POF			OF		
	(1)	(2)	t-test	(1)	(2)	t-test	(1)	(2)	t-test
	Non or incomplete adopters	Full adopters	Difference	Non or incomplete adopters	Full adopters	Difference	Non or incomplete adopters	Full adopters	Difference
Variable	Mean/SE	Mean/SE	(2)-(1)	Mean/SE	Mean/SE	(2)-(1)	Mean/SE	Mean/SE	(2)-(1)
Paddy yield (kg/ha) (follow-up)	1506.545 [189.594]	1897.733 [345.916]	391.188	1509.208 [216.710]	1877.083 [422.779]	367.875	1742.170 [104.960]	2380.032 [335.675]	637.862**
Household size (baseline)	3.222 [0.304]	4.250 [0.559]	1.028	4.294 [0.695]	4.000 [1.528]	−0.294	3.913 [0.199]	4.300 [0.493]	0.387
Head's education (years) (baseline)	3.296 [0.662]	5.875 [1.445]	2.579*	3.529 [0.836]	5.333 [2.906]	1.804	3.913 [0.337]	2.100 [0.794]	−1.813**
Log of assets (baseline)	8.005 [0.261]	8.349 [0.336]	0.345	8.303 [0.371]	7.704 [0.621]	−0.599	7.541 [0.150]	6.690 [0.291]	−0.851**
Total plot area (ha) (baseline)	0.960 [0.191]	0.949 [0.581]	−0.011	0.952 [0.255]	0.233 [0.017]	−0.719	0.583 [0.065]	0.758 [0.326]	0.175
Weather shock in the last rice season (= 1) (baseline)	0.000 [0.000]	0.250 [0.164]	0.250***	0.059 [0.059]	0.000 [0.000]	−0.059	0.163 [0.036]	0.200 [0.092]	0.037

(continued)

Table 6.5 (continued)

	---------LF---------			---------POF---------			---------OF---------		
	(1)	(2)	t-test	(1)	(2)	t-test	(1)	(2)	t-test
	Non or incomplete adopters	Full adopters	Difference	Non or incomplete adopters	Full adopters	Difference	Non or incomplete adopters	Full adopters	Difference
Variable	Mean/SE	Mean/SE	(2)-(1)	Mean/SE	Mean/SE	(2)-(1)	Mean/SE	Mean/SE	(2)-(1)
Weather shock in the last non-rice season (= 1) (baseline)	0.704 [0.090]	0.875 [0.125]	0.171	0.765 [0.106]	1.000 [0.000]	0.235	0.788 [0.040]	0.800 [0.092]	0.012
Proportion of baseline known LF (%) (baseline)	8.708 [1.440]	5.286 [2.282]	−3.422	6.596 [1.017]	2.381 [1.190]	−4.215	8.296 [0.706]	1.964 [0.408]	−6.332***
Proportion of baseline known POF (%) (baseline)	2.783 [0.807]	1.839 [1.361]	−0.944	3.679 [1.036]	4.762 [2.381]	1.083	3.205 [0.339]	3.393 [0.915]	0.188
Proportion of baseline known OF (%) (baseline)	19.162 [3.908]	17.214 [10.451]	−1.948	23.334 [6.256]	42.857 [12.542]	19.523	30.962 [2.649]	19.319 [5.274]	−11.644*

(continued)

Table 6.5 (continued)

Variable	LF			POF			OF		
	(1) Non or incomplete adopters	(2) Full adopters	t-test Difference	(1) Non or incomplete adopters	(2) Full adopters	t-test Difference	(1) Non or incomplete adopters	(2) Full adopters	t-test Difference
	Mean/SE	Mean/SE	(2)-(1)	Mean/SE	Mean/SE	(2)-(1)	Mean/SE	Mean/SE	(2)-(1)
Join all 4 demo farm trainings (follow-up)	0.333 [0.092]	0.750 [0.164]	0.417**	0.118 [0.081]	0.000 [0.000]	−0.118	0.231 [0.042]	0.500 [0.115]	0.269**
Knows at least one full adopter among the baseline members	0.185 [0.076]	0.750 [0.164]	0.565***	0.118 [0.081]	1.000 [0.000]	0.882***	0.327 [0.046]	0.850 [0.082]	0.523***
Existence of known LF from whom respondent learned any 5 practices	0.630 [0.095]	0.125 [0.125]	−0.505**	0.471 [0.125]	0.667 [0.333]	0.196	0.452 [0.049]	0.350 [0.109]	−0.102

(continued)

Table 6.5 (continued)

Variable	LF			POF			OF		
	(1) Non or incomplete adopters	(2) Full adopters	t-test Difference	(1) Non or incomplete adopters	(2) Full adopters	t-test Difference	(1) Non or incomplete adopters	(2) Full adopters	t-test Difference
	Mean/SE	Mean/SE	(2)-(1)	Mean/SE	Mean/SE	(2)-(1)	Mean/SE	Mean/SE	(2)-(1)
Existence of known POF from whom respondent learned any 5 practices	0.444 [0.097]	0.000 [0.000]	-0.444**	0.294 [0.114]	0.000 [0.000]	-0.294	0.106 [0.030]	0.000 [0.000]	-0.106
Existence of known OF from whom respondent learned any 5 practices	0.333 [0.092]	0.000 [0.000]	-0.333*	0.118 [0.081]	0.000 [0.000]	-0.118	0.058 [0.023]	0.000 [0.000]	-0.058
N	27	8		17	3		104	20	
Proportion of full adopters		23%			15%			16%	

(continued)

Table 6.5 (continued)

	--------LF--------			--------POF--------			--------OF--------		
	(1)	(2)	t-test	(1)	(2)	t-test	(1)	(2)	t-test
	Non or incomplete adopters	Full adopters	Difference	Non or incomplete adopters	Full adopters	Difference	Non or incomplete adopters	Full adopters	Difference
Variable	Mean/SE	Mean/SE	(2)-(1)	Mean/SE	Mean/SE	(2)-(1)	Mean/SE	Mean/SE	(2)-(1)
F-test of joint significance (F-stat)			3.227***			65.413***			4.906***
F-test, number of observations			35			20			124

The values displayed for t-tests are the differences in the means across the groups

The values displayed for F-tests are the F-statistics

***, **, and * indicate significance at the 1, 5, and 10% critical levels

Fifth, the project provided five training sessions at the demo plots, and among them, four trainings were relevant to the adoption of the five key practices. Hence, the dummy variable for the participation of all four trainings is created. The table shows that the full adopters were more likely to be the farmers who joined all four trainings, except for the case of POF, where the sample size and therefore the number of full adopters was very small. This indicates the importance of completing the demo farm training for fully adopt the five key practices.

Now we turn to the analysis of social learning or F2FE. The sixth feature is that the full adopters are more likely to have acquaintances who would also lean toward being full adopters (see the variable "Knows at least one full adopter among baseline members"). This feature seems to imply that there are two mechanisms of farmer-to-farmer knowledge dissemination. The first is that similar persons are likely to know each other (a correlated social effect or positive sorting/assortative matching). Second, acquaintances learn new practices from each other (a social learning effect). Statistically disentangling these two effects is difficult unless the researcher can identify the independent variables.

Seventh, in order to obtain insights into the abovementioned identification issue, we created a variable defined as the proportion of known LF, POF, or OF among the members from whom the respondent learned any of the five key practices. For example, in the case of LF, the denominator is the total number of sample members, and the numerator is the number of LF from whom the respondent learned any practices. We compare this proportion between the full adopters and the non- or partial adopters. The table indicates that, in LF's case, learning from the other LF, POF, or OF members was much less for the full adopters and it was statistically significant. Besides, in the POF's and OF's cases, learning was also lower among the full adopters (except in the case of learning by POF from the other LF members), although these figures are not statistically significant. All in all, the results do not demonstrate that social learning or F2FE was a strong channel of diffusion, at least in the duration between the baseline and follow-up survey periods.

Meanwhile, as indicated by the dummy of full training participation, our survey indicated the effectiveness of FFS for full adoption. To investigate this aspect, we constructed Table 6.6, which lists the most important information sources for new practices among adopters in the follow-up season. The sources are classified into six categories: through demonstration plot participation, from extension workers, from other farmers, through observation of the plots of unrecognized farmers, and cases where the practice was already known prior to the training. The results indicate that the demonstration plots or the extension workers were the two key sources where the farmers were exposed to the new practices for the first time, indicating that these two key components of FFS can effectively make farmers aware of these new practices. If this is the case, however, the cost of disseminating new rice production management practices to a large number of rainfed farmers in this country will be high.

Table 6.6 Practice adoption and source of the most important information in the follow-up survey

Practices	Source of information among adopters (%)				
	Demonstration plot participation	Extension workers	From other farmers	Observation	Ever known
(S) Seed test by water	39.39	55.56	4.05	0	0
(N) Nursery bed set up	39.40	55.60	5.05	0	0
(B) Bund construction	44.12	25.49	7.84	4.90	17.65
(L) Leveling	37.62	56.44	0	3.96	1.98
(TP) Straight-row transplanting	33.33	63.89	2.78	0	0
Rice variety (Mamima)	12.82	12.82	0	10.26	64.1
Rice variety (Mocuba)	9.52	68.25	7.94	7.94	6.35
Rice variety (Chupa)	29.82	38.60	3.51	15.79	12.28

Sample size = 257
Source Authors

6.6 Conclusion

This chapter evaluated the RCT of rice farm management training in the rainfed lowlands of Mozambique. Our analyses found a positive impact from the training on the adoption of recommended practices and rice yield. The ITT effect on paddy yield was 447–546 kg/ha (or an increase of between 29 and 36% of the control group average yield). This impact was achieved through the adoption of basic practices alone without modern inputs, indicating that even poor farmers in remote areas can benefit from management practice training.

Our analysis suggests that the full adoption of all five key practices was important for increasing the yield, and FFS was effective in achieving this purpose. Meanwhile, our data did not clearly indicate the dissemination of practices through F2FE or social learning, at least in our survey period. Among many possibilities, one possible reason for the ineffectiveness of F2FE in our survey area can be attributed to the diverse agro-ecological conditions of the rainfed areas. Since plot characteristics are highly heterogeneous among the farmers in rainfed areas, appropriate practices may differ among the plots. Hence, the practices that farmers acquire through social learning may not be appropriate for their own plots, and thus, simply mimicking what they see may not be effective. In the long run, however, as the significant impact of practices becomes well understood (and thus the adopted farmers themselves become good instructors), social learning mechanisms may emerge. Further research on external validity as well as long-run impact assessment would provide a better understanding of the appropriate training design in rainfed field-dominant areas in SSA in general and Mozambique in particular.

Appendix

See Tables 6.7 and 6.8

Table 6.7 Estimation results for the non-attrition probit model

	Non-attrition = 1
Household size	−0.00415
	[0.9165]
Head's education (years)	−0.0335*
	[0.0800]
Head's education squared	0.000253*
	[0.0705]
Log of assets	0.0866*
	[0.0600]
Total plot area (ha)	−0.594
	[0.6500]
Total plot area squared	0.521
	[0.4050]
Proportion of known members (%)	−0.0107*
	[0.0630]
Weather shock in the last rice season (= 1)	−0.341
	[0.1200]
Weather shock in the last non-rice season (= 1)	−0.248
	[0.3970]
Constant	1.382**
	[0.0235]
Observations	311

Wild bootstrap cluster robust p-values in brackets
** $p < 0.05$, * $p < 0.1$
Source Authors

Table 6.8 Full estimation results of ANCOVA model on the impact of training: rice productivity and technology adoption

Variables	(1) Paddy yield	(2) Seed test by water	(3) Nursery bed set up	(4) Bund	(5) Leveling	(6) Straight-row TP	(7) Use all 5	(8) Use mamima	(9) Use Mocuba	(10) Use Chupa
Y0	0.139**	-0.0585	0.214**	0.0530	0.0596	-0.207	N/A	-0.0325	0.141	0.0469
	[0.0250]	[0.1760]	[0.0230]	[0.6040]	[0.2805]	[0.8035]		[0.6185]	[0.1710]	[0.4885]
Demo 1 (treatment)	545.5*	0.570***	0.592***	0.376**	0.609**	0.508*	0.367*	0.0903*	0.0895	0.0899
	[0.0795]	[0.00085]	[0.00005]	[0.0440]	[0.0390]	[0.0750]	[0.06635]	[0.0985]	[0.3710]	[0.6010]
Demo 2 (treatment)	447.5*	0.479*	0.461***	0.326**	0.416**	0.449**	0.200	-0.00583	-0.0568	0.289
	[0.0650]	[0.0710]	[0.0000]	[0.0265]	[0.0400]	[0.0100]	[0.2015]	[0.8485]	[0.7730]	[0.1380]
Household size	-50.13	0.00336	0.0117	-0.0886**	-0.0470	-0.0124	-0.0101	-0.0516	0.0243	0.0241
	[0.7935]	[0.9415]	[0.7195]	[0.0440]	[0.4720]	[0.6835]	[0.6135]	[0.1060]	[0.6995]	[0.5025]
Household size squared	6.379	-1.21e-05	-0.00158	0.00943**	0.00386	0.000529	0.00189	0.00537	-0.00286	-0.00510
	[0.7400]	[0.9980]	[0.6385]	[0.0310]	[0.5495]	[0.8305]	[0.3550]	[0.1430]	[0.5665]	[0.1475]
Head's education (years)	-9.454	-0.00581	-0.0110	-0.00308	6.11e-05	-0.0131	0.00323	-0.00110	-0.00589	0.00546
	[0.5520]	[0.4375]	[0.2720]	[0.6505]	[0.9975]	[0.2320]	[0.6085]	[0.7480]	[0.1570]	[0.5470]
Log of assets	52.09	0.0221	0.000723	0.0210	0.000923	0.0118**	0.00708	-0.00833	0.0223	-0.0249*
	[0.1280]	[0.1715]	[0.9640]	[0.1750]	[0.9670]	[0.0115]	[0.1505]	[0.1940]	[0.1080]	[0.0655]

(continued)

Table 6.8 (continued)

Variables	(1) Paddy yield	(2) Seed test by water	(3) Nursery bed set up	(4) Bund	(5) Leveling	(6) Straight-row TP	(7) Use all 5	(8) Use mamima	(9) Use Mocuba	(10) Use Chupa
Total plot area (ha)	-1,563***	-0.0910	-0.0191	0.0476	-0.201*	-0.0425	-0.141	0.156*	0.0575	0.0456
	[0.00005]	[0.4460]	[0.8800]	[0.5050]	[0.0755]	[0.6575]	[0.1020]	[0.0620]	[0.4120]	[0.3550]
Total plot area squared	240.6***	0.0367	0.0141	-0.00230	0.0574**	0.0246	0.0353*	-0.0295	-0.0123	-0.00121
	[0.0000]	[0.1000]	[0.6005]	[0.8695]	[0.0270]	[0.1820]	[0.0675]	[0.1410]	[0.3420]	[0.9190]
Proportion of known members (%)	3.53	-0.00178	-0.000388	0.00213	0.000296	0.000119	0.000352	-0.000444	0.000147	-0.00120
	[0.1385]	[0.4065]	[0.8235]	[0.1070]	[0.9145]	[0.9570]	[0.9355]	[0.7640]	[0.9340]	[0.3040]
Weather shock rice	130.8	0.0317	-0.0932	0.0340	-0.152	0.0719	0.0852	-0.0312	-0.0490	-0.104
	[0.6210]	[0.6850]	[0.2270]	[0.2275]	[0.3960]	[0.3735]	[0.3490]	[0.6725]	[0.2080]	[0.3905]
Weather shock non-rice	-140.4	-0.136	-0.0140	0.0651	0.125	-0.0616	-0.0155	0.0665*	-0.0308	0.0596
	[0.7535]	[0.1005]	[0.8585]	[0.1620]	[0.1460]	[0.3565]	[0.5815]	[0.0860]	[0.5400]	[0.3820]
Constant	1,735***	0.222	0.266*	0.173	0.0178	0.0317	0.0190	0.0961	0.226	0.316***
	[0.0010]	[0.1255]	[0.0960]	[0.4895]	[0.9205]	[0.8035]	[0.8065]	[0.4015]	[0.4950]	[0.0080]
Local unit FE	Yes	Yes	Yes	Yes	Yes	Yes	Yes	Yes	Yes	Yes
t-test (Demo 1 = Demo 2)	0.8372	0.7911	1.3768	0.5446	1.3559	0.4146	1.7405	2.0472*	1.8529	-1.7335
	[0.5005]	[0.5795]	[0.3330]	[0.7695]	[0.4005]	[0.7605]	[0.1900]	[0.0570]	[0.1910]	[0.2235]
Control mean value	1535	0.141	0.333	0.192	0.038	0.00	NA	0.167	0.231	0.064
Observations	257	257	257	257	257	257	257	257	257	257
R-squared	0.363	0.300	0.413	0.510	0.395	0.380	0.418	0.403	0.525	0.302

Wild bootstrap cluster robust p-values in brackets; Inverse probability weights are used to control for attrition bias (see Table 6.7 for the probit analysis of non-attrition)

N/A: No variation in the baseline observations. *** $p < 0.01$, ** $p < 0.05$, * $p < 0.1$

Source Authors

References

Angrist JD, Pischke J (2009) Mostly harmless econometrics. Princeton University Press, Princeton, NJ

Balasubramanian VM, Sié M, Hijimans RJ, Otsuka K (2007) Increasing rice production in sub-Saharan Africa: challenges and opportunities. Adv Agron 94:55–133

Barker R, Hardt RW (1985) The rice economy of Asia. Resources for the Future, Washington DC

deGraft-Johnson M, Suzuki A, Sakurai T, Otsuka K (2014) On the transferability of the Asian rice green revolution to rainfed areas in sub-Saharan Africa: an assessment of technology intervention in Northern Ghana. Agric Econ 45(5):555–570

FAO (2021) AQUASTAT-FAO's global information system on water and agriculture. https://www.fao.org/aquastat/en/. Accessed 24 May 2021

Hossain M (2006) Rice in Africa. Rice Today 5(1):41

Kajisa K (2015) On the determinants of low productivity of rice farming in Mozambique: pathways to intensification. In: Otsuka K, Larson DF (eds) Pursuit of an African Green Revolution: views from rice and maize farmers' fields. Springer. Heidelberg

Kajisa K, Payongayong E (2011) Potential of and constraints to the rice Green Revolution in Mozambique: a case study of Chokwe irrigation scheme. Food Policy 36:614–625

Kijima Y, Ito Y, Otsuka K (2012) Assessing the impact of training on lowland rice productivity in an African setting: evidence from Uganda. World Dev 40(8):1610–1618

McKenzie D (2012) Beyond baseline and follow-up: the case for more T in experiments. J Dev Econ 99(2):210–221

Ministério da Agricultura e Segurança Alimentar (2015) Anuário de Estatísticas Agrárias 2015. Direcção de Planificação e Cooperação Internacional (DPCI)

Nakano Y, Tanaka Y, Otsuka K (2018) Impact of training on the intensification of rice farming: evidence from rainfed areas in Tanzania. Agric Econ 49(2):193–202

Otsuka K, Larson DF (eds) (2013) An African Green Revolution: finding ways to boost productivity on small farms. Springer, Dordrecht

Otsuka K, Larson DF (2016) In pursuit of an African Green Revolution: views from rice and maize farmers' fields. Springer, Dordrecht

Roodman D, MacKinnon J, Nielsen M, Webb M (2019) Fast and wild: bootstrap inference in Stata using boottest. Stata J 19(1):4–60

USDA (2021) PSD online. https://apps.fas.usda.gov/psdonline/app/index.html#/app/downloads. Accessed 21 April 2021

Wooldridge JM (2010) Econometric analysis of cross section and panel data, 2nd edn. MIT Press, Cambridge

Kei Kajisa is a professor at School of International Politics, Economics and Communication, Aoyama Gakuin University. He worked at the International Rice Research Institute from 2006 to 2012. He obtaned a Ph.D. in Agricultural Economics from Michigan State University in 1999. His area of specialization is international agricultural development.

Trang Thu Vu is a Research Associate at the Asian Development Bank Institute (ADBI), Tokyo, Japan. She previously worked as a Research Assistant at the JICA Ogata Research Institute (JICA-RI). She obtained a Ph.D. degree in Development Economics from the National Graduate Institute for Policy Studies (GRIPS), Japan in 2022. Her current research focus is on impact evaluation, micro econometrics, agriculture, education, health and behavioral development economics.

Part III
Complementary Development Strategies

Chapter 7
Intensification of Rice Farming: The Role of Mechanization and Irrigation

Hiroyuki Takeshima and Yukichi Mano

Abstract PART III of this volume discusses complementary strategies toward the rice Green Revolution in sub-Saharan Africa (SSA), including mechanization, irrigation, and marketing efforts to enhance demand for domestic rice. In Asia, population pressure on land and high rice prices relative to input prices induced farming intensification, while mechanization and irrigation development boosted rice productivity. However, in SSA, increased rice demand has been partly met by rice imports without adequate investments in these important technologies. This chapter reviews the available evidence on the constraints and potentials of these complementary strategies in SSA. In particular, we argue that inadequate understanding of the complementarity of each strategy with rice cultivation training may be an important reason behind low technology adoption. The remaining chapters in PART III attempt to narrow this gap and provide relevant empirical evidence. Intensive land preparation using tractors facilitates the adoption of management-intensive practices, while the provision of rice cultivation training is associated with the improved economic viability of irrigation investment. Increased demand for domestic rice is also expected to raise the returns to rice cultivation training and technology investment. The adoption of modern milling technologies improves the quality of domestic milled rice and promotes rice value-chain transformation. Local traders offer quality-based pricing that incentivizes farmers to supply high-quality paddy, while urban traders and supermarkets sell domestically produced and milled rice with brand names at a premium. In sum, the use of tractors, irrigation investment, and modern milling, combined with improved rice marketing, play critical roles that complement rice cultivation training in achieving the rice Green Revolution in SSA.

H. Takeshima (✉)
International Food Policy Research Institute (IFPRI), Development Strategy and Governance Division, 1201 Eye Street, NW, Washington D.C. 20005, USA
e-mail: H.Takeshima@cgiar.org

Y. Mano
Graduate School of Economics, Hitotsubashi University, 2-1 Naka, Isono Building Room 324, Kunitachi-shi, Tokyo 186-8601, Japan
e-mail: yukichi.mano@r.hit-u.ac.jp

© JICA Ogata Sadako Research Institute for Peace and Development 2023 143
K. Otsuka et al. (eds.), *Rice Green Revolution in Sub-Saharan Africa*, Natural Resource Management and Policy 56, https://doi.org/10.1007/978-981-19-8046-6_7

7.1 Introduction

During the Green Revolution era in Asia, population pressures on land and high rice prices relative to input prices kick-started farming intensification, while mechanization and irrigation investments boosted rice productivity (Barker et al. 1985; David and Otsuka 1994). Between 1950 and 1970, expanded irrigation area and tractor use each increased rice production by 26–30% and 15–16%, respectively, in Asian countries where more than 25% of the farm area was irrigated (Barker et al. 1985, Table 4.19).[1] Despite the significant gains realized during the rice Green Revolution in Asia, investment in these strategies remains low in sub-Saharan Africa (SSA), and rice imports from Asia are required to meet increased rice demand (Chap. 1 of this volume). This low investment is puzzling considering the transferability of improved rice cultivation technologies from Asia to SSA (see Chap. 2 of this volume). It can be hypothesized that some technical constraints interfere with investments in these strategies. Alternatively, the expected returns may be low because of SSA's unique natural and institutional conditions.

This chapter reviews the literature on the constraints and potentials of investment in regard to mechanization and irrigation development for rice production in SSA. We explore mechanization not only for land preparation and harvesting/threshing but also for modern rice milling. We consider properly mechanized rice milling essential because recent studies find that improved rice milling technologies enhance the demand for domestic rice through quality improvement (Tokida et al. 2014; Mano et al. 2022; Chap. 12 of this volume). We focus on demand-side, supply-side, and natural conditions among investment constraints for mechanization and irrigation development. The demand-side factors of mechanization initially include farming intensification due to land pressures (Diao et al. 2020; Boserup 1965; Pingali et al. 1987), and later, increased farm wages (Diao et al. 2016, 2020; Hayami and Ruttan 1970, 1985). The demand-side factors of irrigation development include rice prices (Kikuchi et al. 2021; Chap. 10 of this volume) and the complementary adoption of seed-fertilizer technologies (Otsuka and Larson 2012, p. 18).[2] The supply-side factors of mechanization include the profitability of hiring-out services, constrained by low cropping intensity and few opportunities for multi-functional use (Diao et al. 2020). The supply-side factors of irrigation development include the construction and management cost of irrigation schemes. The natural conditions affecting mechanization include the heavy soil in SSA, while those for irrigation include surface water and groundwater availability. We also explore whether mechanization facilitates farm size expansion and the adoption of improved rice cultivation practices and examine the extent to which irrigation increases rice yield.

---

[1] These countries are the Philippines, Indonesia, Malaysia, Sri Lanka, Pakistan, India, China, and South Korea.

[2] Irrigation demand here refers to farmers' demand for private irrigation systems and demand for public irrigation systems as perceived by irrigation authorities.

Our literature review finds that SSA faces more significant constraints than Asia. Mechanization demand remains low in SSA partly due to farmers' inadequate knowledge of improved cultivation practices. This is a result of the shorter history of rice farming and limited use of animal traction, thereby reducing expected investment returns. Irrigation demand is also limited in SSA partly because of the low global rice prices realized after the Green Revolution in Asia. For the supply side, machinery service provision is constrained by heterogeneous soil types and cropping systems within each locality. These depress utilization rates of machines, a factor that is particularly constraining for large tractors that are popular due to the perception that soil in SSA is heavy. Irrigation development is tricky in SSA because of higher construction costs and improper operation and management. Rice yield improvement in irrigated environments is notable but still modest in contrast to its high potential in SSA.

Several additional issues related to investment in mechanization and irrigation have only recently become apparent and are awaiting more evidence. One such issue is the complementarity of these technologies with rice-farming intensification and improved cultivation practices. For example, proper land preparation using tractors can facilitate the adoption of management-intensive production practices (Mano et al. 2020; Chaps. 8 and 9 of this volume). Providing rice cultivation training is also associated with the improved economic viability of irrigation investment, as suggested in Chap. 10. Another issue is evaluating irrigation investment in a broader framework. We argue that standard cost–benefit analysis of irrigation development fails to consider the spillover (i.e., general equilibrium) effect on the related sectors, including agricultural input suppliers. The discussion also extends to the strategies to improve the demand for local rice, which enhances the investment returns to rice cultivation training, mechanization, and irrigation development. Millers adopt modern milling technologies to improve the quality of milled rice (Mano et al. 2022; Chap. 12 of this volume), while providing farmers with information about paddy quality-enhancing technologies and quality parameters appreciated by the market induces them to supply more aromatic varieties of paddy outside the village and receive a higher sales price (Chap. 13 of this volume). The remaining chapters in PART III examine these emerging issues.

Section 7.2 describes the status of mechanization and irrigation for rice farming in SSA. Section 7.3 discusses the constraints of mechanization for rice production in SSA, while the constraints and potentials of irrigation are considered in Sect. 7.4. Section 7.5 argues that some emerging issues require more empirical evidence. Section 7.6 summarizes this chapter.

7.2 Farm Mechanization, Irrigation, and Rice Milling in SSA

The extent of farm mechanization and irrigation remains low in SSA. Stocks of agricultural machines, like tractors and harvester-threshers, have stagnated in the region (Fig. 7.1) in contrast to the substantial growth in India and the rest of South Asia,

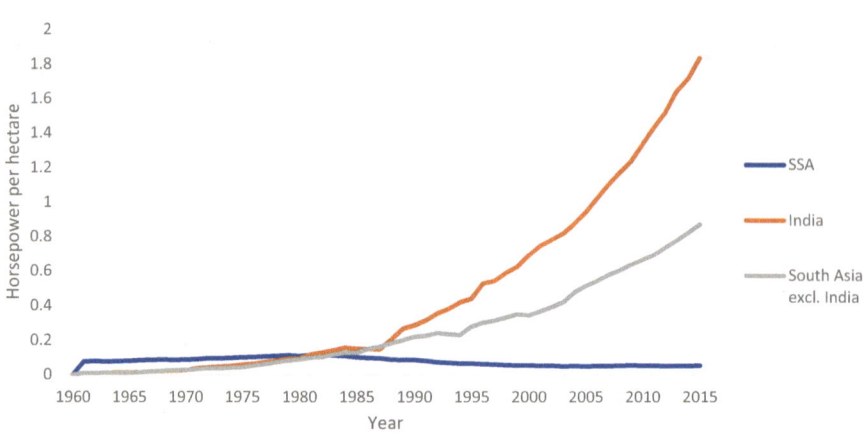

Fig. 7.1 The stock of major machines in use (horsepower per hectare of cultivated areas) in sub-Saharan Africa, India, and the rest of South Asia over time. *Source* USDA (2021). *Note* Horsepower is aggregated over four-wheel tractors, two-wheel tractors, harvester-threshers, and milking machines

where the agro-ecological conditions are similar to SSA. Although animal traction formerly played a significant role in rice-farming intensification in Asia,[3] the rising cost of draft animals combined with a farm wage increase induced tractorization (Binswanger 1978). In contrast, the use of draft animals has been less common in SSA due to the prevalence of sleeping sickness (trypanosomiasis) (Alsan 2015), coupled with deteriorating animal health services and recurring droughts (Takeshima et al. 2015, 2013). These factors have left farmers less familiar with intensive land preparation (Diao et al. 2020). The share of cultivated land under irrigation remained around 3% in SSA from 1973 to 2013. In contrast, it almost doubled from 24 to 46% in South Asia (Fig. 7.2),[4] slightly below the 56% achieved across the whole of Asia by 2005 (Seck et al. 2012).

Limited household-level data suggest that in SSA, mechanization and irrigation development for rice farming are similar to the adoption levels for general agriculture and substantially limited compared to India. For selected countries with readily available nationally representative data, the shares of farm households using tractors were lower in the selected SSA countries than the level in India (90%) in the mid-2010s (Table 7.1). The irrigation use in rice farming was also much lower in the SSA countries than in India.

---

[3] Ox-plows were on a majority of farmland in India in the mid-1960s (Government of India 1977, cited in Delgado and McIntire 1982), and 98% of land was plowed by bullock in Bangladesh in the mid-1980s (IRRI 1986).

[4] The current irrigation adoption rate in SSA may be underestimated (Woodhouse et al. 2017). For example, FAO's AQUASTAT may exclude some form of farmer-devised private irrigation.

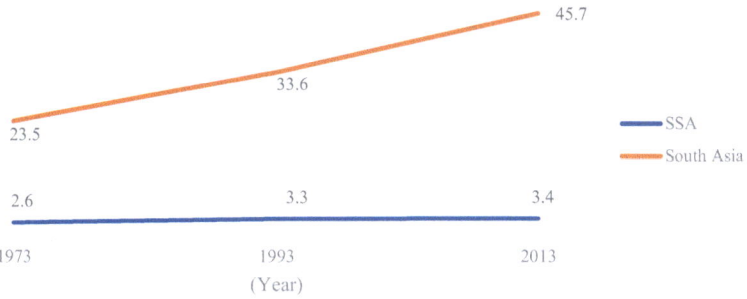

Fig. 7.2 The share of cultivated land under irrigation (%). *Source* FAO (2021)

Table 7.1 Share of farm households using tractors, animal traction, and irrigation for all crops and rice farming in selected SSA countries in the mid-2010s (%)

	Tractors		Animal traction		Irrigation		Data
	All farm	Rice farm	All farm	Rice farm	All farm	Rice farm	
Ghana[a]	13	23	N/A	N/A	2	4	2017 GLSS
Nigeria	10	15	25	39	4	14	2018 LSMS-ISA
Tanzania	9	12	37	41	3	2	2014 LSMS-ISA
Uganda[a]	<1	3	7	20	<1	<1	2018 LSMS-ISA
India[b]	90	90	N/A	N/A	45	57	See table note

Source Authors' calculations. Figures are nationally representative, adjusted using sample weights
[a]"Rice farm" figures for Ghana and Uganda comprise the use of tractors and irrigation among rice-farming households. However, these might not have been used for rice farming and may be used with non-rice crops
[b]Figures for India are area shares based on secondary national-level data. The shares of tractor-using farm areas and tractor-using rice farm areas are obtained from Diao et al. (2020). The shares of the irrigated area among all farm areas and rice farm areas are figures for 2017 from FAOSTAT and AQUASTAT. The figure for "all farm" is the proportion of area fully equipped for irrigation to total arable land. The figure for "rice farm" is the proportion of harvested irrigated rice area to total harvested rice area

As for the types of machinery used, two-wheel tractors (power tillers) have been popular in Asia due to the dominance of lowland ecology (Binswanger 1978),[5] even though four-wheel tractors have also been in use since the 1990s. In SSA, four-wheel tractors are more commonly used in upland areas or sometimes before flooding in irrigated rice farms, while puddling and other operations are implemented using two-wheel tractors in some areas of SSA (Diao et al. 2020). Data are scarce for other power-intensive farm activities, such as harvesting and milling. However, casual observation suggests that harvesting is primarily implemented manually, while large

[5] In the mid-1980s, when mechanization took off in South Asia (Barker et al. 1985, Table 3.1), 87% of rice was grown in lowland areas, while close to 40% of rice growing was in upland areas in SSA in 2010 (Seck et al. 2012).

combine harvesters are used in small pockets of intensive production areas rather than the smaller reapers or smaller combines typically used in Asia.

Similarly, SSA lags behind Asia in the modernization of the milling sector. In SSA, small-scale mills perform most rice milling, with few large-scale mills operating modern milling machines (Soullier et al. 2020). In contrast, in Asia, medium and large rice mills with upgraded milling equipment have grown substantially, improving the quality of milled rice, whereas small mills were forced to exit the market (Reardon et al. 2012). Rice millers in Asia have also often engaged in vertical coordination with farmers (Basu 1983), including interlinkage deals (i.e., contract farming),[6] along with supermarkets, large urban retailers, and wholesalers (Reardon et al. 2012). Most mills in SSA have not adopted modern rice milling machines or do not engage in such vertical coordination (Soullier et al. 2020).[7]

7.3   Roles of Mechanization in SSA

This section explores the reasons behind the low investment in mechanization in SSA through a literature review on the adoption of agricultural machines. Mechanization facilitates farmland expansion and farming intensification, which is particularly relevant now because the land-labor ratio in SSA has reached a level as low as that of tropical Asia in the 1960s, at the beginning of the Asian Green Revolution (Chap. 1 of this volume).[8]

The demand-side factors for mechanization include farming intensification due to population pressure on land and high rice prices (Diao et al. 2020; Boserup 1965; Barker et al. 1985, p. 111; Pingali et al. 1987). The farm wage increase due to urbanization and economic growth also facilitates the substitution of labor with machinery (Diao et al. 2016, 2020; Hayami and Ruttan 1970, 1985). However, mechanization has been limited in SSA. Increased rice demand has been partly met by rice imports in SSA because production has remained unresponsive to increased rice demand (Gyimah-Brempong et al. 2016) and the higher quality of imported rice (Mano et al. 2022; Chaps. 12 and 13 of this volume). The adoption of high-yielding seed-fertilizer technologies also raises returns to intensive tractor use for land preparation (e.g., Takeshima and Liu 2020). However, farmers' knowledge of intensive land preparation and cultivation practices often remains inadequate in SSA (e.g., deGraft-Johnson et al. 2014) because of the short history of modern rice cultivation and limited practice of animal traction (Alsan 2015; Diao et al. 2020).

---

[6] In such contracts, rice mills provide input credit and cultivation advice and receive paddy after harvest. In Latin America, rice mills also provide farmland and irrigation water (Bierlen et al. 1997).

[7] In West Africa, less than 1% of mills engage in vertical integration (Soullier et al. 2020).

[8] Mini tillers raised yield in mid-hills in Nepal (Paudel et al. 2019), which may be applied to hilly areas in SSA. In Tanzania and India's semi-arid region, using combine-harvesters significantly increased yield through reduced crop losses (Diao et al. 2020; Wilson and Lewis 2015). Mechanization also facilitates diversification (i.e., economies of scope) between rice and non-rice crops (Takeshima et al. 2020; Pingali 2007).

Four-wheel tractors are more popular than two-wheel tractors in SSA because they are considered more suitable for heavy soil (Diao et al. 2016).[9] By contrast, two-wheel tractors are ideal for lowland rice farming, where the soil is softened by stored water. Mechanization rates tend to be higher in SSA countries with more active land markets, enabling large tractors to achieve economies of scale (Takeshima et al. 2018; Jayne et al. 2019). However, smallholders remain dominant in the rice sector in SSA due partly to inactive land markets.[10] As in Asia, the development of the machinery service hiring market is critical to realizing the potentially high returns to machinery investment (Binswanger 1986; Diao et al. 2020).

The supply-side factors of mechanization include constraints on service providers to capitalize on a scale economy and recover investment costs. Heterogeneous soil types and cropping systems within each locality prevent machinery service coordination and provision in SSA (Diao et al. 2020). Compared to Asia, the machinery utilization rate is also lower in SSA due to there being fewer production seasons because of the dominance of rainfed agriculture and fewer opportunities for multifunctional use of machinery, such as its use as a power source to run irrigation pumps (Binswanger 1986; Diao et al. 2020). Poor road infrastructure in SSA also prevents service providers from reaching break-even utilization rates through extensive migratory services. Furthermore, inadequate, unstable paddy production also constrains investments in modern rice mills, which often require sufficient utilization rates to be profitable (Gyimah-Brempong et al. 2016; Takeshima et al. 2017).

7.4 Roles of Irrigation in SSA

In the twentieth century, significant irrigation expansion was driven by the development of large-scale public irrigation schemes (Inocencio et al. 2007). In SSA, however, the long-term decline in global rice prices brought about by the Asian Green Revolution has made it difficult to justify new large-scale irrigation projects, even though the population growth and urbanization have increased the demand for rice. Indeed such projects had nearly disappeared by the late 1990s (Kikuchi et al. 2021; Chap. 10 of this volume).

Despite the popular belief that SSA has constraints on irrigation development due to the paucity of surface and groundwater endowments, SSA has considerable irrigation expansion and groundwater potentials (World Bank 2007, Box 2.5). Potential irrigation areas in SSA are as many as 22 million hectares compared with 7 million

---

[9] Diao et al. (2020) provide some indicative evidence that more SSA countries belong to land areas categorized as having greater soil workability constraints, for which larger, powerful tractors are arguably more suitable. However, it is still unclear whether the differences are substantial enough to limit the adoption of smaller tractors like power tillers in SSA today. Besides this, soil data for regions like SSA remain poor and require further improvement.

[10] Scale economy in rice production also emerged in Japan as large tractors were introduced in the 1980s (Hayami and Kawagoe 1989), whereas the inactive land rental market has prevented the further structural transformation of Japanese agriculture (Hayami and Godo 2001).

hectares already realized as of 2010 (You et al. 2011), including 6–14 million hectares in drylands (Xie et al. 2018). Groundwater potentials exploitable for irrigation expansion are also tremendous (Cobbing and Hiller 2019), including more than 10,000 km^3 found in Nigeria, Ethiopia, Angola, Botswana, South Africa, and Kenya (Cassman and Grassini 2013). However, there is heterogeneity in water availability because of different evapotranspiration rates (Burney et al. 2013) and few major rivers (Otsuka and Kalirajan 2006), as well as growing water scarcity due to population pressure[11] and climate change (World Bank 2007, p. 65 and Focus F).

The higher construction cost of public irrigation schemes than elsewhere has also prevented irrigation investment in SSA (Inocencio et al. 2007; World Bank 2007). Investment costs per hectare irrigated in West Africa have been three times higher than in Asia because poorly managed water supplies enabled only single cropping, reducing the effective irrigated area (Schoengold and Zilberman 2007, p. 2970). Furthermore, rice farming requires more significant infrastructure investments, such as dams and reservoirs, than non-rice crops, which require less water (Inocencio et al. 2007). Irrigation management is also critical for realizing the potential irrigation returns in SSA,[12] and some regional-level governance structures have been criticized for low efficiency (e.g., Barrow 1998). In Mali, for example, irrigation management reforms involving greater farmer participation quadrupled rice yield and increased rice income between 1982 and 2002 (World Bank 2007). Successful management transfer depends on legal frameworks defining the responsibilities of water user groups (World Bank 2007) and requires significant technical and managerial support for entities newly tasked with water management (Cambaza et al. 2020), which is inadequate in many SSA countries.

However, these practical difficulties do not mean that SSA should abandon the development of irrigation schemes. Inocencio et al. (2007) suggest that "successful schemes" in SSA are comparable with irrigation schemes elsewhere. While some studies indicate that small irrigation systems may be more sustainable (Burney et al. 2013), efficient expansion of irrigation schemes is possible in SSA with scale economy, improved management, and the right mix of system components (Inocencio et al. 2007; Faltermeier and Abdulai 2009; Akpoti et al. 2021). In expanding irrigation infrastructure for rice in SSA, it is crucial to identify appropriate areas based on agro-ecological and market conditions where rice irrigation can bring higher returns than other high-value crops (World Bank 2007; Herdt 2010).[13]

---

[11] A country is considered water-scarce if annual internal renewable water resources are less than 1000 m^3 per capita (Rosegrant and Perez 1997). As recently experienced in South Asia, intensifying water scarcity may lower the net returns to rice relative to other crops that require less irrigation water, like vegetables (Shah et al. 2009).

[12] In Asia, communal management of irrigation systems sustained the initial success of the Green Revolution (Otsuka and Kijima 2010), and public irrigation schemes also benefited from management decentralization (Alaerts 2020).

[13] There has been growing competition over water between rice and non-rice crops in SSA (e.g., Qadir et al. 2003). Because of intensifying water competition across sectors, arsenic-contaminated water used for irrigation in South Asia accumulates in produce, especially rice (Balasubramanya and Stifel 2020).

Table 7.2 Agro-climatic potential yield for irrigated and rainfed rice in SSA and elsewhere (lowland rice and upland rice)

Regions	Lowland rice (in areas where irrigation is at least marginally suitable)		Upland rice (in areas where irrigation is at least marginally suitable)
	Irrigated	Rainfed	Rainfed
Sub-Saharan Africa	7.6	5.1	3.2
Rest of the world	7.8	5.5	3.2

Source Authors' calculation based on FAO and IIASA (2021)
Notes These simulation exercises assume production with "high-input levels" (FAO and IIASA 2021), adopting high-yielding varieties, fully mechanized with low labor intensity, and the application of proper nutrients and pest, disease, and weed control chemicals. The results should therefore be interpreted as potentials arising in the long term. Although the simulation assumes the RCP 8.5 climate scenario, results are similar under other RCP scenarios. While FAO and IIASA (2021) limit the analysis to irrigated conditions in certain agro-ecological areas, results for rainfed conditions are also presented exclusively for the same agro-ecological areas for comparison purposes

Irrigation is expected to enhance rice productivity (Nakano et al. 2012), reduce production risk in SSA (Gebretsadik and Romstad 2020),[14] and reduce rice prices, thereby promoting food security and poverty alleviation (e.g., Hanjra et al. 2009; Dillon 2011; Gyimah-Brempong et al. 2016). A secure water supply raises the productivity of modern varieties and chemical fertilizers (Otsuka and Larson 2012, p. 18), and extension efforts are likely to be more effective in irrigated areas (Fuwa et al. 2007; Kijima et al. 2012; Chap. 2 of this volume). While globally, rice yield under irrigated conditions attained 5.3 tons per hectare in 2000 (Dobermann 2000), irrigation in SSA has agro-climatic potential of lowland rice yields that is comparable to that in Asia. Simulation models, such as FAO and IIASA (2021), suggest that irrigation can improve the agro-climatically potential lowland rice yields in SSA compared to rainfed conditions in a similar way as the rest of the world (Table 7.2). Despite such high potentials, farmers attain 2–5 tons per hectare in most irrigated areas in SSA due to irregular irrigation, inadequate input use, and poor cultivation practices (Miézan and Sié 1997; Riddell et al. 2006; Balasubramanian et al. 2007).[15]

---

[14] Reduction in production risk due to irrigation has also been reported in China (Wang et al. 2018) and Indonesia (Gatti et al. 2021).

[15] Although yields in Mali and Senegal are comparable to those in South and Southeast Asia (Cassman and Grassini 2013), they are lower elsewhere in SSA (e.g., Borgia et al. 2012; Gyimah-Brempong et al. 2016). Even within an irrigation scheme, there are variations in irrigation facility utilization rate and rice yield fluctuations over the medium-to-long term caused by soil-nutrient depletion (Dembele et al. 2012; Borgia et al. 2012).

7.5 Emerging Issues

In addition to a somewhat established understanding of constraints on mechanization and irrigation development in SSA, several issues have been increasingly recognized as important, including the complementarity between investments in farm mechanization and irrigation and improved rice cultivation practices. The complementarity between the increased demand for domestic rice through the adoption of modern milling machines and the investments in cultivation training, farm mechanization, and irrigation development has also been increasingly recognized. These issues are important for consideration because ignoring this complex complementarity would underestimate the true potential of investments in farm mechanization, irrigation, and modern rice milling facilities. As shown in the details below, we highlight emerging evidence that intensive land preparation using tractors facilitates the adoption of improved rice cultivation practices, and rice cultivation training is more effective in irrigated areas. Furthermore, improved milling technologies when combined with rice value-chain modernization enhance the quality and price of domestic rice. They also contribute to increased returns to rice cultivation training for timely harvesting and appropriate drying, as well as promoting investment in farm mechanization and irrigation development. We thus argue that insufficient understanding of these issues may reduce expected investment returns and prevent farm mechanization, irrigation development, and modernization of rice value chains. These issues are discussed in the remaining chapters in Part III.

7.5.1 Complementarity with Improved Cultivation Practices

As emphasized in Part II, rice-farming intensification is crucial for increasing rice productivity in SSA. Emerging literature further finds suggestive evidence on the complementary roles of investments in mechanization and irrigation and rice-farming intensification, consistent with the conceptual framework illustrated in Fig. 1.8 in Chap. 1.

In Asia, intensive land preparation enabled by animal traction facilitated farming intensification and the adoption of improved rice cultivation practices, with tractors only substituted for animal traction later (Pingali et al. 1987; Binswanger 1978). By contrast, animal traction has been less common in SSA (Alsan 2015), and intensive land preparation using tractors plays a key role in facilitating rice-farming intensification and the adoption of improved rice cultivation practices, as suggested by emerging studies from Benin, Cote d'Ivoire, and Tanzania (Tanaka et al. 2013; Mano et al. 2020; Chaps. 8 and 9 of this volume). In particular, two-wheel tractors used for intensive land preparation enable even water depth that prevents the overgrowth of weeds and promotes straight-row transplanting, among other improved rice cultivation practices (Baudron et al. 2015). These effects are stronger than four-wheel tractors, the dominant types of tractors in SSA, which may be less suitable because of the difficulty in maneuvering them in heavy muddy fields.

Irrigation is also complementary to rice-farming intensification and improved cultivation practices. Kajisa and Payongayong (2011) find complementary roles of irrigation with fertilizer application and yield enhancement in Mozambique. Based on the literature review, Chap. 2 of this volume finds that rice cultivation training effectively improves rice yield in irrigated and favorable rainfed areas (see Chaps. 3–6 of this volume for evidence from SSA). Furthermore, although the estimated returns to irrigation projects are typically modest in SSA, Chap. 10 of this volume argues that the economic viability of irrigation investment can be enhanced through the provision of proper rice cultivation training.

Increased demand for domestic rice through post-harvest value-chain modernization is also expected to raise the investment returns to mechanization and irrigation development, as well as to rice cultivation training. Recent studies discuss strategies to increase the demand for domestic rice by improving its quality during the milling process (Soullier et al. 2020). The introduction of improved milling machines,[16] including a destoner component, enhances the quality and price of milled rice in Tanzania (Kapalata and Sakurai 2020), Uganda (Tokida et al. 2014), and Kenya (Mano et al. 2022; Chap. 12 of this volume).[17] In the Kenyan case, a significant rice value-chain transformation is observed, and millers increasingly source paddy from local traders and sell milled rice to urban supermarkets and urban traders and consumers. Traders can facilitate farmers to improve paddy quality by informing them about paddy quality-enhancing technologies and introducing quality-based pricing in Ghana (see Chap. 13 of this volume).[18] Nonetheless, few millers engage in contract farming or provide cultivation advice in SSA due to limited paddy availability (Soullier et al. 2020), suggesting a complementarity between strategies to increase rice production and improve rice quality.

7.5.2 Technological Knowledge and Operating Skills

Maximizing the complementarity mentioned above between technologies and improved rice production practices also requires mastering these technologies themselves. Insufficient machinery knowledge and operating skills, including the advantages of different tractor types, machine operations, maintenance, and repairs, may remain key constraints in mechanization. In Nigeria, for example, machine-hiring service providers promoted by the government lacked proper knowledge of machine

---

[16] Emerging modern rice mills reduced milling costs in Ghana's Kpong Irrigation Scheme to a level comparable to Asia (Takeshima et al. 2017).

[17] Large modern rice millers have also voluntarily fortified rice with micronutrients to raise market premiums in countries like Colombia (Tsang et al. 2016).

[18] Bernard et al. (2017) also find that quality-based pricing improves the quality and price of onions in Senegal.

types compared to experienced informal service providers and failed to keep their businesses viable (Diao et al. 2020). In a successful case, after introducing modern milling machines from China, entrepreneurial rice millers in Kenya learned machine operations and maintenance from Chinese suppliers, thereby improving the quality of milled rice to compete with imported Asian rice (Mano et al. 2022; Chap. 12 of this volume). Late adopters also learned such technological information from the early adopters operating in the neighborhood and chose smaller, more improved milling machines than before, a practice that is not yet widespread across SSA.

Knowledge and operating skills of irrigation are also insufficient among farmers in SSA. A review of irrigation challenges in SSA highlights inadequate knowledge and skills in using irrigation water at the farmers' level (Nakawuka et al. 2018). Scientific knowledge of irrigated agriculture may not be adequately transferred from research institutions and universities to farmers. Inadequate knowledge of proper irrigation use, scheduling methods, and benefits from irrigation technologies often results in water wastage and high fuel and labor costs.

Although some essential knowledge can be acquired and diffused by profit-seeking individuals, the question of whether governmental interventions, such as providing training, is necessary for accelerating knowledge acquisition can be an important area for future research.

7.5.3   Other Issues that Affect Irrigated Rice Production in SSA

Several other issues also remain relevant in assessing irrigation's potential contribution to rice-sector growth in SSA. These issues require closer investigation, including the spillover (i.e., general equilibrium) effects of irrigation development to related activities, the environmental externality of irrigation schemes, and financing for irrigation development.[19]

The general equilibrium effects of large irrigation infrastructure development on the related rice-sector activities are strongly felt by input suppliers expanding business with rice farmers cultivating in the irrigation scheme (see Chaps. 10 and 12 of this volume). These effects are usually neglected in the standard cost–benefit analysis of large-scale irrigation investments, which restricts attention to direct benefits to the farmers on the irrigation scheme and local and urban traders and rice millers. However, these spillover effects may bring about significant social benefits triggered by farming intensification and productivity improvement directly associated with the irrigation infrastructure.

The effect of environmental externalities is another issue to be investigated as rice irrigation expands in SSA. Large irrigation infrastructure, such as dams, improves

---

[19] Higher waterborne disease rates in SSA have also been linked to irrigation, including greater malaria incidence through mosquitoes (Malabo Montpellier Panel 2019; Mutero et al. 2004; Schoengold and Zilberman 2007; Asenso-Okyere et al. 2011).

water supply stability in downstream basins while incurring external costs in more upstream basins that experience displacement (e.g., Strobi and Strobi 2011; Duflo and Pande 2007). Dam construction often leaves native populations worse off (Schoengold and Zilberman 2007). SSA needs to deal with these social costs seriously when developing irrigation infrastructure with dams.[20]

Financing irrigation expansion in SSA requires a significant increase in public investments. In Zambia, however, only 3% of the government budget went to irrigation development and other rural infrastructure (World Bank 2007, p. 116, Box 4.8). In Nigeria, public investments in irrigation development used to be comparable to several Asian countries but have substantially declined since the 1990s (Gyimah-Brempong et al. 2016). Countries like Tanzania experimented with providing supplementary funds on a competitive basis to local governments to finance medium-scale irrigation schemes and focused national public spending on inducing private irrigation investment (World Bank 2007, p. 243).

7.6 Summary

Mechanization and irrigation development played significant roles in the Asian Green Revolution, but investments in and updates of these technologies have remained low in SSA. Our literature review finds that such underutilization of these technologies occurs partly because SSA faces greater constraints and heterogeneous potentials. Machinery service provision is constrained by heterogeneous soil types and cropping systems within each locality, while large tractors, typically adopted in SSA, require a higher utilization rate to break even. Irrigation development is constrained by higher construction costs in SSA, while its potential is heterogeneous because of heterogeneity in water availability and irrigation management quality.

However, recent studies also suggest that there is some scope to enhance the returns to these technologies in SSA by exploiting their complementarity with rice-farming intensification and improved cultivation practices. An inadequate understanding of this complementarity and underappreciation of achievable potentials are likely to comprise some of the reasons behind the low investment in and adoptions of these technologies for rice production in SSA. PART III of this volume fills this gap by providing evidence for these complementary strategies, including efforts to promote the supply of high-quality domestic rice through the improvement of rice milling activities toward the Green Revolution in SSA. Chapters 8 and 9 find that land preparation using two-wheel tractors improves productivity and profitability by facilitating fertilizer application and the adoption of improved cultivation practices in Cote d'Ivoire and Tanzania. Chapters 10 and 11 find that the high economic viability of irrigation construction is associated with farming intensification and improved rice cultivation practices in Kenya and Senegal. Chapter 12 examines a case in

---

[20] In the early 2000s, the number of large dams reached 6,575, 4,291, and 2,675 in USA, India, and Japan, respectively, compared to, for example, less than 100 in Nigeria (WCD 2000).

which millers enhanced the quality of milled rice and demand for domestic rice by adopting improved milling technologies in Kenya. This technological improvement in the milling sector also induces rice value-chain transformation and likely increases the demand for high-quality paddy. Chapter 13 analyzes the case in which traders incentivize farmers to provide high-quality paddy by offering quality-based pricing in Ghana.

References

Akpoti K, Dossou-Yovo ER, Zwart SJ, Kiepe P (2021) The potential for expansion of irrigated rice under alternate wetting and drying in Burkina Faso. Agric Water Manag 247:106758

Alaerts GJ (2020) Adaptive policy implementation: process and impact of Indonesia's national irrigation reform 1999–2018. World Dev 129:104880

Alsan M (2015) The effect of the tsetse fly on African development. Am Ec Rev 105(1):382–410

Asenso-Okyere K, Asante FA, Tarekegn J, Andam KS (2011) A review of the economic impact of malaria in agricultural development. Agric Econ 42(3):293–304

Balasubramanya S, Stifel D (2020) Water, agriculture and poverty in an era of climate change: why do we know so little? Food Policy 93:101905

Balasubramanian V, Sie M, Hijmans RJ, Otsuka K (2007) Increasing rice production in sub-Saharan Africa: challenges and opportunities. Adv Agron 94:55–133

Barker R, Herdt RW, Rose B (1985) The rice economy of Asia. RFF, Washington DC

Barrow CJ (1998) River basin development planning and management: a critical review. World Dev 26(1):171–186

Basu K (1983) The emergence of isolation and interlinkage in rural markets. Oxf Econ Pap 35:262–280

Baudron F, Sims B, Justice S, Kahan DG, Rose R, Mkomwa S, Kaumbutho P, Sariah J, Nazare R, Moges G, Gérard B (2015) Re-examining appropriate mechanization in Eastern and Southern Africa: two-wheel tractors, conservation agriculture, and private sector involvement. Food Secur 7(4):1–16

Bernard T, de Janvry A, Mbaye S, Sadoulet E (2017) Expected product market reforms and technology adoption by Senegalese onion producers. Am J Agric Econ 99(4):1096–1115

Bierlen R, Wailes EJ, Crammer GL (1997) The MERCOSUR rice economy (No. 954). Arkansas Agricultural Experiment Station

Binswanger H (1978) Economics of tractors in South Asia: an analytical review. Agric Dev Council, New York

Binswanger H (1986) Agricultural mechanization: a comparative historical perspective. World Bank Res Obs 1:27–56

Borgia C, García-Bolaños M, Mateos L (2012) Patterns of variability in large-scale irrigation schemes in Mauritania. Agric Water Manag 112:1–12

Boserup E (1965) The conditions of agricultural growth: the economics of agrarian change under population pressure. Transaction Pub, New Brunswick, NJ

Burney JA, Naylor RL, Postel SL (2013) The case for distributed irrigation as a development priority in sub-Saharan Africa. PNAS 110(31):12513–12517

Cambaza C, Hoogesteger J, Veldwisch GJ (2020) Irrigation management transfer in sub-Saharan Africa: an analysis of policy implementation across scales. Water Int 45(1):3–19

Cassman K, Grassini P (2013) Can there be a green revolution in Sub-Saharan Africa without large expansion of irrigated crop production? Glob Food Sec 2:203–209

Cobbing J, Hiller B (2019) Waking a sleeping giant: realizing the potential of groundwater in Sub-Saharan Africa. World Dev 122:597–613

David CC, Otsuka K (1994) Modern rice technology and income distribution in Asia. Lynne Rienner, Boulder

deGraft-Johnson M, Sakurai SA, T, Otsuka K, (2014) On the transferability of the Asian rice green revolution to rainfed areas in sub-Saharan Africa: an assessment of technology intervention in Northern Ghana. Agric Econ 45(5):555–570

Delgado CL, McIntire J (1982) Constraints on oxen cultivation in the Sahel. Am J Agr Econ 64(2):188–196

Dembele Y, Yacouba H, Keïta A, Sally H (2012) Assessment of irrigation system performance in south-western Burkina Faso. Irrig Drain 61(3):306–315

Diao X, Silver J, Takeshima H (2016) Agricultural mechanization and agricultural transformation. Background paper for African economic transformation report. Submitted to the African Center of Economic Transformation

Diao X, Takeshima H, Zhang X (2020) An evolving paradigm of agricultural mechanization development: how much can Africa learn from Asia? IFPRI, Washington DC

Dillon A (2011) Do differences in the scale of irrigation projects generate different impacts on poverty and production? J Agr Econ 62(2):474–492

Dobermann A (2000) Future intensification of irrigated rice systems. In: Sheehy JE, Mitchell PL and Hardy B (eds) Redesigning rice photosynthesis to increase yield. International Rice Research Institute, Elsevier Science, Makati City (Philippines), Amsterdam, pp 229–247

Duflo E, Pande R (2007) Dams. Q J Econ 122(2):601–646

Faltermeier L, Abdulai A (2009) The impact of water conservation and intensification technologies: empirical evidence for rice farmers in Ghana. Agric Econ 40:365–379

FAO (2021) FAOSTAT. https://www.fao.org/faostat/

FAO, IIASA (2021) Global Agro Ecological Zones version 4 (GAEZ v4). http://www.fao.org/gaez/

Fuwa N, Edmonds C, Banik P (2007) Are small-scale rice farmers in eastern India really inefficient? examining the effects of microtopography on technical efficiency estimates. Agric Econ 36(3):335–346

Gatti N, Baylis K, Crost B (2021) Can irrigation infrastructure mitigate the effect of rainfall shocks on conflict? Evidence from Indonesia. Am J Agr Econ 103(1):211–231

Gebretsadik KA, Romstad E (2020) Climate and farmers' willingness to pay for improved irrigation water supply. World Dev Perspect 20:100233

Government of India, Ministry of Agriculture and Irrigation (1977) Indian agriculture in brief, 16th edn. New Delhi

Gyimah-Brempong K, Johnson M, Takeshima H (2016) The Nigerian rice economy: policy options for transporting production, marketing, and trade. University of Pennsylvania Press, Philadelphia

Hanjra M, Ferede T, Gutta D (2009) Reducing poverty in sub-Saharan Africa through investments in water and other priorities. Agric Water Manag 96(7):1062–1070

Hayami Y, Godo Y (2001) Nougyo Keizai Ron: Shin-ban (Agricultural Economics: New edition). Iwanami Shoten, Tokyo

Hayami Y, Kawagoe T (1989) Farm mechanization, scale economies, and polarization: the Japanese experience. J Dev Econ 31(2):221–239

Hayami Y, Ruttan V (1970) Factor prices and technical change in agricultural development: the United States and Japan. J Polit Econ 78(5):1115–1141

Hayami Y, Ruttan V (1985) Agricultural development: an international perspective. Johns Hopkins University Press, Baltimore

Herdt R (2010) Development aid and agriculture. In: Pingali P, Evenson R (eds) Handbook of agricultural economics. Elsevier, pp 3253–3304

Inocencio A, Kikuchi M, Tonosaki M, Maruyama A (2007) Costs and performance of irrigation projects: a comparison of Sub-Saharan Africa and other developing regions. IWMI Research Report 109

IRRI (1986) Small farm equipment for developing countries. IRRI, Los Banos, the Philippines

Jayne TS, Muyanga M, Wineman A, Ghebru H, Stevens C, Stickler M, Chapoto A, Anseeuw W, van der Westhuizen D, Nyange D (2019) Are medium-scale farms driving agricultural transformation in sub-Saharan Africa? Agric Econ 50:75–95

Kajisa K, Payongayong E (2011) Potential of and constraints to the rice Green Revolution in Mozambique: a case study of the Chokwe irrigation scheme. Food Policy 35:615–626

Kapalata D, Sakurai T (2020) Adoption of quality-improving rice milling technologies and its impacts on millers' performance in Morogoro region, Tanzania. J Agric Econ 22:101–105

Kijima Y, Ito Y, Otsuka K (2012) Assessing the impact of training on lowland rice productivity in an African setting: evidence from Uganda. World Dev 40(8):1610–1618

Kikuchi M, Mano Y, Njagi TN, Merrey D, Otsuka K (2021) Economic viability of large-scale irrigation construction in Sub-Saharan Africa: what if Mwea irrigation scheme were constructed as a brand-new scheme? J Dev Stud 57(5):772–789

Malabo Montpellier Panel (2019) WATER-WISE: Smart irrigation strategies for Africa. Malabo Montpellier Panel Report

Mano Y, Takahashi K, Otsuka K (2020) Mechanization in land preparation and agricultural intensification: the case of rice farming in the Cote d'Ivoire. Agric Econ 51(6):899–908

Mano Y, Njagi T, Otsuka K (2022) An inquiry into the process of upgrading rice milling service: the case of Mwea Irrigation Scheme in Kenya. Food Policy 106

Miézan KM, Sié M (1997) Varietal improvement for irrigated rice in the Sahel. Irrigated rice in Sahel: prospect for sustainable development. West African Rice Development Association (WARDA), pp 443–455

Mutero CM, Kabutha C, Kimani V, Kabuage L, Gitau G, Ssennyonga J, Gighure J, Muthami L, Kaida A, Musyoka L, Klarie E, Oganda M (2004) A transdisciplinary perspective on the links between malaria and agroecosystems in Kenya. Acta Trop 89:171–186

Nakano Y, Bamba I, Diagne A, Otsuka K, Kajisa K (2012) The possibility of a Rice Green Revolution in large-scale irrigation schemes in Sub-Saharan Africa. In: Otsuka K and Larson D (eds) An African Green Revolution: finding ways to boost productivity on small farms. Springer Science & Business Media, Berlin, Germany

Nakawuka P, Langan S, Schmitter P, Barron J (2018) A review of trends, constraints and opportunities of smallholder irrigation in East Africa. Glob Food Sec 17:196–212

Otsuka K, Kalirajan K (2006) Rice green revolution in Asia and its transferability to Africa: an introduction. Dev Econ 44(2):107–122

Otsuka K, Kijima Y (2010) Technology policies for a Green Revolution and agricultural transformation in Africa. J Afr Econ 19(2):ii60–ii76

Otsuka K, Larson D (eds) (2012) An African Green Revolution: finding ways to boost productivity on small farms. Springer Science & Business Media, Berlin

Paudel GP, Kc DB, Rahut DB, Justice SE, McDonald AJ (2019) Scale-appropriate mechanization impacts on productivity among smallholders: evidence from rice systems in the mid-hills of Nepal. Land Use Policy 85:104–113

Pingali P (2007) Agricultural mechanization: adoption patterns and economic impact. In Evenson R, Pingali P (eds) Handbook of agricultural economics. Elsevier, Amsterdam

Pingali P, Bigot Y, Binswanger H (1987) Agricultural mechanization and evolution of farming systems in sub-Saharan Africa. Johns Hopkins University Press, Baltimore

Qadir M, Boers TM, Schubert S, Ghafoor A, Murtaza G (2003) Agricultural water management in water-starved countries: challenges and opportunities. Agric Water Manag 62(3):165–185

Reardon T, Chen KZ, Minten B, Adriano L (2012) The quiet revolution in staple food value chains in Asia: enter the dragon, the elephant, and the tiger. Asian Development Bank and IFPRI, November

Riddell PJ, Westlake MJ, Burke J (2006) Demand for products of irrigated agriculture in sub-Saharan Africa. FAO, Rome

Rosegrant M, Perez ND (1997) Water resources development in Africa: a review and synthesis of issues, potentials, and strategies for the future. EPTD Discussion Paper 28. IFPRI, Washington DC

Schoengold K, Zilberman D (2007) The economics of water, irrigation, and development. Handb Agric Econ 3:2933–2977

Seck PA, Diagne A, Mohanty S (2012) Crops that feed the world 7: Rice. Food Secur 4:7–24

Shah T, Hassan M, Khattak MZ et al (2009) Is irrigation water free? A reality check in the Indo-Gangetic Basin. World Dev 37(2):422–434

Soullier G, Demont M, Arouna A, Lançon F, del Villar PM (2020) The state of rice value chain upgrading in West Africa. Glob Food Secur 25:100365

Strobi E, Strobi RO (2011) The distributional impact of large dams: evidence from cropland productivity in Africa. J Dev Econ 96(2):432–450

Takeshima H, Edeh HO, Lawal AO (2015) Characteristics of private-sector tractor service provisions: insights from Nigeria. Dev Econ 53:188–217

Takeshima H, Agandin J, Kolavalli S (2017) Growth of modern service providers for the African agricultural sector: an insight from a public irrigation scheme in Ghana. IFPRI Discussion Paper 01678

Takeshima H, Houssou N, Diao X (2018) Effects of tractor ownership on agricultural returns-to-scale in household maize production: evidence from Ghana. Food Policy 77:33–49

Takeshima H, Hatzenbuehler P, Edeh H (2020) Effects of agricultural mechanization on economies of scope in crop production in Nigeria. Agric Syst 177:102691

Takeshima H, Liu Y (2020) Smallholder mechanization induced by yield-enhancing biological technologies: Evidence from Nepal and Ghana. Agric Syst 184:102914

Takeshima H, Nin-Pratt A, Diao X (2013) Mechanization, agricultural technology evolution, and agricultural intensification in Sub-Saharan Africa: typology of agricultural mechanization in Nigeria. Am J Agric Econ 95:1230–1236

Tanaka A, Saito K, Azoma K, Kobayashi K (2013) Factors affecting variation in farm yields of irrigated lowland rice in southern-central Benin. Eur J Agron 44:46–53

Tokida K, Haneishi Y, Tsuboi T, Asea G, Kikuchi M (2014) Evolution and prospects of the rice mill industry in Uganda. Afr J Agric Res 9(33):2560–2573. https://doi.org/10.5897/AJAR2014.8837

Tsang BL, Moreno R, Dabestani N, Pachón H, Spohrer R, Milani P (2016) Public and private sector dynamics in scaling up rice fortification: the Colombian experience and its lessons. Food Nutr Bull 37(3):317–328

USDA (United States Department of Agriculture) (2021) Economic Research Service Agricultural Productivity Project. https://www.ers.usda.gov/data-products/international-agricultural-productivity/. Accessed 1 Dec 2021

Wang Y, Huang J, Wang J, Findlay C (2018) Mitigating rice production risks from drought through improving irrigation infrastructure and management in China. Aust J Agric Resour Econ 62(1):161–176

Wilson RT, Lewis I (2015) The rice value chain in Tanzania: a report from the Southern Highlands Food Systems Programme. FAO, Rome

World Bank (WB) (2007) World development report 2008: Agriculture for development. World Bank

World Commission on Dams (WCD) (2000) Dams and development: a new framework for decision-making. Report of the World Commission on Dams. Earthscan Publications, London

Woodhouse P, Veldwisch GJ, Venot JP, Brockington D, Komakech H, Manjichi A (2017) African farmer-led irrigation development: re-framing agricultural policy and investment? J Peasant Stud 44(1):213–233

Xie H, Perez N, Anderson W, Ringer C, You L (2018) Can Sub-Saharan Africa feed itself? The role of irrigation development in the region's drylands for food security. Water Int 43(6):796–814

You L, Ringler C, Wood-Sichrabet U et al (2011) What is the irrigation potential for Africa? a combined biophysical and socioeconomic approach. Food Pol 36(6):770–782

Hiroyuki Takeshima is a Senior Research Fellow of Development Strategy and Governance Division, International Food Policy Research Institute, Washington D.C., USA. He received Ph.D in Agriculture & Consumer Economics from University of Illinois in 2008. His area of specialization is Agricultural Economics and International Development.

Yukichi Mano is a professor at Hitotsubashi University, Japan, and is a fellow at Tokyo Center for Economic Research (TCER). He received Ph.D. in Economics from the University of Chicago in 2007. His scholarly interests include agricultural technology adoption, horticulture and high-value crop production, business and management training (KAIZEN), human capital investment, migration and remittance, and universal health coverage in Asia and sub-Saharan Africa.

Chapter 8
Mechanization in Cote d'Ivoire: Impacts of Tractorization on Agricultural Intensification

Yukichi Mano, Kazushi Takahashi, and Keijiro Otsuka

Abstract It is critically important to intensify farming systems in sub-Saharan Africa by disseminating improved agronomic practices and increasing the application of modern inputs (Chaps. 1 and 2 of this volume). One of the region's challenges is that proper land preparation is difficult due to the scarcity of draft animals and the underdevelopment of the tractor rental market (Chap. 7). Our analysis of rice production in Cote d'Ivoire reveals that farmers who use two-wheel tractors in land preparation are more likely to adopt proper, labor-intensive rice cultivation practices and apply fertilizer more intensively, thereby raising productivity. Thus, the diffusion of two-wheel tractors appears to be critical to the intensification of rice-farming systems in sub-Saharan Africa.

8.1 Introduction

In view of the need to modernize agriculture in Africa, where productivity has been stagnant for an extended period, farm mechanization has attracted considerable attention from economists and policymakers (Chap. 7 of this volume). Unlike tropical Asia, the use of draft animals has been limited in this region due to the prevalence

---

This chapter draws heavily on Mano et al. (2020).

Y. Mano (✉)
Graduate School of Economics, Hitotsubashi University, 2-1 Naka, Isono Building Room 324, Kunitachi-shi, Tokyo 186-8601, Japan
e-mail: yukichi.mano@r.hit-u.ac.jp

K. Takahashi
Graduate School of Policy Studies, National Graduate Institute for Policy Studies (GRIPS), 7-22-1, Room 1211, Minato-ku, Roppongi, Tokyo 106-8677, Japan
e-mail: kaz-takahashi@grips.ac.jp

K. Otsuka
Graduate School of Economics, Kobe University, Fourth Academic Building, 2-1 Rokkodai-cho, Nada-ku, 5th Floor, Room 504, Kobe 657-8501, Hyogo, Japan
e-mail: otsuka@econ.kobe-u.ac.jp

© JICA Ogata Sadako Research Institute for Peace and Development 2023
K. Otsuka et al. (eds.), *Rice Green Revolution in Sub-Saharan Africa*, Natural Resource Management and Policy 56, https://doi.org/10.1007/978-981-19-8046-6_8

of trypanosomiasis, or sleeping sickness (Alsan 2015), coupled with deteriorating animal health services and recurring droughts (Mrema et al. 2008; Takeshima et al. 2013, 2015; Takeshima 2015). The public sector made substantial efforts to promote mechanization by distributing tractors[1] at subsidized prices in sub-Saharan Africa (SSA) in the 1960s and 1970s (Pingali 2007; Adu-Baffour et al. 2019). However, these government-sponsored attempts often failed due to governance challenges such as lack of political interest and elite capture (Daum and Birner 2017), as well as low demand for tractor services among local farmers, leaving tractors idle or scrapped (Pingali et al. 1987). Along with intensified concerns about increased unemployment in the face of growing populations, agricultural mechanization lost its momentum in SSA after the 1970s (Diao et al. 2014).

Renewed interest in agricultural mechanization has recently emerged with an emphasis on the role of the private sector in the provision of tractor services (Mrema et al. 2008; Diao et al. 2014; Daum and Birner 2017; Adu-Bauffour et al. 2019). Although machines are expensive and indivisible, they are purely private goods, and governmental intervention may not be justifiable if there are sizeable private service providers that are not credit-constrained. Proper land preparation by tractors is expected to improve farmers' agronomic practices, especially by allowing for better tillage, weed control, and water management. Whether this private sector model is successful crucially depends on its profitability for farmers.

While tractor use has been considered a substitute for labor and draft animals in Asia, the association between tractors and labor use is ambiguous in SSA. If the land is thoroughly plowed by tractor in place of manual labor and if this facilitates the adoption of input- and labor-intensive agronomic practices, the adoption of proper rice cultivation practices and the application of modern inputs may increase along with the use of tractors. Such intensification, accompanied by mechanization, would occur if proper land preparation and agronomic practices were complementary. However, we know little about whether farmers who use tractors in land preparation are also more likely to choose farming intensification and how this affects the productivity and profitability of crop production in SSA. This study fills these research gaps by revisiting the association between agricultural mechanization and intensification in farming systems as well as productivity and profitability improvements in SSA. We do this by drawing on a case study of rice farming in Cote d'Ivoire.

This focus on rice production is suitable for our research purpose because the productivity of rice, especially lowland rice, depends on the use of modern inputs and improved rice cultivation practices, such as bund construction, leveling, and transplanting (see Chap. 2 of this volume). Therefore, whether farmers employ these recommended practices and apply appropriate modern input levels in conjunction with their choices concerning tractor use for land preparation is an important issue impacting the realization of the Green Revolution in SSA (Chap. 7).

---

[1] Broadly speaking, there are two types of tractors: Power tillers (also called 2-wheel tractors) and riding tractors (also called 4-wheel tractors) (Takeshima 2015). Power tillers were prevalent in Asia until recently, and are popular in irrigated areas in SSA, where the soil is moist and soft. Conversely, riding tractors are particularly common in rainfed areas in SSA. In this study, a "tractor" refers to a power tiller, which is common in our study sites.

To examine the association between tractor use and farming intensification, we use plot-level data collected from 111 farmers in ten villages in 2015. We apply a regression framework with village fixed effects and a doubly robust (DR) method to address the inherent differences between the plots plowed with tractors and those cultivated with hand hoes.

We find that the farmers who use tractors for land preparation more intensively adopt proper labor-intensive rice cultivation practices and apply modern inputs, thereby improving their productivity. The results are robust regardless of empirical specifications and estimation methods, providing evidence that mechanization in land preparation will be complimentary to input- and labor-intensive production systems in SSA. This finding is consistent with the emerging literature (Chap. 9 of this volume; Takeshima et al. 2013) and suggests that promoting mechanization in land preparation does not lead to increased rural unemployment in SSA (Binswanger 1986; Panin 1995; Pingali 2007; Adu-Bauffour et al. 2019). Our result contrasts with the experiences of the US since the 1940s, Europe and Japan since the 1950s, and tropical Asia since the 1960s, where rising wage rates have induced the substitution of capital for labor and draft animals (e.g., David and Otsuka 1994; Wang et al. 2016). The current chapter illustrates the potential of a unique agricultural development path in SSA in the absence of draft animals and underdevelopment of the tractor rental market, which is different from the case of Tanzania examined in Chap. 9, where draft animals co-exist with agricultural machines.

The rest of this chapter is organized as follows: Sect. 8.2 explains the study setting and discusses the descriptive statistics. The study's empirical strategy is presented in Sect. 8.3. Section 8.4 discusses the estimation results regarding the association between mechanization in land preparation and the rice-farming intensification. Section 8.5 provides some concluding remarks and recommendations for future studies.

8.2 Data

8.2.1 Study Setting

Our study site is the Yamoussoukro District in Cote d'Ivoire. Located between the country's business center, Abidjan, and the headquarters of AfricaRice (formerly the West Africa Rice Development Association, or WARDA) in Bouake, it has good access to markets and technical information about agronomic practices. Our site is unique in that, unlike those in other SSA countries, virtually all the sample rice farmers grow WITA 9, a modern variety of rice developed by AfricaRice, and apply significant amounts of chemical fertilizer under the influence of training programs provided by local governments, Japan International Cooperate Agency (JICA), and international organizations such as the World Bank (Takahashi et al. 2019). Some

non-governmental organizations may have also learned improved agronomic practices from AfricaRice's publications and website and then disseminated them to
farmers.

However, as in many countries in SSA, proper land preparation is difficult due to
the scarcity of draft animals and the underdevelopment of the tractor rental market.
According to our interviews with farmers, only a limited number of tractors are
available in the region. Furthermore, the supply of tractors is dominated by Chinese
products that frequently break down while spare parts are lacking and the maintenance system for repairs remains underdeveloped. Farmers usually contact the tractor
owners directly before a crop season. The tractor owners are not usually farmers but
wealthy entrepreneurs living in towns who arrange tractors and operators to provide
plowing services across villages and regions. In 2013, a local agricultural company
was also established to offer tractor services and modern inputs on credit via contract
farming.

At the beginning of the main cultivation season, roughly from July to December
each year, the tractor demand surges against their limited supply, making it difficult
to secure them. Manual land preparation is possible but laborious without sufficient
irrigation water, which provides a soft soil mass but is often inadequate. Due to
constraints on plowing because of the absence of draft animals and tractors, it is
difficult to intensify land use through improved agronomic practices such as bund
construction, leveling, transplanting, weeding, and water control.[2]

8.2.2 Sampling Framework

A household survey of 111 households was conducted in early 2015. This survey
comprised a census of all rice farmers in the ten villages that the local agricultural
company targeted to operate contract farming.[3] However, according to our interviews
with local farmers, this agricultural company was neither familiar with the local
production environment nor well organized in contract management. Since the impact
of contract farming was limited, this study focuses on the effects of tractor use in
land preparation on the adoption of input- and labor-intensive agronomic practices
by using the pooled sample comprising farmers who engaged in contract farming
and those who did not.[4]

Our sample covered the main cropping season, July to December 2014. Since
some farmers cultivate more than one plot, 136 rice plots were surveyed. All the

---

[2] Farming intensification may be constrained by different factors in different environments. Emerick
et al. (2016) recently documented how improved rice varieties (i.e., flood tolerance) reduced the
production risk—which promoted the intensive application of inputs—and thereby increased the
adoption of agronomic practices and improved agricultural productivity in eastern India.

[3] We omit one rice plot used for seed production, which is different from ordinary rice production,
from our analysis.

[4] See Mano et al. (2017) for details of the impact of contract farming.

rice plots were under irrigation. The collected data pertain to household socioeconomic characteristics, detailed rice production (including input use, the adoption of agronomic practices, and output), labor time, water access, and past rice cultivation training experience.

8.2.3 Descriptive Analysis

Table 8.1 presents the basic characteristics of the rice plots by land preparation method, either tractor or manual. A total of 56 plots were plowed using tractors, and 80 plots were plowed manually. The household head was typically a male in his mid-40 s with minimal education, and the average household consisted of around ten members. Family members had previously received agricultural training on several agronomic practices provided by the government and international organizations such as the World Bank. We conducted a *t*-test of the equality of means between the plots plowed with tractors and those manually. We did not find significant differences in most key variables between the tractor plots and the manual plots except for education, assets, and plot size.[5] Although we do not have detailed data, agricultural experts knowledgeable about the local rice farming also confirm that plot conditions, including soil quality and plot slope, are largely homogeneous within each irrigation scheme.

Table 8.2 illustrates the application of production factors and the amount of applied chemical fertilizer by the land preparation method. Family and hired labor, machinery, and chemical fertilizer were more intensively applied in the plots plowed with tractors. On average, about 298 kg of chemical fertilizer per hectare were used in the tractor plots, compared with 181 kg per hectare in the manual plots.[6] This highly intensive fertilizer application in Cote d'Ivoire is likely related to the technical training and campaigns provided by the government and international agencies through local extension services and a well-functioning chemical fertilizer market. When tractors were available and land preparation was adequately done, farmers also intensively applied chemical fertilizer.

Table 8.3 illustrates the adoption of agronomic practices using the land preparation method. We pay special attention to the rice cultivation practices commonly adopted in rice production in Asia, which led to success in the rice-based Green Revolution (Chap. 2 of this volume; David and Otsuka 1994; Otsuka and Larson 2013, 2016). These agronomic practices, known to be highly complementary, include leveling, bund construction, canal construction, seed selection, seed incubation using paper or straw, and transplanting.

---

[5] We confirmed this observation through regression estimation.

[6] Njeru et al. (2016) summarize the FAO data showing that farmers in Indonesia and Kenya apply almost 150 kg of chemical fertilizer per hectare compared to the much lower amounts applied in other countries in Southeast and South Asia and sub-Saharan Africa.

Table 8.1 Basic characteristics of rice plots by land preparation method

	Tractor	Manual	Diff
Head female (=1)	0.01	0.02	−0.007
	(0.13)	(0.15)	
Head age	46.32	44.65	1.67
	(9.62)	(12.12)	
Head any school (=1)	0.66	0.52	0.14*
	(0.47)	(0.50)	
Family size	10.0	9.45	0.55
	(6.25)	(5.93)	
No. techniques trained in past	7.14	6.54	0.59
	(4.02)	(4.20)	
Asset (000 FCFA)	144.53	80.01	64.51*
	(282.2)	(143.7)	
Plot size (ha)	0.90	0.71	0.19**
	(0.48)	(0.44)	
Land rent (000 FCFA/ha)	7.58	8.33	−0.74
	(29.25)	(15.12)	
Inadequate water (=1)	0.01	0.03	−0.02
	(0.13)	(0.19)	
Contract farming (=1)	0.19	0.26	−0.06
	(0.40)	(0.44)	
Obs.	56	80	

Notes The unit of observation is the plot. Standard deviations are in parentheses. ** and * indicate statistical significance at the 5% and 10% levels, respectively

Table 8.2 The application of production factors by land preparation method

Outcomes	Tractor	Manual	Diff.
Family labor (000FCFA /ha)	153.7	76.71	77.03**
	(236.6)	(99.40)	
Hired labor (000FCFA/ha)	92.22	24.86	67.36***
	(93.35)	(55.92)	
Machine (000FCFA/ha)	59.41	5.36	54.04***
	(44.94)	(30.00)	
Chemical fertilizer (kg/ha)	298.3	181.6	116.7***
	(196.9)	(158.7)	
Obs.	56	80	

Note Standard deviations are in parentheses. The t-test is between "Tractor" and "Manual." *** and ** indicate statistical significance at the 1% and 5% levels, respectively

Table 8.3 Adoption of agronomic practices and fertilizer application by land preparation method

	Tractor	Manual	Diff.
No. of adopted agronomic practices[a]	4.21	3.70	0.51**
	(1.23)	(1.21)	
Canal construction (=1)	0.82	0.68	0.13*
	(0.38)	(0.46)	
Bund construction (=1)	0.73	0.43	0.29***
	(0.44)	(0.49)	
Seed selection (=1)	0.83	0.86	−0.02
	(0.37)	(0.34)	
Leveling (=1)	0.66	0.68	−0.02
	(0.47)	(0.46)	
Seed incubation (=1)	0.55	0.45	0.10
	(0.06)	(0.05)	
Transplanting (=1)	0.60	0.57	0.03
	(0.49)	(0.49)	
Obs.	56	80	

Note Standard deviations are in parentheses. The *t*-test is between "Tractor" and "Manual." ***, ** and * indicate a statistical significance levels of 1%, 5% and 10%, respectively

[a]The practices refer to leveling, bund construction, canal construction, seed selection, seed incubation, and transplanting

The plots adequately plowed with tractors recorded a significantly greater number of adopted agronomic practices (Table 8.3). In particular, the tractor plots were associated with the adoption of improved land preparation practices, such as bund construction and canal construction. As Tables 8.2 and 8.3 indicate, farmers significantly increased labor use to apply more chemical fertilizer and adopt improved rice cultivation practices when tractors were available and land preparation was adequately executed. More precisely, the use of tractors saves labor in land preparation, whereas it was also positively associated with adopting labor-intensive agronomic practices such as more thorough canal and bund construction. This proper land preparation enabled better water control, which improved the effectiveness of fertilizer and the productivity of careful transplanting. Overall, both family and hired labor were used significantly more intensively on the tractor plots than on the manual plots for crop care and harvesting (Mano et al. 2020).[7]

We now turn to the economic performance of rice farming by tractor use (Table 8.4). Income was defined as the value of production minus the paid-out cost, and profit was income minus the imputed cost of family labor.[8] We observe higher productivity

---

[7] Crop care consists of weeding, fertilizer application, pesticide application, and water control. Harvesting includes threshing and drying.

[8] We compute profit without deducting the labor cost of bird-scaring because this activity is often carried out by children at play, whose market wage is not available.

Table 8.4 Rice yield, income, and profit by land preparation method

	Tractor	Manual	Diff.
Rice yield (t/ha)	4.71	3.58	1.12***
	(2.34)	(1.52)	
Rice income (ha) (000FCFA/ha)	554.3	513.1	41.22
	(395.3)	(300.2)	
Rice profit (ha) (000FCFA/ha)	429.7	435.1	−5.3
	(429.8)	(320.9)	
Total rice income per plot (000FCFA)	470.6	321.2	149.3***
	(364.1)	(289.2)	
Total rice profit per plot (000FCFA)	395.8	280.9	114.9**
	(381.3)	(283.5)	
Obs.	56	80	

Note Standard deviations are in parentheses. The t-test is between "Tractor" and "Manual." *** and
** indicate statistical significance at the 1% and 5% levels, respectively

and profitability of the tractor plots compared with the manual plots. In particular, rice yield per hectare, total rice income per plot, and total rice profit per plot were significantly higher on the tractor plots.[9]

8.3   Empirical Strategy

We conduct regression analysis to examine the impact of tractor use on rice-cultivation performance. Estimating the causal effect of tractor use on farming system choice, while desirable, is difficult because of a lack of plausible instruments and panel data. Moreover, at the outset of the season, farmers may decide to employ input- and labor-intensive farming systems in conjunction with their choice of whether to use a tractor to prepare the land. These considerations illustrate endogenous technology choice, suggesting that we must refrain from a causal inference regarding tractor use per se (Mundlak et al. 1999, 2012; Larson and Leon 2006). However, we will compare plots endowed with similar attributes but with and without tractor use to isolate the effect of tractor use.

Consider the following cross-sectional regression function:

$$Y_i = \beta_0 + \beta_1 M_i + X_i' \beta_X + \varepsilon_i \tag{8.1}$$

---

[9] Notice that rice yield is around 4 tons per hectare. This exceeds the average yield of 2.4 tons per hectare in SSA and is comparable to the average yield in irrigated areas in SSA in the recent period and tropical Asia in the late 1980s (Njeru et al. 2016).

where: Y_i is the outcome variable of plot i, such as the input application, the adoption of rice-cultivation practices, productivity, and profitability in rice farming; M_i is a dummy variable for tractor use (or mechanization) in land preparation; X_i is the vector of the basic characteristics of plot i and the cultivator; β s are the regression parameters to be estimated, where β_1 is assumed to capture the statistical association between tractor use and the outcome variables, which is our primary interest; and ε_i is a random error term.

Outcome variable Y_i is represented by the input application, consisting of (A1) the imputed cost of family labor (000FCFA/ha), (A2) the cost of hired labor (000FCFA/ha), (A3) the cost of machinery (000FCFA/ha), (A4) the application of chemical fertilizer (kg/ha); the adoption of proper rice-cultivation practices, consisting of (B1) the number of adopted practices, (B2) canal construction, (B3) bund construction, (B4) leveling, (B5) seed selection, (B6) seed incubation, (B7) transplanting; and (C) rice-farming performance, consisting of (C1) rice yield (t/ha), (C2) rice income per hectare (000FCFA/ha), (C3) rice profit per hectare (000FCFA/ha), (C4) total rice income per plot (000FCFA), and (C5) total rice profit per plot (000FCFA).

Covariate X_i is the vector of the basic characteristics of plot i and its cultivator. These basic characteristics consist of (1) a dummy variable for female-headed households, (2) the age of the household head and its square term, (3) a dummy variable for the household head who received any schooling, (4) the number of household members, (5) the number of agronomic practices that the farmers learned in formal agricultural training programs in the past, (6) the value of agricultural assets,[10] (7) the plot size, (8) the rent or rental value of the cropland, (9) whether the farmer perceived that the water was not adequate, and (10) whether the plot was under contract farming. We selected these variables because they are considered exogenous and are likely to affect outcomes; thus, they are often used as controls in the literature (see Table 8.1 for the summary statistics of these variables). The covariates regarding market access are omitted due to their small variation. The rice plots are concentrated within each sample village, and the villages are located within 30 min of the capital city. To make farmers with and without tractors in land preparation more comparable, we include village fixed effects in the regression model to address the possible correlation in outcomes due to the common production environments within the villages.

Another estimation strategy we use is to explicitly control the selection on observables and match plots with similar characteristics. A range of estimation methods exists, such as propensity score matching, inverse probability weighting, and doubly robust (DR) estimator. The use of these methods is common in the literature on new agricultural technologies and institutions, including contract farming, when there are no plausible instruments or panel data (Takahashi and Barrett 2014; Bellemare and Novak 2016; Mishra et al. 2016; Khan et al. 2019). To address the effect of selection on partially observable characteristics, we apply the DR estimation, or more precisely, inverse-probability-weighted regression adjustment, which combines the

---

[10] We asked the farmers about their willingness to pay for each specific agricultural asset and calculated the total value.

regression and propensity score weighting (Wooldridge 2007, 2010, Sect. 21.3.4).[11] The DR method is more robust than the propensity score matching estimator and the inverse-probability-weighting estimator because it can provide a consistent estimate as long as either the propensity score for tractor use or the regression function of outcomes Y_i in terms of covariates X_i is correctly specified (Wooldridge 2010).[12] More specifically, we first estimate the probability of tractor use by using a logit model with a set of covariates, including the basic characteristics of the plot and the cultivator as described above. Then, each expected outcome with and without a tractor is computed by:

$$\widehat{E(Y_1)} = \frac{1}{n} \sum_{i=1}^{n} \left[\frac{M_i}{p(X_i)} Y_{1i} + \left(1 - \frac{M_i}{p(X_i)} \right) \widehat{Y_{1i}} \right] \text{and}$$

$$\widehat{E(Y_0)} = \frac{1}{n} \sum_{i=1}^{n} \left[\frac{1 - M_i}{1 - p(X_i)} Y_{0i} + \left(1 - \frac{1 - M_i}{1 - p(X_i)} \right) \widehat{Y_{0i}} \right],$$

where variable $p(X)$ is the estimated probability of tractor use, $\widehat{Y_{1i}}$ and $\widehat{Y_{oi}}$ are the predicted values from estimated regression Eq. (8.1) with and without tractors ($M_i = 1$ and $M_i = 0$), respectively, and n is the number of sample plots. Taking the difference between the two estimators above, $\widehat{E(Y_1)} - \widehat{E(Y_0)} = \widehat{E(Y_1 - Y_0)}$, we can obtain an unbiased estimate of the statistical association between tractor use and outcomes.

8.4   Estimation Results

Tables 8.5, 8.6, and 8.7 present the estimated impact of tractor use on input application, the adoption of proper rice-cultivation practices, and rice-farming performance. The estimation results regarding the adoption of agronomic practices obtained with these two methods are similar (see Tables 8.5, 8.6, and 8.7).[13] Because the central aim of this study was to explore the impact of tractor use on rice-farming intensification while controlling for plot and farmer characteristics, we used the DR estimator

---

[11] We used STATA command *teffects ipwra* to implement the DR method.

[12] Whether the DR estimator is more or less biased when both the outcome and the treatment model are misspecified is still a matter of debate in the literature (Kang and Schafer 2007; Robins et al. 2007; Tan 2010).

[13] To address concerns about omitted variable bias, we apply Oster's (2019) methodologies to test the robustness of significantly estimated coefficient on tractor use β_1 to unobservables, assuming the proportional selection relationship on observed and unobserved variables. We used STATA command *psacalc* to implement Oster's (2019) robustness tests and confirmed the robustness of regression estimates to unobservables.

Table 8.5 Association between tractor use in land preparation and the application of production factors

Outcomes	Village FE	DR
Family labor (000FCFA/ha)	99.35**	101.79***
	(40.11)	(34.58)
Hired labor (000FCFA/ha)	45.38**	44.39***
	(19.00)	(16.60)
Machine (000FCFA/ha)	55.50***	53.40***
	(11.03)	(7.01)
Chemical fertilizer (kg/ha)	86.73*	127.37***
	(44.86)	(41.50)

Note Robust standard errors are in parentheses. We also controlled for all the characteristics presented in Table 8.1. ***, **, and * indicate statistical significance at the 1%, 5%, and 10% levels, respectively

Table 8.6 Association between tractor use in land preparation and the adoption of agronomic practices

Outcomes	Village FE	DR
No. adopted practices	0.63**	0.47**
	(0.29)	(0.25)
Canal construction (=1)	0.12	0.12
	(0.10)	(0.07)
Bund construction (=1)	0.34***	0.29***
	(0.11)	(0.08)
Leveling (=1)	0.02	–0.09
	(0.11)	(0.08)
Seed selection (=1)	–0.02	–0.02
	(0.08)	(0.06)
Seed incubation (=1)	0.24**	0.23**
	(0.11)	(0.09)
Transplanting (=1)	–0.07	–0.05
	(0.10)	(0.08)

Note Robust standard errors are in parentheses. We also controlled for all the characteristics presented in Table 8.1. *** and ** indicate statistical significance at the 1% and 5% levels, respectively

in the following analysis, which directly addresses endogenous tractor use based on selection on observables.[14]

The use of tractors in land preparation had positive associations with intensified input and labor application per hectare of land, including chemical fertilizer, family

---

[14] To examine whether the failed contract farming influenced the results, we also conducted the analyses using the subsample of farmers who did not participate in the contract farming (non-CF farmers). The findings essentially remain unaffected (see Appendix Tables 2–4 of Mano et al. 2020).

Table 8.7 Association between tractor use in land preparation and productivity and profitability

Outcomes	Village FE	DR
Rice yield (ton/ha)	1.15***	1.22***
	(0.49)	(0.52)
Rice income (ha) (000FCFA/ha)	84.46	67.06
	(97.97)	(87.85)
Rice profit (ha) (000FCFA/ha)	30.82	26.51
	(102.97)	(91.47)
Total rice income per plot (000FCFA)	41.30	25.96
	(59.52)	(62.06)
Total rice profit per plot (000FCFA)	11.05	-3.23
	(60.56)	(52.02)

Note Robust standard errors are in parentheses. We also controlled for all the characteristics presented in Table 8.1. *** indicates statistical significance at the 1% level

labor, and hired labor (see Table 8.5). Tractor use was significantly associated with the number of proper rice-cultivation practices adopted, specifically bund construction and seed incubation (Table 8.6). Bund construction enabled effective water control and increased the effectiveness of fertilizer use.[15] Furthermore, while the use of tractors was not significantly associated with the use of either family or hired labor in land preparation, it increased the application of family labor in crop care and harvesting, as well as hired labor in crop establishment, crop care, and harvesting (see Table 6 of Mano et al. 2020).[16]

Table 8.7 presents the association of tractor use with rice yields, as well as incomes and profits from rice farming. Tractor use is significantly and positively associated with rice yields, consistent with the greater application of inputs and labor to more carefully implement improved rice-cultivation practices (see Tables 8.5 and 8.6). Given the mean values of rice yields in the case of manual land preparation, the increase in rice yields associated with tractor use is 39.6%. However, tractor use was not significantly associated with income or profit, perhaps because of the increased cost of the labor input, including family and hired laborers. Another possibility is that the rental price of the tractor service is adjusted to make the profit indifferent between the manual plots and tractor plots, given the limited tractor service availability. These results align with the recent evidence presented by Benin (2015) and Adu-Baffour et al. (2019).

---

[15] Regarding modern inputs, all the sample farmers grow the same improved seeds and apply chemical fertilizer.

[16] While crop establishment consists of seeding and transplanting, crop care consists of weeding, fertilizer application, pesticide application, and water control. Harvesting includes threshing and drying.

Overall, we confirmed that farmers who use the tractor for land preparation also intensively apply modern inputs, use more labor to implement proper rice-cultivation practices, and improve productivity.

8.5 Conclusions

Farming intensification is becoming critically important for improving food security in SSA, where agricultural productivity has long been stagnant. This chapter analyzed the statistical association between tractor use in land preparation and the adoption of intensive farming methods. We used primary data drawn from Cote d'Ivoire and studied farmers with good access to water, markets, and skills in improved agronomic practices. We found that tractor use in land preparation is positively associated with intensively applying family and hired labor and chemical fertilizer and the number of adopted proper rice-cultivation practices, specifically bund construction and seed incubation. Tractor use is also found to increase paddy yield per hectare. Mano et al. (2020) also found tractor use was associated with more careful implementation of crop establishment, such as seed preparation and transplanting, and crop care such as weeding, fertilizer application, and water control.

As exemplified by induced innovation theory, the conventional view of agricultural development assumes that capital substitutes for labor as wages increase due, for example, to the development of the non-farm sector (Hayami and Ruttan 1985). However, our analysis of rice farming in Cote d'Ivoire suggests that the availability of tractor services is *positively* associated with the application of labor and more intensive implementation of improved agronomic practices, suggesting a potentially complementary role for capital and labor, as discussed in Adu-Baffour et al. (2019), Takeshima et al. (2013), and Pingali (2007), among others. Tractor use may have saved labor in land preparation, but more importantly, it also intensified the farming system and increased labor application. This offsets, or more than offsets, the potential reduction of labor use in land preparation. Thus, a complementarity was found between tractor use in land preparation and input- and labor-intensive farming methods. Tractor use appears to serve as a substitute for labor when it replaces draft animals, as in tropical Asia (Binswanger 1978). However, tractor use may complement labor when it replaces manual labor in land preparation, which makes it effective in adopting proper rice-cultivation practices, as in our case.

The generalizability of our results is questionable at this stage, given that our sample is drawn from areas with favorable access to water, markets, and technological information. However, our findings support the emerging literature suggesting complementarity between mechanization in land preparation and the adoption of intensive farming systems in SSA, including Chap. 9 of this volume discussing the case of Tanzania (Takeshima et al. 2013, 2015; Pingali 2007).

The government may be able to promote the private supply of high-quality tractors by establishing a public quality inspection and certification system and promoting the development of a maintenance and repair system (Daum and Birner 2017). It is also vital to train tractor owners and operators to encourage careful maintenance and provide a public-sector extension service because knowledge of improved cultivation practices, including tractor use, is a local public good.

Therefore, future studies should rigorously evaluate whether the following policies will promote the intensification of rice-farming systems in SSA: (1) building an extension system that promotes both the adoption of proper rice-cultivation practices and tractor use, (2) helping develop a tractor service market by providing information on tractor quality through inspections, and (3) training tractor owners and operators in the careful maintenance of tractors and building the expertise of mechanics in tractor-repair services.

References

Adu-Baffou F, Daum T, Birner R (2019) Can small farms benefit from big companies' initiatives to promote mechanization in Africa? A case study from Zambia. Food Policy 84:133–145

Alsan M (2015) The effect of the tsetse fly on African development. Am Econ Rev 105(1):382–410

Bellemare M, Novak L (2016) Contract farming and food security. Am J Agric Econ 99(2):357–378

Benin S (2015) Impact of Ghana's agricultural mechanization services center program. Agric Econ 46:103–117

Binswanger H (1978) Economics of tractors in South Asia: an analytical review. Agricultural Development Council, New York

Binswanger H (1986) Agricultural mechanization: a comparative historical perspective. World Bank Res Obs 1(1):27–56

David CC, Otsuka K (1994) Modern rice technology and income distribution in Asia. Riennerand International Rice Research Institute, Boulder

Daum T, Birner R (2017) The neglected governance challenges of agricultural mechanisation in Africa–insights from Ghana. Food Secur 9:959–979

Diao X, Cossar F, Hossou N, Kolavalli S (2014) Mechanization in Ghana: emerging demand, and the search for alternative supply models. Food Policy 48:168–181

Emerick K, de Janvry A, Sadoulet E, Dar MH (2016) Technological innovations, downside risk, and the modernization of agriculture. Am Econ Rev 106(6):1537–1561

Hayami Y, Ruttan VW (1985) Agricultural development: an international perspective, Revised and Expanded Edn. John Hopkins University Press, Baltimore and London

Kang JDY, Schafer JL (2007) Demystifying double robustness: a comparison of alternative strategies for estimating a population mean from incomplete data. Stat Sci 22(4):523–539

Khan MF, Nakano Y, Kurosaki T (2019) Impact of contract farming on land productivity and income of maize and potato growers in Pakistan. Food Policy 85:28–39

Larson DF, Leon M (2006) How endowments, accumulations, and choice determine the geography of agricultural productivity in Ecuador. World Bank Econ Rev 20:449–471

Mano Y, Takahashi K, Otsuka K (2017) Contract farming, farm mechanization, and agricultural intensification: the case of rice farming in Cote d'Ivoire. JICA Research Institute Working Paper, 157

Mano Y, Takahashi K, Otsuka K (2020) Mechanization in land preparation and agricultural intensification: the case of rice farming in Cote d'Ivoire. Agric Eco 51(6):899–908

Mishra AK, Kumar A, Joshi PK, D'souza A (2016) Impact of contracts in high yielding varieties seed production on profits and yield: the case of Nepal. Food Policy 62:110–121

Mrema CG, Baker D, Karlan D (2008) Agricultural mechanization in sub-Saharan Africa: time for a new look. Rome. Agricultural Management, Marketing and Finance Occasional Paper 22

Mundlak Y, Larson DF, Butzer R (1999) Rehinking within and between regressions: the case of agricultural production functions. Annales d'Economie et de Statistique, No 55/56, Econometrie des Donnees de Panel/Panel Data Econometrics, pp 475–501

Mundlak Y, Butzer R, Larson DF (2012) Heterogeneous technology and panel data: the case of the agricultural production function. J Dev Econ 99(1):139–149

Njeru TN, Mano Y, Otsuka K (2016) Role of access to credit in rice production in sub-Saharan Africa: the case of the Mwea Irrigation Scheme in Kenya. J Afr Econ 25(2):300–21

Oster E (2019) Unobservable selection and coefficient stability: theory and evidence. J Bus Econ Stat 37(2):187–204

Otsuka K, Larson D (2013) An African Green Revolution: finding ways to boost productivity on small farms. Springer, Dordrecht, Netherlands

Otsuka K, Larson D (2016) In pursuit of an African Green Revolution: views from rice and maize farmers' fields. Springer, Dordrecht, Netherlands

Panin A (1995) Empirical evidence of mechanization effects on smallholder crop production systems in Botswana. Agric Syst 47:199–210

Pingali PL (2007) Agricultural mechanization: adoption patterns and economic impact. In: Evenson R, Pingali PL (eds) Handbook of agricultural economics, vol 3. Elsevier/North-Holland, Amsterdam, pp 2779–2805

Pingali PL, Bigot Y, Binswanger H (1987) Agricultural mechanization and the evolution of farming systems in sub-Saharan Africa. Johns Hopkins University Press, Baltimore

Robins JM, Sued M, Lei-Gomez Q, Rotnitzky A (2007) Comment: Performance of double-robust estimators when "inverse probability" weights are highly variable. Stat Sci 22(4):544–559

Takahashi K, Barrett CB (2014) The system of rice intensification and its impacts on household income and child schooling: evidence from rural Indonesia. Am J Agric Econ 96:269–289

Takahashi K, Mano Y, Otsuka K (2019) Learning from experts and peer farmers about rice production: Experimental evidence from Cote d'Ivoire. World Dev 122:157–169

Takeshima H, (2015) Market imperfections for tractor service provision in Nigeria. IFPRI Discussion Paper 1424, Washington, DC

Takeshima H, Edeh HO, Lawal AO, Isiaka MA (2015) Characteristics of private-sector tractor service provisions: insights from Nigeria. Dev Econ 53(3):188–217

Takeshima H, Nin-Pratt A, Diao X (2013) Mechanization, agricultural technology evolution, and agricultural intensification in sub-Saharan Africa: typology of agricultural mechanization in Nigeria. Am J Agric Econ 95(5):1230–1236

Tan Z (2010) Bounded, efficient and doubly robust estimation with inverse weighting. Biometrika 97(3):661–682

Wang X, Yamauchi F, Huang J (2016) Rising wages, mechanization, and the substitution between capital and labor: evidence from small scale farm system in China. Agric Econ 47(3):309–317

Wooldridge JM (2007) Inverse probability weighted estimation for general missing data problems. J Econom 141(2):1281–1301

Wooldridge JM (2010) Econometric analysis of cross section and panel data, 2nd edn. MIT Press, Cambridge, MA

Yukichi Mano is a professor at Hitotsubashi University, Japan, and is a fellow at Tokyo Center for Economic Research (TCER). He received Ph.D. in Economics from the University of Chicago in 2007. His scholarly interests include agricultural technology adoption, horticulture and high-value crop production, business and management training (KAIZEN), human capital investment, migration and remittance, and universal health coverage in Asia and sub-Saharan Africa.

Kazushi Takahashi is a professor at the National Graduate Institute for Policy Studies (GRIPS), and is a director of the Global Governance Program at GRIPS, Japan. He received Ph.D. in Development Economics from GRIPS. His scholarly interests include agricultural technology adoption, rural poverty dynamics, microfinance, human capital investment, and aid effectiveness in Asia and sub-Saharan African countries.

Keijiro Otsuka is a professor of development economics at the Graduate School of Economics, Kobe University and a chief senior researcher at the Institute of Developing Economies in Chiba, Japan since 2016. He received Ph.D. in economics from the University of Chicago in 1979. He majors in Green Revolution, land tenure and land tenancy, natural resource management, poverty reduction, and industrial development in Asia and sub-Saharan African countries.

Chapter 9
Mechanization in Tanzania: Impact of Tractorization on Intensification and Extensification of Rice Farming

Eustadius Francis Magezi, Yuko Nakano, and Takeshi Sakurai

Abstract In this chapter, we explore the effects of mechanized tillage among rice farmers in Tanzania. We use two-year panel data to examine the expansion of the area under rice cultivation (extensification) and the intensive use of yield-enhancing rice technologies (intensification). We categorize farmers based on the implements they use to prepare their rice plots, namely four-wheel tractors (4WTs), two-wheel tractors or power tillers (2WTs), draft animals (DAs), and the use of manually operated tools such as hand hoes (HTs). We then examine how the use of each implement is associated with intensification and extensification. We find, among other things, that 2WTs are highly associated with high adoption rates of improved rice technology, resulting in high paddy yield. We do not find any strong evidence that 4WTs or 2WTs significantly affect extensification compared to DAs. This chapter's findings suggest that, since 2WTs can play a role in the intensification of rice farming in SSA, policies to promote the use of this mechanical technology should be implemented.

E. F. Magezi (✉)
Department of Agricultural Economics, Graduate School of Agricultural Science, Tohoku University, N211, 468-1 Aramaki Aza Aoba, Aoba-ku, Sendai, Miyagi 980-8572, Japan
e-mail: magezi@tohoku.ac.jp

Y. Nakano
Faculty of Humanities and Social Sciences, University of Tsukuba, Jinsya Building A309, 1-1-1 Tennodai, Tsukuba, Ibaraki 305-8577, Japan
e-mail: nakano.yuko.fn@u.tsukuba.ac.jp

T. Sakurai
Department of Agricultural and Resource Economics, Graduate School of Agricultural and Life Sciences, University of Tokyo, Yayoi Campus, 1-1-1 Yayoi, Bunkyo-ku, Tokyo 113-8657, Japan
e-mail: takeshi-sakurai@g.ecc.u-tokyo.ac.jp

© JICA Ogata Sadako Research Institute for Peace and Development 2023
K. Otsuka et al. (eds.), *Rice Green Revolution in Sub-Saharan Africa*, Natural Resource Management and Policy 56, https://doi.org/10.1007/978-981-19-8046-6_9

9.1 Introduction

Sub-Saharan Africa (SSA) has continued to rely heavily on human power in agriculture, and the use of farm machinery remains the lowest in the world (Diao et al. 2020). There have been some attempts to promote agricultural mechanization in SSA, notably during the 1960s and 1970s when several countries in the region, including Tanzania, launched state-sponsored mechanization projects (Pingali 2007).[1] Although these projects contributed to an increase in the number of potentially available tractors, due to poor management, most failed to realize the objective of widespread tractor use. As a result, the farming system in the region remained far from intensive, and sustainable demand for mechanized tillage was not generated (Pingali et al. 1987).

During the 1970s and 1980s, agricultural experts and development agencies shifted their attention to supporting the development of draft animal-powered implements and low-power mechanical equipment, which were seen as appropriate labor-saving technologies for smallholder farmers in SSA (Binswanger and McIntire 1987; Binswanger and Pingali 1989; Ruthenberg 1980). However, there was no widespread adoption of these technologies. By the 1990s, the number of tractors, as well as interest in investment in draft animal traction, had declined significantly in the SSA. This was in contrast to other developing regions, such as Southeast Asia and South Asia, which saw a significant increase in mechanization over the same period (Aryal et al. 2019; Belton et al. 2021; Paudel et al. 2019).

In the past three decades, however, two important trends have emerged related to mechanization in SSA. First, there has been renewed interest in agricultural mechanization in SSA, driven by wage growth, high food demand due to urbanization, and growth in the rural non-farm economy as discussed in Chap. 7. In particular, the use of tractors to perform power-intensive tillage activities is becoming common in some parts of SSA (Diao et al. 2020). In some areas, custom tractor hire services have emerged, where tractor owners, mostly large- and medium-scale farmers, provide tillage services to smallholder farmers (Daum et al. 2021). Some governments consider custom tractor hire services as an essential tool in the mechanization of small farms and have stepped up their efforts to reduce bottlenecks in the dissemination of tractors and spare parts (Adu-Baffour et al. 2019; Diao et al. 2014; Takeshima et al. 2013).

Second, while tractor use shows an upward trend, the number of draft animals in SSA has been declining since the late 1990s (Baudron et al. 2015; Kirui and von Braun 2018). Although draft animals are concentrated in a few areas where the prevalence of the tsetse fly (a vector of trypanosomiasis) is low, they have been

---

[1] In this chapter, we define mechanization as the process by which machine-powered technologies—that is, four-wheel tractors (4WTs) and two-wheel tractors, also known as power tillers (2WTs)—are applied to perform land preparation activities. In doing so, they provide a substitute for: (i) manual technologies using simple hand tools with human labor as a power source (HTs); and/or (ii) draft animal powered implements (DAP) and their associated implements. The term tractors is used when we collectively refer to 4WTs and 2WTs.

playing a significant role in tillage and transport (Sims and Kienzle 2017).[2] The decline in draft animals is mainly caused by frequent epidemics of animal diseases and severe drought, which has reduced the availability of feed grasses (Mrema et al. 2008). Other factors include the shrinking of communal grazing areas, which leads to increases in the cost of keeping draft animals (Mrema et al. 2020; Tegebu et al. 2012). Recently, there has been a call to gradually transition from draft animals to mechanized tillage (Kormawa et al. 2018).

Despite these trends, the question of how tractors can be beneficial to the majority of smallholder farmers in SSA remains. Generally, tractors have been known for facilitating the expansion of cultivation areas by converting the fallow land to cultivable land (extensification) and improving labor productivity by reducing the labor required for crop production per unit area (Pingali 2007; Pingali et al. 1987). Although some positive effects of tractor use on farm extensification have been reported in recent years in SSA (Adu-Baffour et al. 2019; Jayne et al. 2019), the effects tend to vary with the farming systems (Takeshima et al. 2013). One strand of empirical evidence also suggests that the mechanization of land preparation can induce an increased application of labor-intensive, yield-enhancing agricultural technology (intensification), resulting in higher land productivity (Baudron et al. 2019; Mano et al. 2020; Nin-Pratt and McBride 2014; Takeshima and Liu 2020). However, some studies did not find distinguishable effects of tractor use on land productivity (Benin 2015; Houssou and Chapoto 2015). It is also notable that studies comparing different types of tools, such as large four-wheel tractors (4WT), small two-wheel tractors (2WT), and draft animal power (DA), remain scarce. Given the diverse agroecological conditions and farming systems in SSA, rigorous analyses of the effects of tractor use are crucial in identifying and implementing appropriate mechanization policies (Kormawa et al. 2018).

In this chapter, we use two-year panel data collected from rice farmers in Tanzania, one of the major rice producers in SSA, to discuss the effects of tractor use on the intensification and extensification of rice farming. Tanzania was among the SSA countries with comparably high use of tractors, as we will discuss in Sect. 9.2. Farmers can access different types of tractors, mainly through custom tractor hire services provided by private machinery operators. This practice is becoming common, especially in maize and rice farming areas. This allows us to examine the differential effects of 2WTs and 4WTs, DAs, and handheld tools (HT) such as hand hoes. We find that the use of 2WTs for land preparation significantly increases the adoption of transplanting in rows, improved modern varieties, and chemical fertilizers, resulting in an increase in paddy yield compared to the use of DAs. The positive effects of the 2WTs on rice productivity might be due to their effectiveness in puddling, thereby increasing the plant's ability to absorb nutrients from the soil (Sharma and de Datta 1985). On the other hand, we found little evidence that 4WT use contributes to the intensification or extensification of rice farming. Our findings contribute to the

---

[2] Tsetse fly free areas include the lake Victoria region, Zambia, Zimbabwe, Malawi, Eritrea, Ethiopia, Burkina Faso and Mali. In addition, some cattle breeds that are resistant to trypanosomiasis have been developed, and they can remain unaffected even after being exposed to the disease.

growing mechanization literature in SSA by showing that two types of tractors might play different roles in rice cultivation.

The rest of the chapter is organized as follows. Section 9.2 offers general information about rice cultivation and tractor use in Tanzania, whereas Sect. 9.3 presents details about the study site and data collection method. Section 9.4 presents descriptive analyses, and Sect. 9.5 explains estimation methods. We discuss our estimation results in Sect. 9.6 and offer some conclusions in Sect. 9.7.

9.2 Rice Production and Tractor Use in Tanzania

9.2.1 Rice Production

Rice production in Tanzania has increased from 1.76 million tons of milled rice equivalent in 2010 to 3.03 million tons in 2020, overtaking Madagascar to become the second largest rice producer in SSA after Nigeria (FAO 2022). The increase is partly in response to the growing demand for rice, especially in urban areas, where consumers prefer rice to other traditional staples. Increasing rice production in Tanzania is considered important because the country has a high per capita rice consumption. There is a strong preference for domestically produced rice over imported rice, which is mostly of low quality (Lazaro et al. 2017). Furthermore, rice produced in Tanzania is exported to neighboring countries, including Rwanda, Uganda, Kenya, Zambia, Malawi, and the Democratic Republic of Congo (Sekiya et al. 2020).

Rice in Tanzania is produced in three agroecological zones: the Lake Zone located in the northwestern part of the country, the Eastern Zone, and the Southern Highlands zone. About 70% of the total production in the country is undertaken in five administrative regions, namely Morogoro and Pwani in the Eastern Zone, Mbeya in the Southern Highlands Zone, and Tabora and Shinyanga in the Lake Zone. About 70% of the land suitable for rice cultivation is situated in rainfed lowlands ecosystems, with the remainder either in the rainfed uplands or irrigated ecosystems. Production techniques tend to vary across the country as farmers adapt to the agroecological environment around them (Sekiya et al. 2020).

9.2.2 The Trend of Tractor Use in Tanzania

Tractors were first introduced to Tanzania in the 1940s by the colonial government as part of economic recovery programs following World War II (Pingali et al. 1987). The statistics from FAO in Fig. 9.1 show that, by the time Tanzania became independent in 1961, there were 16,550 operational tractors.

When the government implemented a socialist and self-reliance policy, the number of tractors declined significantly from 17,000 units in 1970 to about 8,000 units in

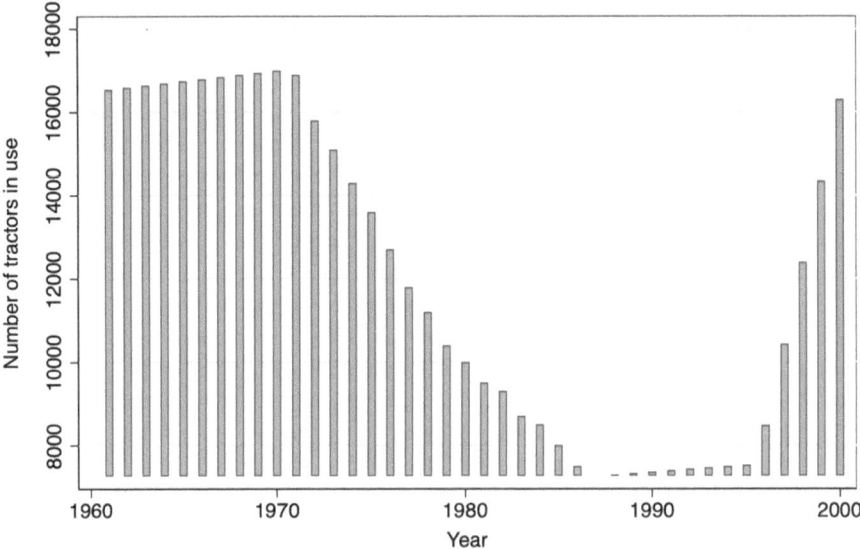

Fig. 9.1 Changes in number of tractors in use in Tanzania[3] (*Source* FAO 2021)

1985 (FAO 2022).[4] Under the policy, private enterprises, including large-scale private farms, were nationalized, and millions of rural residents were relocated to communal settlements (Owens 2014). Incentives to increase production in these farms were low mainly due to centralized price control and poor management of communal production activities. By the early 1980s, it was becoming clear that most state-sponsored projects, including mechanization, were costly and unsustainable, and the government was unable to support them due to budget deficits. In 1986, Tanzania accepted an offer to transform its economic policy toward economic liberalization under the Structural Adjustment Program (SAP).

Although tractor usage remained low in the early years after joining the SAP, policies adopted during that time laid the foundation for subsequent growth in tractorization rates. Among others, the liberalization of the financial sector in 1991, Village Land Act of 1990, and new investment policies were crucial in attracting private investment in agriculture and contributed to the increase in the number of tractors after the mid-1990s.

Since then, there has been an upward trend in tractorization in Tanzania, partly due to conducive conditions for investing in agricultural sector and the emergence of manufacturers of affordable tractors—particularly 2WTs. The 2WTs were first introduced in Tanzania in the early 2000s as part of a policy aiming to encourage

---

[3] FAO data on tractors in use in Tanzania are available up to 2002. They are based on data reported by the country in FAO questionnaires, official government reports as well as FAO estimates.

[4] Although the FAO statistics do not distinguish tractors by type, it is considered that the reported number of tractors refers to 4WTs only, since 2WTs were not introduced in Tanzania until the early 2000s.

the use of appropriate mechanization (Agyei-Holmes 2016). Since then, the number of 2WTs grew rapidly, making Tanzania one of the countries with the largest 2WT fleets in SSA.[5] 4WTs have also been increasing, but at a slower pace compared to 2WTs (Mrema et al. 2020). Tractors operating in Tanzania are mostly imported from Europe and Asia by private entrepreneurs and well-off farmers, but the process of importing only a few tractors at a time is considered to be costly. For new tractors, the government has intervened periodically by importing them in bulk and then selling them to private owners at subsidized prices, while second-hand tractors are traded in private markets without subsidies.

The use of 4WT for tillage is common in maize and rice farming system, while 2WTs are commonly used in rice farming as they are not sufficiently powerful to use on upland fields, where the soil is harder than in paddy fields. Smallholder farmers can hire a tractor (4WT or 2WT) from private service providers, and contracts are made based on the market-determined piece rate per unit of land. Although studies on the supply side of mechanization in Tanzania remain rare, our field observations tell us that tractor hire service providers move from one area to another as cultivation seasons of rice and other crops, such as sunflowers and cowpeas, tend to differ across the locations.

9.3 Study Site and Data

To examine the impact of machinery on rice farming, we use a part of a Rice Extensive Survey conducted in three administrative regions, namely Morogoro (Eastern Zone), Mbeya (Southern Highland Zone), and Shinyanga (Lake Zone). In each region, we selected two rice-growing districts, Kilombero and Mvomero in Morogoro Region, Kahama and Shinyanga Rural in Shinyanga Region, and Mbarali and Kyela in Mbeya Region. We used information from the 2002–2003 agricultural census to determine the number of villages to be covered in each district and randomly selected a total of 76 villages in all six districts. In each village, we randomly selected 10 rice-growing households, making a total sample size at the baseline survey of 760 households.

During the interview, we observed that farmers grow rice on multiple plots. We asked each household to identify the rice production plot that they considered to be their most important (hereafter referred to as a sample plot) so that we could collect information on one plot per household. We gathered detailed information about technology adoption, production costs, and rice productivity in the sample plot. In addition to the household-level survey, interviews were conducted with village leaders in all 76 villages, in which information about rice cultivation, mechanization, and access to public services and markets was collected. We revisited and interviewed the same households in 2018. For this second round of surveys, we interviewed a

---

[5] In fact, three countries—Tanzania, Madagascar and South Africa—account for about 70% of all 2WTs in SSA (Mrema et al. 2018).

randomly selected replacement household if the household interviewed in 2009 could not be traced.

Although we intended to analyze the impact of mechanization using the total sample, we excluded observations from Kahama, Shinyanga Rural, and Kyela districts after finding that about 90% of surveyed farmers used draft animals to prepare their rice plots, and the use of tractors in these districts was low. One of the reasons for this is that most households are agro-pastoralists, who grow rice and other crops during the main rain season and graze their livestock herds in plains near the settlements. Therefore, we use data only from Kilombero, Mvomero, and Mbarali. After dropping outliers, our data becomes unbalanced two-year household-level panel data with a total of 662 observations.

Since we could not reinterview a significant number of our original sample households in the endline survey in 2018, our results may suffer from attrition bias. To examine this, we estimate the attrition probit model using the 2009 observations, where the dependent variable is the dummy variable taking one if the household is attrited in the endline, and the independent variables are baseline household characteristics. We find that there are no household characteristics that significantly influence the probability of being attritted, suggesting that the attrition could be considered random to some extent. We also find that farmers in Kilombero and Mvomero districts were significantly less likely to be attritted than those in Mbarali district. Given that our sample includes replacement households, however, handling attrition is not simple. Although we admit that attrition is not fully controlled, we mitigate this problem by adding district-fixed effects.

In our study sites, there are two cultivation seasons; the main season from October to June and the dry season from July to September. During the main season, farmers grow rice in the rainfed and irrigated lowland plots and other crops such as maize in the upland plots. During the dry season, lowland rainfed plots are usually left to fallow due to water scarcity, and irrigated plots are used to grow rice, vegetables, and other crops depending on water availability. As only a few plots, if any, are used for rice cultivation during the dry season, our analysis focuses on rice cultivated in the sample plot during the main season.

Farmers in our study sites use 4WTs, 2WTs, DAs, or HTs to perform land preparation activities, including plowing, harrowing, and puddling.[6] The choice of implements is determined by several factors, including accessibility, soil characteristics, and farmers' socioeconomic characteristics. 4WTs, for example, are often used for plowing and harrowing in rainfed lowland plots before they are flooded. They are also preferred for plowing and harrowing in plots with heavy clay soil since their engines are powerful. 2WTs are commonly used for puddling in standing water conditions in irrigated or rainfed rice plots surrounded by bunds. 4WTs fitted with special implements can also be used to perform puddling in standing water conditions, although in many areas, this is rare due to bogging or sinking in muddy fields. Thus, 2WTs, which are much lighter, have the advantage in puddling.

---

[6] Puddling is the process of mixing soil and water in a paddy field to make soft structureless mud before direct seeding or transplanting rice.

As some rice farmers in our sample utilized more than one implement for preparing their plots, we generate four mutually exclusive dummy variables—namely, 4WTs, 2WTs, DAs, and HTs—to simplify our analyses. Out of 662 observations in our sample, 22 farmers used 4WTs along with 2WTs, 16 used 4WTs with DAs, and 49 used 4WTs with HTs. After close examination, we categorize these farmers as 4WT users. Even when we categorize the 22 farmers who used 4WTs and 2WTs as 2WT users, our estimation results remain largely the same. Similarly, the 35 farmers who used 2WTs with DAs and 49 with HTs are categorized under 2WTs, and 73 farmers who used DAs with HTs are categorized under DAs. Note, however, that the results of our main analyses are largely the same, even if we allow for the use of multiple instruments by one household.

Tractors (4WTs and 2WTs) are usually hired from private operators, where agreements between the two parties depend on the size and condition of the plot. Payment is made before or immediately after the work is completed. The hiring of DAs for plowing also follows such an agreement, although most farmers use their own cattle. During the study period, there were no cases of farmers in our sample who used their own 4WTs, and just 13% of those who use 2WTs (about 2% of the active sample of 662 observations) own them.

9.4 Descriptive Analysis

Table 9.1 shows changes in farm appliances used by farmers in our sample to prepare rice plots between 2009 and 2018 (Panel A) and other variables related to tractor access and related village-level variables. The data show that, in general, the use of tractors for land preparation has increased over time, while the use of DAs and HTs has decreased. The percentage of farmers who used 4WTs to prepare their rice plots increased from 34.2% in 2009 to 46.4% in 2018, while users of 2WTs increased from 7.6 to 22.3% during the same period. The increase is in line with data presented by Mrema et al. (2020), showing that 2WTs have contributed substantially to the trend toward increased tractor use in Tanzania. Conversely, the use of DAs decreased somewhat, declining from 18.5% to 14.2%, and the use of HTs fell from about 39.7% to 17.17%.

Panel B reports a significant increase in the number of tractors owned by villagers, particularly 2WTs. In 2009, the average number of 4WTs and 2WTs per village was merely 2.3 units and 0.7 units, respectively. In 2018, the average number of 4WTs per village increased slightly to 3.4 units, while the number of 2WTs increased to 8.2 units. The increase in the number of tractors and the decline in machinery hire fees suggest that tractor hire services are becoming more accessible and affordable. The fees for hiring DAs also decreased to a large extent, but their use also declined. Despite the fact that DAs are cheaper than machinery, farmers may prefer machinery because they are more time efficient than DAs. Panel B also shows that the village-level population density also increased from 142 to 165 persons per square kilometer

Table 9.1 Accessibility and use of tractor hire services and related variables

	(1)	(2)
Variables	2009	2018
Panel A: Farm appliances used by the household to prepare rice plots		
Four-wheeled tractors (4WTs: %)	34.24	46.39
Two-wheeled tractors (2WTs: %)	7.58	22.29
Traction animal power (DAs: %)	18.48	14.16
Hand hoe (HT: %)	39.70	17.17
Panel B: Machinery access and population density at the village-level		
Number of four-wheeled tractors in the village	2.27	3.37
Number of two-wheeled tractors in the village	0.74	8.15
4WT hire fees per acre ('000 TShs)	40.73	25.22
2WT hire fees per acre ('000 TShs)	43.16	32.14
DA hire fees per acre ('000 TShs)	38.12	22.72
Village population density ('00 people/km^2)	1.42	1.65
Observation (households)	330	332

Notes (i) 4WTs, 2WTs, DAs, and HTs stand for four-wheeled tractors, four-wheeled tractors, and draft animal power, and handheld tools respectively. (ii) All the monetary values are adjusted for inflation using the 2009 value of Tanzanian Shillings (TShs). (iii) Since not all villages have machinery hire markets, 4WT and 2WT hire rates are based on villages where markets exists

between 2009 and 2018, suggesting increased availability of labor and demand for food in rural areas.

To fully understand how farmers transit from one source of power to another, we conducted a further descriptive investigation. To make it easier to identify the transition path, we use 236 households with balanced panel household-level observations. Table 9.2 shows how farmers changed the implements they used for land preparation between 2009 and 2018. Bolded numbers show farmers who used the same implements between the two periods. Out of the farmers who used DAs in 2009, only about 36% remained in the same category in 2018, while others shifted to 2WTs (35.7%), 4WTs (19.1%), and HTs (9.5%). Similarly, 36% of HTs users in 2009 remained in the same category, while others shifted to 4WTs (33%), 2WTs (16.5%,) and DAs (14.3%). These trends indicate a shift away from labor-intensive and time-consuming implements to tractors.

In Table 9.3, we compare rice cultivation based on farm implements used to prepare the rice plots. We stratify our sample on whether the sample rice plot was irrigated (Panel A) or rainfed (Panel B). As indicators for extensification, we focus on the total area under rice cultivation at the household level and the size of the area under rice cultivation within the sample plot. Farmers sometimes cultivate only a part of the plot and leave the remaining part under fallow for various reasons, including labor shortages, the use of labor-intensive cultivation methods, and lack of sufficient water. Therefore, the actual area under rice cultivation within a plot can be smaller

Table 9.2 Power substitutions over time from 2009 to 2018

	2018: 4WTs	2018: 2WTs	2018: DAs	2018: HTs	2009: Total
2009: 4WTs	**80.7**	6.8	6.8	5.7	100
2009: 2WTs	0.0	**80.0**	6.7	13.3	100
2009: DAs	19.1	35.7	**35.7**	9.5	100
2009: HTs	33.0	16.5	14.3	**36.3**	100

Note The table now shows how farmers transitioned from the implement they used in 2009 to the one used in 2018. The bolded numbers show farmers who remained in the same category. For example, out of all farmers who used 4WTs in 2009, about 81% continued to use them in 2018, while the others transitioned to 2WTs (6.8%), DAs (6.8%), and HTs (5.7%)

than the size of the plot. In such environments, the use of tractors is considered to help farmers prepare a larger area of each plot within a short time. Regarding the area under rice cultivation at the household level, it is important to note that we collected data on machinery use only in the sample plot, and we do not have similar data for other rice plots. Therefore, when the area is measured at the household level, our machinery use variables are prone to measurement error.

As indicators for intensification, we focus on the adoption of yield-enhancing technologies and paddy yield. The key technologies include transplanting in rows and the use of modern rice varieties (MVs), chemical fertilizer, and insecticides and herbicides.[7] The modern variety widely available in Tanzania is TXD 306, commonly known as SARO 5. This variety has high yield potential, particularly in irrigated ecosystems, and has some characteristics that are preferred by consumers, such as the aroma inherited from traditional parental varieties (Nakano et al. 2018; Sekiya et al. 2020).

We conduct a *t*-test for comparisons between 4WTs (Column 1), 2WTs (Column 2), and HTs (Column 3), against the reference category, DAs (Column 4). In order to compare the effectiveness between machinery and draft animal power, which can be substitutable without substantially affecting paddy yield —as discussed in Chap. 7— we keep the DAs as a base category. Regarding the irrigated lowlands, we find that the areas cultivated at the household level and within the sample plot by 4WT users do not significantly differ from DA users. 4WT users, however, have significantly higher adoption rates of the modern rice variety, and they apply more chemical fertilizers, insecticides, and herbicides than DA users. Users of 4WTs achieve a paddy yield of 4.0 tons per hectare, which is higher than the 3.3 tons per hectare of DA users. Similar to 4WT users, farmers who use 2WTs also do not cultivate significantly larger areas than DA users, but they have significantly higher adoption rates of all key rice technologies than DA users. Moreover, the users of 2WTs achieve an average paddy yield as high as 5.1 tons per hectare, which is significantly higher than all other farmers. Regarding the HT users, we find that they have high adoption rates of key technologies and cultivate relatively small areas at both the household

---

[7] See Chap. 4 for a detailed explanation of these technologies.

Table 9.3 Cultivated area, technology adoption, and paddy yield by farm implements used for land preparation

	(1)	(2)	(3)	(4)
Variables	4WTs	2WTs	HTs	DAs
Panel A: Irrigated Plots				
Area under rice cultivation at HH level (ha)	1.77	1.86	0.97[a]	1.57
Cultivated area within the plot (ha)	1.31	1.02	0.53[a]	1.06
Plots transplanted in rows (%)	22.58	50.00[b]	48.91[a]	27.59
Plots using modern varieties (%)	41.94[a]	54.55[a]	79.35[a]	5.17
Chemical fertilizer use (kg/ha)	66.40[b]	97.77[a]	59.12[b]	29.46
Insecticide and herbicide use (liter/ha)	1.95[a]	1.68[a]	1.86[a]	0.41
Paddy yield in sample plot (tons/ha)	3.97[c]	5.07[a]	3.87[c]	3.32
Sample size (Households)	31	44	92	58
Panel B: Rainfed Plots				
Area under rice cultivation at HH level (ha)	1.60	2.18	1.20[b]	1.68
Cultivated area within the plot (ha)	1.22	1.41	0.83[a]	1.27
Plots transplanted in rows (%)	3.39[a]	38.18[a]	10.42	14.00
Plots using modern varieties (%)	22.88	47.27[a]	38.54[a]	14.00
Chemical fertilizer use (kg/ha)	24.10	91.09[a]	16.04	28.00
Insecticide and herbicide use (liter/ha)	2.52[b]	2.19[b]	1.17[a]	3.42
Paddy yield in sample plot (tons/ha)	2.32	4.34[a]	2.67	2.64
Sample size (Households)	236	55	96	50

Note [a]Denotes significant at 1%, [b]Significant at 5%, and [c]Significant at 10% in *t*-test comparison between HTs, 4WT, and 2WTs against the reference category (DAs)

level and within the sample plot, achieving an average paddy yield of about 3.8 tons per hectare.

In the rainfed lowlands, there is no significant difference between 4WT and DA users regarding cultivated areas, adoption rates of MVs, and chemical fertilizer application. Although users of 4WTs have a lower adoption rate of transplanting in rows and application of insecticides and herbicides than DA users, we find no significant difference in paddy yield between the two. This finding supports the hypothesis of Alsan (2015) that the unavailability of DAs is a major constraint on the intensification of farming in many areas in SSA. On the other hand, we find that farmers who use 2WTs have a higher adoption rate of transplanting in rows, the use of MVs, and chemical fertilizer use than DA users, achieving a high paddy yield of 4.3 tons per hectare under rainfed conditions. This is remarkable because this yield is comparable to the average yield in tropical Asia as well as in India (see Chap. 1). We also find that HT users cultivate small areas and have low adoption rates of MVs and application of insecticides and herbicides, but there is no significant difference in paddy yield between HT users and DA users.

Our descriptive analyses suggest that, although the use of tractors, in general, may not demonstrate a clear advantage over DAs in the extensification process, the use of 2WTs is likely to be more beneficial than DAs in enhancing technology adoption and paddy yield.

9.5 Estimation Methods

To examine the effects of mechanization of rice cultivation, we employ the fixed effects (FE) model specified as follows for the household i in year t:

$$y_{it} = \alpha + \beta_1 HT_{it} + \beta_2 2WT_{it} + \beta_3 4WT_{it} + \theta_1 T_t + X_{it}\eta + c_i + u_{it}, \qquad (9.1)$$

where y_{it} denotes the aggregate area under rice cultivation at the household level, the area under rice cultivation within the sample plot, technology adoption rates (including transplanting in rows, modern varieties, chemical fertilizers, insecticides, and herbicides), and paddy yield. The main explanatory variables of interest are HT_{it}, $2WT_{it}$ and $4WT_{it}$ which are binary variables that respectively indicate whether the household used HTs, 2WTs, or 4WTs in year t. As discussed above, we define our machinery use variable to be mutually exclusive and we collected the data on one sample plot for each household. Thus, although we measure yield, technology adoption, and machinery use at the plot level (i.e., in the sample plot), we put subscript i (household) for these variables.

We control for several time-varying plot-, household-, and village-level characteristics (denoted by a vector X_{it}), including the number of working-age adults, years of schooling of household head, female-headed household (dummy), age of household head, total landholdings (ha), the value of non-farm household assets (million TShs), amount of credit received by the household ('00,000 TShs), size of the sample plot (ha), dummy variables indicating whether the sample plot is irrigated, has clay soil, or has bunds, as well as a time dummy variable T_t, which takes the value of 1 if year is 2018. β_1, β_2, β_3, θ_1, and η are parameters to be estimated, while c_i and u_{it} respectively denote unobserved time-invariant household characteristics and the error term. The key parameters of interest are β_1, β_2, and β_3, respectively, denoting the effects of using HTs, 2WTs, and 4WTs relative to the reference category of DA.

For a robustness check, we employ the correlated random effects approach (CRE). The CRE, which is based on the random effects estimator, adjusts for time-invariant unobserved heterogeneity by including averages of time-varying household-level explanatory variables (referred to as MC devices) as additional regressors in the random effect model (Wooldridge 2019). We use similar sets of dependent and independent variables as in the FE model and include district-fixed effects and the MC devices instead of household FE.

9.6 Estimation Results and Discussion

Table 9.4 presents the estimation results for the impact of mechanization. Panel A shows the results of FE estimation followed by that of the CRE model in Panel B. In each estimation, we report the robust standard errors clustered at the village level in parentheses. To shorten the presentation of our findings, we report only the estimation results of key coefficients and exclude those of the control variables. We obtain similar results for both FE and CRE estimations, suggesting the robustness of our results. In addition, although our analytical framework assumes that each farmer's choice of mechanization is mutually exclusive, some farmers use multiple implements as mentioned earlier. We therefore conducted further analysis to examine whether the estimation results would change if we allowed the choice of multiple types of mechanization in one household. For this purpose, we estimate the FE and CRE models using the dummy variables that take one as long as a farmer uses each instrument. Although the results are not shown, the estimates are consistent with the results presented in this chapter.

Results of CRE estimation show that 4WT users cultivate a significantly larger area than DA users. 4WT users are significantly more likely to use MVs but less likely to adopt transplanting in rows. These significant effects, however, are not observed in FE estimation. We do not find any significant effects of HT or 2WT use on the areas of cultivation at the household level or within the sample plot relative to DA use. These results suggest that neither 2WTs nor 4WTs have a strong advantage in the extensification of rice cultivation compared to DAs. However, regarding the adoption of yield-enhancing technologies, our estimation results of both FE and CRE modes show that the use of 2WTs is associated with an increased adoption rate of transplanting in rows, modern varieties, and the application of chemical fertilizer. Specifically, the use of 2WTs significantly increases the adoption of transplanting in rows by 21 percentage points, modern varieties by 31 percentage points, and the application of chemical fertilizer by about 39 kg per hectare relative to the use of DAs. The use of 2WTs is also positively associated with an increase in paddy yield of about 1.1 tons per hectare, possibly due to the high adoption rates of yield-enhancing technologies. In addition, the use of HTs is associated with an increase in modern varieties and a decrease in the use of insecticides and herbicides. The use of HTs also increases the paddy yield by 1.0 tons per hectare compared to the use of DAs. In contrast, we do not find any significant effect of using 4WTs on rice technology variables.

Although the exact mechanism by which the use of 2WTs leads to input intensification and paddy yield remains unclear, it may be linked to the effectiveness of 2WTs in puddling in muddy paddy fields. As we discussed earlier, although any implement can be used to perform this activity, 2WTs are considered the most effective for puddling partly because they are lightweight and easily maneuverable in small paddy plots. Effective puddling by 2WTs could increase nutrient uptake by plants, and thus, increase the performance of yield-enhancing technology (Sharma and de Datta 1985). This agronomic observation is consistent with our findings that

Table 9.4 Impacts of tractorization on cultivated area, technology adoption, and paddy yield

	(1)	(2)	(3)	(4)	(5)	(6)	(7)
Variables	Area under rice cultivation at HH level (ha)	Cultivated area within the plot (ha)	Adoption of transplanting in rows (%)	Adoption of modern varieties (%)	Chemical fertilizer use (kg/ha)	Insecticide and herbicide use (liter/ha)	Paddy yield in sample plot (tons/ha)
Panel A: FE model							
4WTs (dummy)	0.30 (0.323)	0.22 (0.185)	−12.59 (8.049)	10.56 (11.527)	1.78 (18.727)	−0.47 (0.677)	0.43 (0.368)
2WTs (dummy)	0.97 (0.733)	0.34 (0.304)	20.59[b] (9.056)	31.00[b] (13.326)	38.86[c] (21.137)	−0.40 (0.526)	1.13[a] (0.406)
HTs (dummy)	0.25 (0.473)	0.07 (0.212)	−0.61 (7.707)	34.57[a] (11.242)	18.42 (22.043)	−1.02[c] (0.536)	1.06[a] (0.361)
Control variables	Yes	Yes	Yes	Yes	Yes	Yes	Yes
Year FE	Yes	Yes	Yes	Yes	Yes	Yes	Yes
Constant	−0.08 (0.770)	0.54 (0.359)	22.74 (27.063)	34.57 (25.468)	46.45 (37.195)	0.22 (1.765)	2.90[a] (0.581)
Panel B: CRE model							
4WTs (dummy)	0.16 (0.228)	0.30[c] (0.161)	−20.67[a] (7.584)	11.20[b] (5.137)	7.70 (10.898)	0.01 (0.278)	0.41 (0.275)
2WTs (dummy)	0.40 (0.285)	0.21 (0.169)	11.77[c] (6.832)	31.47[a] (6.738)	43.01[a] (13.535)	−0.07 (0.246)	1.26[a] (0.357)
HTs (dummy)	0.07 (0.241)	0.11 (0.151)	−3.52 (6.640)	37.91[a] (5.802)	18.01 (12.856)	−0.34 (0.300)	0.86[a] (0.232)
Control variables	Yes	Yes	Yes	Yes	Yes	Yes	Yes
District and Year FE	Yes	Yes	Yes	Yes	Yes	Yes	Yes
MC Devise	Yes	Yes	Yes	Yes	Yes	Yes	Yes
Constant	0.27 (0.761)	−0.14 (0.538)	−75.54[a] (29.262)	10.75 (25.523)	−58.47 (78.078)	−0.42 (1.713)	2.57[c] (1.413)
Observations	662	662	662	662	662	662	662

Notes (i) Robust standard errors clustered at the village level in parentheses. (ii) The reference category is the dummy variable for the use of DA to prepare rice plot (iii) [a]Denotes significant at 1%, [b]Significant at 5%, and [c]Significant at 10%. (iv) We control for the number of working-age adults, years of schooling of household head, female-headed household (dummy), age of household head, total landholdings (ha), the value of non-farm household assets (million TShs), amount of credit received by the household ('00,000 TShs), size of the sample plot (ha), dummy variables indicating whether the sample plot is irrigated, has clay soil, or has bunds, as well as year dummies. (v) The MC device and district FE are included the CRE model only

almost all 2WT users used them for puddling. It must also be emphasized that 2WTs are as efficient as 4WTs—and probably more efficient than DAs—in plowing and harrowing if the soil is not very hard. In sum, our analysis suggests that 2WTs are the appropriate technology for rice farming intensification in many areas in Tanzania. This is consistent with the rapidly increasing adoption of 2WTs reported in Table 9.1.

9.7   Conclusion

In this chapter, we examined the effects of mechanization on rice production and productivity using two-year panel data collected in Tanzania, one of the major rice producers in SSA. Specifically, we investigated whether mechanization of land preparation activities using 2WTs or 4WTs results in the expansion of the cultivated area and increasing use of yield-enhancing technologies compared to Das. Conducting this study in Tanzania, where DAs are widely used in rice cultivation, allows us to compare all four types of implements used to prepare the rice plots, including HTs. We estimated the effects of using HTs, 2WTs, and 4WTs on rice cultivation practices and yield using FE and CRE estimation methods.

Overall, we found that the adoption of 2WTs contributes to the adoption of transplanting in rows, modern rice varieties, and chemical fertilizer application, resulting in high paddy yield compared to the use of DAs. On the other hand, we find that the use of 4WTs does not have significantly different effects on rice cultivation compared with the use of DAs. Results of our analyses suggest that the effects of tractorization on rice intensification may differ depending on the type of tractors used. Our results are partly consistent with the case of Côte d'Ivoire reported in Chap. 8, which demonstrates the positive relationship between the use of 2WTs and intensification.

Our estimation results showing that 2WT tractors play a significant role in the intensification of rice farming in SSA are consistent with observations in some Asian countries, including Bangladesh, India, and Nepal (Aryal et al. 2019; Belton et al. 2021; Paudel et al. 2019). Regarding the 4WTs, our estimation results indicate that they do not have a significant advantage over DAs concerning extensification, adoption of rice technologies, and paddy yield. The heavier weight and lower maneuverability of large machines in small muddy patches of paddy fields are a disadvantage for use in rice production. This is partly consistent with Pingali (2007), who argued that 4WTs may be beneficial over draft animals, but only if they contribute to a significant reduction in the hours of labor use required for land preparation.

Our results provide supportive evidence for the recent trend toward the promotion of small-scale mechanization, especially for the intensification of rice farming in SSA. Note, however, that our results should not be interpreted as "small is beautiful." The results may be location-specific because of the different functions that large- and small-scale machinery (i.e. 4WTs and 2WTs) play in rice cultivation and because of soil conditions (Daum et al. 2022). For example, 4WTs are used for plowing and a pair of oxen are used for puddling in Mwea Irrigation Scheme in Kenya, where the soil is vertisol, which is particularly hard. Further investigation is needed to identify the conditions under which 2WTs are particularly useful for intensifying rice farming in SSA.

References

Adu-Baffour F, Daum T, Birner R (2019) Can small farms benefit from big companies' initiatives to promote mechanization in Africa? A case study from Zambia. Food Policy 84:133–145

Agyei-Holmes A (2016) Technology transfer and agricultural mechanization in Tanzania: Institutional adjustments to accommodate emerging economy innovations. Innov Dev 6(2):195–211

Alsan M (2015) The effect of the tsetse fly on African development. Am Econ Rev 105(1):382–410

Aryal JP, Rahut DB, Maharjan S, Erenstein O (2019) Understanding factors associated with agricultural mechanization: a Bangladesh case. World Dev Perspect 13:1–9

Baudron F, Misiko M, Getnet B, Nazare R, Sariah J, Kaumbutho P (2019) A farm-level assessment of labor and mechanization in Eastern and Southern Africa. Agron Sustain Dev 39(2):1–13

Baudron F, Sims B, Justice S, Kahan DG, Rose R, Mkomwa S, Kaumbutho P, Sariah J, Nazare R, Moges G, Gérard B (2015) Re-examining appropriate mechanization in Eastern and Southern Africa: two-wheel tractors, conservation agriculture, and private sector involvement. Food Secur 7(4):889–904

Belton B, Win MT, Zhang X, Filipski M (2021) The rapid rise of agricultural mechanization in Myanmar. Food Policy 101:102095

Benin S (2015) Impact of Ghana's agricultural mechanization services center program. Agric Econ 46(S1):103–117

Binswanger HP, McIntire J (1987) Behavioral and material determinants of production relations in land-abundant tropical agriculture. Econ Dev Cult Chang 36(1):73–99

Binswanger H, Pingali P (1989) Technological priorities for farming in sub-Saharan Africa. J Int Dev 1(1):46–65

Daum T, Villalba R, Anidi O, Mayienga SM, Gupta S, Birner R (2021) Uber for tractors? Opportunities and challenges of digital tools for tractor hire in India and Nigeria. World Dev 144:105480

Daum T, Seidel A, Getnet B, Birner R (2022) Animal traction, two-wheel tractors, or four-wheel tractors? A best-fit approach to guide farm mechanization in Africa. Hohenheim Working Papers on Social and Institutional Change in Agricultural Development. https://doi.org/10.2139/ssrn.4092687

Diao X, Cossar F, Houssou N, Kolavalli S (2014) Mechanization in Ghana: emerging demand, and the search for alternative supply models. Food Policy 48:168–181

Diao X, Takeshima H, Zhang X (2020) An evolving paradigm of agricultural mechanization development: how much can Africa learn from Asia? International Food Policy Research Institute, Washington DC

FAO (2022) FAOSTAT. License: CC BY-NC-SA 3.0 IGO. Extracted from: https://www.fao.org/faostat/en/#data/QCL. Accessed 30 July 2022

FAO (2021) FAOSTAT. License: CC BY-NC-SA 3.0 IGO. Extracted from: https://www.fao.org/faostat/en/#data/RM. Accessed 13 Apr 2021

Houssou N, Chapoto A (2015) Adoption of farm mechanization, cropland expansion, and intensification in Ghana. Agriculture in an Interconnected World. In: Conference of the International Association of Agricultural Economists. Milan, Italy, 9–14 Aug 2015. https://doi.org/10.22004/AG.ECON.211744

Jayne TS, Muyanga M, Wineman A, Ghebru H, Stevens C, Stickler M, Chapoto A, Anseeuw W, van der Westhuizen D, Nyange D (2019) Are medium-scale farms driving agricultural transformation in sub-Saharan Africa? Agric Econ 50(S1):75–95

Kirui O, von Braun J (2018) Mechanization in African agriculture: a continental overview on patterns and dynamics. SSRN Electron J. https://doi.org/10.2139/ssrn.3194466

Kormawa P, Mrema G, Mhlanga N, Fynn MK, Kienzle J, Mpagalile J (2018) Sustainable agricultural mechanization: a framework for Africa. Sustainable agricultural mechanization: a framework for Africa. FAO, Rome; African Union Commission, Addis Ababa

Lazaro E, Sam AG, Thompson SR (2017) Rice demand in Tanzania: an empirical analysis. Agric Econ 48(2):187–196

Mano Y, Takahashi K, Otsuka K (2020) Mechanization in land preparation and agricultural intensification: the case of rice farming in the Cote d'Ivoire. Agric Econ (United Kingdom) 51(6):899–908

Mrema GC, Baker D, Kahan D (eds) (2008) Agricultural mechanization in sub-Saharan Africa: time for a new look. Agricultural Management, Marketing and Finance Service Occasional Paper No. 22. Rural Infrastructure and Agro-industries Division, FAO, Rome

Mrema GC, Kienzle J, Mpagalile J (2018) Current status and future prospects of agricultural mechanization in sub-Saharan Africa (SSA). Agric Mech Asia Afr Lat Am 49(2):13–30

Mrema GC, Kahan DG, Agyei-Holmes A (2020) Agricultural mechanization in Tanzania. In: Diao X, Takeshima H, Zhang X (eds) An evolving paradigm of agricultural mechanization development: how much can Africa learn from Asia? International Food Policy Research Institute (IFPRI), p 457–496

Nakano Y, Tsusaka T, Aida T, Pede V (2018) Is farmer-to-farmer extension effective? The impact of training on technology adoption and rice farming productivity in Tanzania. World Dev 105:336–351

Nin-Pratt A, McBride L (2014) Agricultural intensification in Ghana: evaluating the optimist's case for a Green Revolution. Food Policy 48:153–167

Owens GR (2014) From collective villages to private ownership. J Anthrop Res 70(2):207–231

Paudel GP, Kc DB, Rahut DB, Khanal NP, Justice SE, McDonald AJ (2019) Smallholder farmers' willingness to pay for scale-appropriate farm mechanization: evidence from the mid-hills of Nepal. Technol Soc 59:101196

Pingali P (2007) Chapter 54 Agricultural mechanization: adoption patterns and economic impact. Handbook of agricultural economics, vol 3. Elsevier, Amsterdam, pp 2779–2805

Pingali P, Bigot Y, Binswanger HP (1987) Agricultural mechanization and the evolution of farming systems in sub-Saharan Africa. Johns Hopkins University Press, Baltimore

Ruthenberg H (1980) Farming systems in the tropics, 3rd edn. Clarendon Press, Oxford, UK

Sekiya N, Oizumi N, Kessy TT, Fimbo KM, Tomitaka M, Katsura K, Araki H (2020) Importance of market-oriented research for rice production in Tanzania: a review. Agron Sustain Dev 40(1):1–16

Sharma PK, de Datta SK (1985) Puddling influence on soil, rice development, and yield. Soil Sci Soc Am J 49(6):1451–1457

Sims B, Kienzle J (2017) Sustainable agricultural mechanization for smallholders: What is it and how can we implement it? Agriculture 7(6):50

Takeshima H, Liu Y (2020) Smallholder mechanization induced by yield-enhancing biological technologies: Evidence from Nepal and Ghana. Agric Syst 184:102914

Takeshima H, Nin-Pratt A, Diao X (2013) Mechanization and agricultural technology evolution, agricultural intensification in sub-Saharan Africa: typology of agricultural mechanization in Nigeria. Am J Agric Econ 95(5):1230–1236

Tegebu FN, Mathijs E, Deckers J, Haile M, Nyssen J, Tollens E (2012) Rural livestock asset portfolio in northern Ethiopia: a microeconomic analysis of choice and accumulation. Trop Anim Health Prod 44(1):133–144

Wooldridge JM (2019) Correlated random effects models with unbalanced panels. J Econom 211(1):137–150

Eustadius Francis Magezi is an assistant professor at Graduate School of Agricultural Science and Faculty of Agriculture, Tohoku University, Japan. He obtained a Ph.D. from the University of Tokyo, Japan in 2021. His area of specialization is agricultural and resource economics.

Yuko Nakano is associate professor at Faculty of Humanities and Social Science, University of Tsukuba, Japan. She received Ph.D. degree in Development Economics in 2009 from National Graduate Institute for Policy Studies, Japan. Her specialization is development and agricultural economics.

Takeshi Sakurai is a professor of the University of Tokyo since 2014. He obtained a Ph.D. in agricultural economics from Michigan State University in 1995. His expertise is in agricultural economics and development economics.

Chapter 10
Irrigation in Kenya: Economic Viability of Large-Scale Irrigation Construction

Masao Kikuchi, Yukichi Mano, Timothy N. Njagi, Douglas J. Merrey, and Keijiro Otsuka

Abstract During the past decade, investments in large-scale irrigation development in sub-Saharan Africa (SSA) have re-emerged. Given past experiences, this revival is not without controversy. This chapter examines whether large-scale irrigation construction in SSA is economically viable by estimating how much it would cost if the Mwea Irrigation Scheme in Kenya, one of the best-performing irrigation schemes in SSA, were to be constructed today. The results show that constructing the Mwea Scheme today would be economically viable except in a situation where (1) the shadow price of modern rice varieties falls as low as the world price that prevailed during the late twentieth century, i.e., in 1986–2004, when large-scale irrigation projects mostly disappeared at any project cost level; or (2) the shadow price is at the medium level prevailing in 2014–2018 for a high project cost. There is undoubtedly untapped physical potential in SSA for large-scale irrigation development, but the economically viable potential remains limited. International donor agencies and

---

This chapter draws heavily on Kikuchi et al. (2021).

M. Kikuchi (✉)
Center for Environment, Health and Field Sciences, Chiba University, 6-2-1, Kashiwanoha, Kashiwa 277-0882, Chiba, Japan
e-mail: m.kikuchi@faculty.chiba-u.jp

Y. Mano
Graduate School of Economics, Hitotsubashi University, 2-1 Naka, Isono Building Room 324, Kunitachi-shi, Tokyo 186-8601, Japan
e-mail: yukichi.mano@r.hit-u.ac.jp

T. N. Njagi
Tegemeo Institute of Agricultural Policy and Development, Tegemeo Institute, Egerton University, Kindaruma lane, orr Ngong road, P.O. Box 20498, Nairobi 00200, Kenya
e-mail: tnjagi@tegemeo.org

D. J. Merrey
SW 50th Blvd, Gainesville, FL 2565, 32608, USA

K. Otsuka
Graduate School of Economics, Kobe University, Fourth Academic Building, 2-1 Rokkodai-cho, Nada-ku, 5th floor, Room 504, Kobe 657-8501, Hyogo, Japan
e-mail: otsuka@econ.kobe-u.ac.jp

national governments wanting to construct large-scale irrigation projects are recommended to assess whether their plan is likely to be economically profitable. In addition to proper operation and maintenance, Mwea's success also points to the importance of adopting modern inputs and improved rice cultivation practices, facilitated by thorough land preparation using tractors and oxen, and improving returns from irrigation investments.

10.1 Introduction

To enhance food security and reduce rural poverty, sub-Saharan Africa (SSA) has long been awaiting a Green Revolution (Chap. 1 of this volume; Diao et al. 2008; Ejeta 2010; Sanchez et al. 2009; Otsuka and Larson 2012). Three technological innovations made the Asian Green Revolution possible: high-yielding varieties, chemical fertilizers, and irrigation (Diao et al. 2008; Estudillo and Otsuka 2012). Among these, irrigation is by far the most critical, as an assured water supply is a prerequisite for effective fertilizer application, without which the high-yielding potential of modern seeds is not fully exploited. Among the world's developing regions, SSA's irrigation is the least developed (Chap. 7 of this volume; Balasubramanian et al. 2007), despite its rich endowment of fresh-water resources (You et al. 2010; Zwart 2013). However, there have been serious debates in the last few decades regarding what types of irrigation investments are the most profitable and sustainable.

An important mode of irrigation development during the twentieth century comprised large-scale projects to construct or rehabilitate irrigation infrastructure, funded by international donors, implemented by the governments of recipient countries, and operated and maintained by national irrigation agencies (Jones 1995; Inocencio et al. 2007). By the late 1990s, large-scale irrigation projects had nearly disappeared from the global agricultural development agenda for good reasons. One of them was the success of the Green Revolution, which brought about historic low prices of cereal crops. For rice, the world price in 2000 was as low in real terms as 25% of the level prevailing during the pre-Green-Revolution period (Fig. 10.1).[1] Such low crop prices made it virtually impossible to justify costly large-scale irrigation projects (Inocencio et al. 2007).

A more serious reason was that many large-scale irrigation projects implemented in the latter half of the last century were characterized by many problems and defects; inadequate designs, faulty construction, less-than-satisfactory achievements, and poor operation and maintenance (O&M) (Plusquellec 2019). When evaluated at the time of construction completion, one-third of the large-scale irrigation projects were found to be 'failure' projects (assessed as 'failure' if the ex-post internal rate

---

[1] In this study, Thai 5% broken or Thai A1 super (broken rice) is taken as representative rice of ordinary quality in the world rice market, used to obtain widely applicable implications. However, farmers in Mwea cultivate high-quality Basmati 370 for sales and a small amount of high-yield variety BW196 only for home consumption.

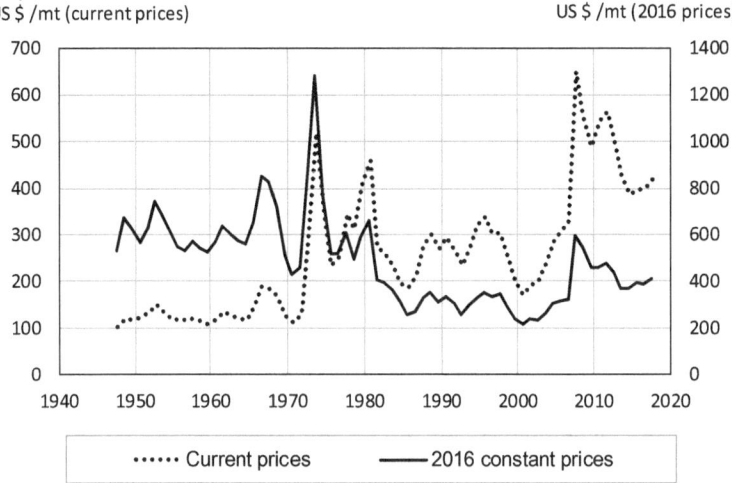

US $ /mt (current prices)   US $ /mt (2016 prices)

Fig. 10.1 World rice price (Thai 5% broken FOB Bangkok), 1948–2018. For 1960–2018, World Bank Pink Sheet 1960–2018 (World Bank 2019) for both current and constant prices; for before 1960, the current price series, compiled by using data from IRRI (2000) and Barker et al. (1985), is linked at 1960 with the World Bank series then deflated by the IMF World Export Price Index (1948–1960) and the world GDP implicit deflator (1960–2017) (World Bank 2020a)

of return [IRR] was less than 10%) (Belli et al. 1998; Inocencio et al. 2007), and the risk of 'failure' increased to 50% when evaluated six to eight years after completion (World Commission on Dams [WCD] 2000). The mode of O&M of these schemes was so institutionally defective that many new and rehabilitated irrigation schemes, even non-failure projects, rapidly deteriorated (Adams 1990; Ostrom 1992; World Bank 2005; Borgia et al. 2013). Moreover, the implementation of large-scale irrigation rehabilitation projects created the 'build-neglect-rebuild' syndrome, depriving national irrigation agencies of incentives to maintain their irrigation systems well (Huppert et al. 2003; Suhardiman and Giordano 2014). Higginbottom et al. (2021) report that more than 80% of large-scale irrigation projects in SSA implemented in the six decades since 1945 had failed to deliver the promised benefits, with no improvement in the performance throughout the period.

Finally, growing environmental and social concerns worked against large-scale new construction projects involving the construction of large dams and the relocation of inhabitants. Both the WCD and the World Bank proposed alternative agricultural development options, such as improving the performance and productivity of existing irrigation schemes through institutional reforms for O&M, developing small-scale irrigation schemes, and investing in micro-irrigation technology and in-field rainwater management rather than resorting to large-scale irrigation projects (WCD 2000; World Bank 2005).

This virtual 'ban' on large-scale irrigation projects was most effective in SSA, where the twentieth century Green Revolution had not taken root, and the irrigation sector was characterized by more handicaps than any other developing region.

Moris (1986) and Biswas (1986) pointed out that large-scale irrigation development in SSA was always problematic because of poor design, lack of understanding of grass-root conditions, inadequate technology choice, and inefficient bureaucratic O&M. Olivares (1989), Jones (1995), and Inocencio et al. (2007) criticized large-scale irrigation projects for their high costs and low performance. Higginbottom et al. (2021) attribute the poor performance of large-scale irrigation projects in SSA in the six decades since 1945 to political and institutional factors and highlight the need for greater learning from past investment outcomes. Moigne and Barghouti (1990) stated that new large-scale irrigation projects should not be considered unless lower-cost technologies or production systems with higher returns were identified.

The shift of focus from large-scale projects led by donors and governments to small-scale projects has been apparent since early in the 2000s—from a nearly single-minded focus on physical infrastructure to a greater emphasis on strengthening institutions, and more recently, from promoting public investments in small gravity and pump irrigation to policy reforms to encourage farmer-led investment in micro-irrigation technology (NEPAD 2003; Rockström et al. 2007; World Bank 2007; Burney et al. 2013; de Fraiture and Giordano 2014). By 2017 or so, 'farmer-led irrigation' had become the dominant focus of efforts to expand irrigation in SSA (Woodhouse et al. 2017; Lefore et al. 2019).

However, parallel to this development, large-scale irrigation projects have also returned to center stage. For example, a loan agreement was signed in 2007 between Kenya and the Kuwait Fund for Arab Economic Development for financing the Bura Irrigation and Settlement Scheme Rehabilitation Project (Reliefweb 2007; National Irrigation Board [NIB] 2018); a loan agreement was signed in 2010 between Kenya and the Japan International Cooperation Agency (JICA) to finance the Mwea Irrigation Rehabilitation Project (JICA 2010); and in 2017, the World Bank approved a loan for funding the Shire Valley Transformation Program in Malawi (World Bank 2017). Irrigation development in Shire Valley was first envisaged in the 1940s, and its implementation has been considered but abandoned several times since then because the construction costs were considered too high (Harrison 2018).

Why have these types of irrigation investments been resurrected? One possible reason could be the food crisis in 2008, which raised food prices sharply. The world rice price soared in 2008 to a historic high level in current prices, nearly four times as high as the 2001 price (Fig. 10.1). This surge in food prices may have reminded policymakers in SSA and international donors of the vulnerability of the world's food production and the need to enhance food security by increasing domestic food production. This has prompted them to bring back large-scale irrigation projects to increase food production (Lankford et al. 2016).[2]

Another reason could be recent advances in yield-increasing technology and its dissemination through proper rice cultivation training (see Chaps. 1 and 2 of this volume for overviews). The adoption of modern inputs and improved rice cultivation

---

[2] There has been a recent surge in food prices, driven by persistent conflict, pre-existing and COVID-19-related economic shocks, weather extremes, and the war in Ukraine (WFP 2022; FAO and WFP 2022). But these are too recent to have affected investment decisions so far.

practices are facilitated by thorough land preparation using tractors and oxen (Chap. 7 of this volume). For rice, the present technology gives a yield of 6 t/ha/season, or even higher, if grown under excellent conditions; and farmers in a few large-scale irrigation schemes in SSA are attaining that yield level for two crops per year (Nakano et al. 2012; Bartier et al. 2014).[3] The availability of such technologies also improves the ex-ante economic performance of large-scale irrigation projects, particularly when coupled with higher crop prices. It is further enhanced by adopting improved rice milling technologies and proper post-harvest rice management practices encouraged by quality-based pricing (Chaps. 7, 12 and 13 of this volume); hence, policymakers could be encouraged to promote such projects.

The recent re-emergence of large-scale irrigation projects has evoked many heated reactions, mostly critical (Burney et al. 2013; Lankford et al. 2016; Merrey and Sally 2017; Woodhouse et al. 2017; Harrison 2018). These studies share the same basic question, raised explicitly by Crow-Miller et al. (2017, 195): 'do these new projects have different justifications from those of the past?' The mode of large-scale irrigation development in the latter half of the last century was so defective that many projects failed to attain their planned level of performance. Unless national governments and international donors are confident that they have found effective remedies for previous defects, it is not clear why they would invest in new large-scale projects. The recent story of the Bura Rehabilitation Project in Kenya reported by *Business Today* (KNA 2018) is alarming: the current rehabilitation project, which commenced in 2013, was only 30% complete as of 2018, 38 months behind schedule.[4] The problems are the same as those found in many project completion reports of failed large-scale irrigation projects implemented 20 to 40 years ago. The original Bura Irrigation Settlement Project, implemented in 1979–1987, was the most infamous dam project in SSA following its disastrous failure; however, the failure had been anticipated before the project began (Chambers 1969; Moris 1973), and the details of the failure were widely reported (Moris 1986; Adams 1990; World Bank 1990). The recent Bura Project could be another example of 'informed amnesia [...], where the major actors involved in irrigation development tend to ignore past mistakes, despite ample proof of the futility of their efforts' (Veldwisch et al. 2009, 21).

In this paper, we approach this problem with a basic economic question: has the construction of large-scale irrigation schemes in SSA, once abandoned because of high investment costs and low profitability, become economically justifiable in this century? To answer this question, we examine if it would be economically justifiable today to construct the Mwea Irrigation Scheme (abbreviated hereafter as Mwea Scheme), an existing well-performing successful irrigation scheme in Kenya, by estimating how much it would cost if the scheme were newly constructed as it is now.

---

[3] This is in sharp contrast to ordinary large schemes in SSA, where rice yield ranges from 2 to 5 t/ha/season with cropping intensity generally less than 2.0 (Balasubramanian et al. 2007, 81–85).

[4] Long delays between signing the agreement and project implementation are common: the Bura project agreement was signed in 2007, work started in 2013. Similarly, the JICA Mwea rehabilitation project agreement was signed in 2010; work began in 2017.

In the next section, we give an overview of large-scale irrigation projects implemented during the last four decades of the twentieth century, mostly financed by the World Bank, with special reference to the cost structure of these projects. In the third section, we present the estimated costs of a modern-day Mwea Scheme construction project. The fourth section examines the project's economic viability compared with recent large-scale projects under implementation, appraisal, or study. We assess some of the broader policy implications in our conclusion in Sect. 10.5.

10.2 Irrigation Projects in the Twentieth Century and Their Cost Structure

Before proceeding to the cost estimation of the Mwea Scheme, we first review large-scale irrigation projects implemented during the last four decades of the twentieth century, with a particular focus on their cost structure. Large-scale irrigation projects generally have a specific cost structure that has been molded partly by the nature of the project as a public construction project and partly by the history of project implementation during irrigation development in the latter half of the twentieth century, which few studies have documented explicitly. We document the cost structure using the project completion reports of 182 large-scale irrigation projects implemented in developing regions worldwide during the latter half of the twentieth century.[5] Although this dataset includes only 19 SSA projects, these projects share the same salient features of irrigation projects in SSA found by Inocencio et al. (2007): compared to other developing regions in the world, (1) the project size, measured by the total irrigated area, was smaller; (2) the unit cost was higher; and (3) the risk of failure was higher.

The costs of large-scale irrigation projects consist of both direct construction costs and various indirect, overhead costs. In this study, the cost structure of public irrigation projects is explored by classifying the project costs into four cost groups:

(1) Costs for civil works directly related to constructing the infrastructure, including materials and equipment used, and indirect construction costs such as field administration and supervision, safety control, and contractor's profit (henceforth referred to as 'Civil-work' costs);

(2) Overhead costs for management, including preparatory surveys and studies, system designing, engineering management and supervision during the implementation, and general project administration and management ('Management' costs);

---

[5] These irrigation projects, for which project costs are reported with appropriate breakdown, are selected from 314 irrigation projects in the database prepared by Inocencio et al. (2007). Their study analyzed many aspects of large-scale irrigation projects in SSA in comparison with those in other developing regions in the world but did not touch the cost structure of irrigation projects.

Table 10.1 Cost structure of twentieth century large-scale irrigation projects in developing regions[a]

	SSA %	Non-SSA
Civil-work cost	61	77
Management cost	27	14
Ag-support cost	8	4
Other-overhead cost	4	6
Total project cost	100	100

[a] Data are from 182 large-scale irrigation projects, a subset of the projects studied by Inocencio et al. (2007), which reported the project cost with appropriate cost-breakdown

(3) Overhead costs for agricultural support, O&M equipment, O&M planning, and training of irrigation officials, water users' groups, and farmers ('Ag-support' costs); and

(4) Other-overhead costs, such as land acquisition/compensation, relocation, settlement construction, other social infrastructure, and environmental measures ('Other-overhead' costs).

'Management' costs and 'Civil-work' costs are ordinary costs in all construction projects, including irrigation. As we have emphasized the importance of rice cultivation training in Chap. 2 of this volume, 'Ag-support' costs as well as 'Other-overhead' costs, which are indispensable to the success of the projects, are largely specific to large-scale irrigation projects.

Table 10.1 shows the cost structure of twentieth-century irrigation projects in terms of the percentage share of the component costs in the total project cost. At the mean of the sample projects, the share of overhead costs (i.e., 'Management' cost, 'Ag-support' cost, and 'Other-overhead' cost) in SSA averaged 39%, more than in other regions. The difference between SSA and other regions is particularly large for 'Management' and 'Ag-support' costs.[6]

The unit total project cost and four unit-component-costs are all correlated negatively with project size (Fig. 10.2). The strong scale economy of irrigation project construction costs was pointed out by Inocencio et al. (2007) for the total project cost and by Fujiie et al. (2011) for some overhead costs. This study reveals that the 'Civil-work' cost, which includes indivisible elements, such as dams, headworks, and heavy construction equipment, also had a strong scale economy.

Using the data in Fig. 10.2 together with some sample-specific characteristics, regression equations are estimated for the unit total project cost and four unit-component-costs (Table 10.2). The project size has a highly significant negative coefficient for all the unit costs, suggesting that the scale economy exists in all the component costs. 'Ag-support' cost has the highest size-elasticity, closely followed

---

[6] For more details, see SP-M Table I-6 in Sub-section I-2–2 of the Supplementary Material (hereafter referred to as SP-M) of Kikuchi et al. (2021).

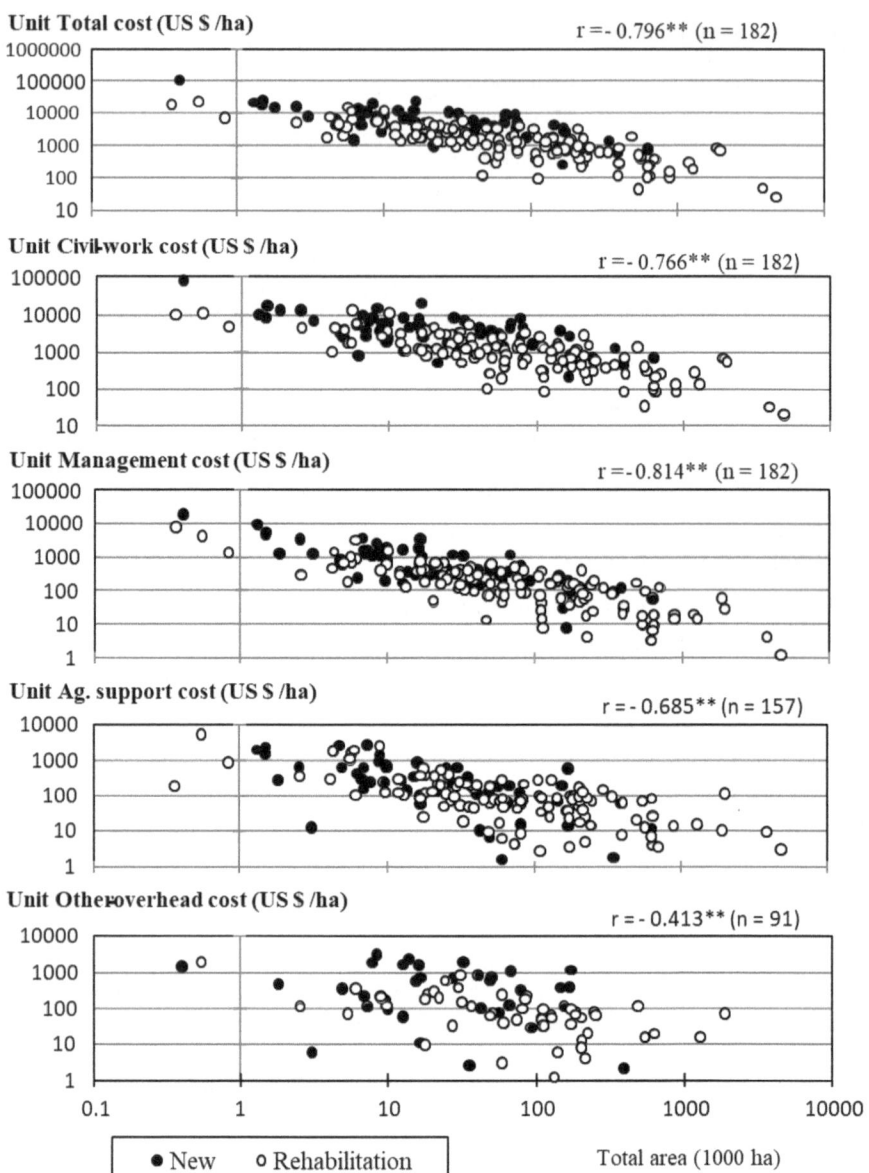

Fig. 10.2 20th-century large-scale irrigation projects: Correlation between project size (total area) and unit costs

by 'Management' cost. The unit total project cost of rehabilitation projects is significantly lower than that of new construction projects, brought about by lower 'Civil-work,' 'Management,' and 'Other-overhead' costs. 'Failure' projects have higher unit total project costs than 'successful' ones due to higher 'Civil-work' costs. There is a tendency that newer projects have lower unit 'Civil-work' and 'Management' costs, reducing the total project cost. This suggests that the performance of twentieth-century irrigation projects improved as project experience accumulated. The most important result of the regression analyses is that the SSA regional dummy is not statistically significant in all the regression equations, implying that the higher unit project costs required to develop irrigation infrastructure in SSA were due mainly to the small size of the irrigation projects, not for SSA-specific reasons.

We try to estimate the construction costs of the Mwea Scheme by using the four component costs.

10.3   Estimation of Project Costs of the Mwea Scheme

10.3.1   The Mwea Irrigation Scheme

Mwea Irrigation Scheme is situated 65 km south of Mount Kenya, 90 km northeast of Nairobi, and 650 km northwest of Mombasa. It is a river-diversion surface irrigation scheme, taking water from two tributaries in the Upper Tana basin on the heavily watered south-eastern slopes of Mt. Kenya. This favorable water potential, coupled with a gently sloping terrain and fertile black soil of volcanic origin, makes the Mwea plain an ideal physical environment for constructing an irrigation scheme (Moris 1973). Construction began in 1954 as a settlement scheme to provide farmland to landless people (Chambers 1973).

The abundant water sources in the area have made it possible to expand the irrigated area rapidly. Starting from 2,000 ha in the 1950s, the Scheme's net irrigable area increased to 6,000 ha by the late 1980s (Table 10.3). A modernization-rehabilitation project, implemented in 1989–1992 with assistance from JICA (henceforth referred to as Mwea Project 1990), eventually expanded the Scheme's irrigable area to 8,500 ha. Another modernization-rehabilitation project by JICA completed in 2022 (henceforth referred to as Mwea Project 2022) expands the irrigable area to 12,400 ha, including three out-grower sections previously developed by farmers themselves with World Bank assistance.

The favorable water and soil conditions have made the Mwea Scheme one of the most effective rice irrigation schemes in SSA and indeed the world. The average farmer's rice yield at Mwea from 1961 to 1971 was 6.4 t/ha/season (Chambers 1973). This was an exceptionally high-yield level for irrigated rice in the twentieth century among developing countries. Even after the rice Green Revolution, 4 t/ha/season was the target yield of many irrigation projects but was rarely attained. The Mwea

Table 10.2 Results of regression analysis, regressing unit total project cost and unit-component-costs (US $/ha; in logarithm) on total project area (in logarithm) and some project-specific variables

	Ln Total project cost		Ln Civil work		Ln Management		Ln Ag.support		Ln Other overhead	
	Coef	p-value	Coef	p-value	Coef	p-value	Coef	p-value	Coef	p-value
Ln Total area (1000 ha)	− 0.512	2.01E-29	− 0.486	6E-26	− 0.650	6.1E-31	− 0.653	1.2E-15	− 0.331	0.022
Rehabilitation[a]	− 0.688	8E-08	− 0.684	3E-07	− 0.559	0.000	− 0.163	0.496	− 1.082	0.006
Failure[b]	0.420	0.001	0.458	7E-04	0.239	0.126	− 0.125	0.610	0.246	0.566
Year started[c]	− 0.026	0.003	− 0.030	9E-04	− 0.024	0.023	− 0.001	0.96497	− 0.012	0.709
Sub-Saharan Africa[d]	− 0.070	0.712	− 0.229	0.249	0.201	0.390	0.560	0.118	− 0.474	0.561
Intercept	60.788	4E-04	68.331	1E-04	54.994	0.008	8.579	0.798	29.800	0.635
R^2	0.731		0.701		0.707		0.481		0.254	
No. of observations	182		182		182		157		91	

[a] A dummy variable that takes 1 if the project is a rehabilitation project and 0 if it is a new construction project
[b] A dummy variable that takes 1 if the internal rate of returns of the project is less than 10% and 0 if 10% or higher
[c] The year the project started
[d] A dummy variable that takes 1 if the project is of sub-Saharan Africa and 0 if otherwise

Table 10.3 Brief history of Mwea Scheme development

Year	Events	Sources
1954–1960	Constructed by the Kenyan government with assistance from the British and the US governments. Settler farmers (tenants) given 4 ac of land	Chambers (1968, 1973)
1960	Irrigated paddy area = 2,000 ha; Rice yield = 6.4 t/ha (average for 1961–1971; variety = Sindano); 1 crop/year	Chambers (1973, p. 68), Veen (1973, p. 107)
1960–1972	Step-by-step extensions of irrigation units with assistance from Kenyan government, UK Freedom from Hunger, and German aid agency of KFW	Chambers (1973)
1972	Irrigated paddy area = 4,800 ha	Chambers (1973, p. 71)
1988	Irrigated paddy area = 5,900 ha; Rice yield = 4.8 t/ha (average over varieties planted in 1988), (Sindano (37%) 5.0 t/ha, Basmati (55%) 4.5 t/ha, BW 196 (8%, from IRRI) 6.0 t/ha); 1 crop/year	JICA (1988, pp. 2–5)
1989–1992	Modernization-rehabilitation project (JICA project) implemented (New Nyamindi Headworks and Link Canals I and II newly constructed)	JICA (1989)
1997	Irrigated paddy area = 6,000 ha; Rice yield = 4.6 t/ha (average for 1991–96) (Sindano (33%) 6.0 t/ha, Basmati (67%) 3.9 t/ha); 1 crop/year	JICA (1997, pp. 18–19)
1998	Management of the Scheme, formerly run by NIB (National Irrigation Board), taken up by two farmers' cooperatives	Kabutha and Mutero (2002)
2003	Joint management between NIB (from head works to the secondary distribution systems) and WUA (water-users' association; tertiary distribution system and below) established	NIB (2010), Baldwin et al. (2015)
2009	Irrigated paddy area = 7,900 ha; Rice yield in 2008 = Basmati 3.6 t/ha; ratooning 1.4 t/ha, Sindano 5.0 t/ha: 1.7 crops/year (1.0 for first-planting and 0.7 for ratooning)	JICA internal data
2011	Rice yield in 2011 = 5.0 t/ha (for all varieties)	Njeru et al. (2016, Table 3)

(continued)

Table 10.3 (continued)

Year	Events	Sources
2007–2013	Water Management Development Project of World Bank implemented (Paddy-fields in Nderewa North and Marura Out-growers sections developed)	JICA Research Institute & Nippon Koei (2018)
2017	Irrigated paddy area = 8,500 ha; Rice yield = 6.2 t/ha (improved Basmati; average of two years); 2.0 crops/year (1.1 for first-harvesting and 0.9 for ratooning)	Our field surveys in 2016 and 2018
2017–	Ongoing modernization-rehabilitation project (JICA project) (New Thiba dam and Link Canal III to be constructed, Mutithi East area to be expanded, along with rehabilitation and improvements for other components of the scheme. After completion; irrigated area = 8,910 ha including out-growers' sections (12,410 ha according to our survey in 2017–2018)	JICA (2010) JICA internal data our surveys in 2017–2018

Scheme was the only successful large irrigation scheme in East Africa (Chambers and Moris 1973) or, by some accounts, even in Africa (Biswas 1986).

The scheme experienced radical changes in its O&M institutional framework at the turn of the century; the management of the Scheme by the National Irrigation Board (NIB) was taken over by farmers' groups in 1998 (Kabutha and Mutero 2002). The mode of O&M was further reformed in 2003 to a joint-management arrangement between NIB and farmers' groups (water-users associations) (Baldwin et al. 2015). The farmers' takeover of Scheme management resulted from their protests against the NIB management under which they had been treated as quasi-slave tenants with virtually no discretion as to their rice production and marketing.

This joint management by the NIB and farmers seems to have been successful, having had little effect on the Scheme's yield performance.[7] Rice yields slightly declined from 4.8 t/ha in 1988 to 4.6 t/ha in 1997 (Table 10.3) because of the shift in rice variety from traditional Sindano, which recorded an average yield of 6 t/ha over six years, to Basmati, which had higher quality but lower yield. The average rice yields were 5.0 t/ha in 2011 and 6.2 t/ha in 2017 (averaged over two years) with recently developed high-yielding Basmati varieties. The cropping intensity has improved from earlier 1.0/year to nearly 2.0/year by 2017, mainly resulting from the introduction of water rotations and rice ratoon harvesting. It is important to note that the rice market has been well developed in Kenya since the liberalization

---

[7] Under the Irrigation Act No. 14 2019, NIB was replaced by a new National Irrigation Authority with a broader mandate than NIB had.

in 1999, facilitating farmers to continue adopting improved rice varieties, modern agricultural inputs, and proper rice cultivation practices such as land preparation by tractors (Mano et al. 2022). These were in place even before the liberalization under the strict control by the irrigation bureaucracy but have come under farmers' own discretion since then. Farmers in Mwea have received agricultural training since 1991, when it was initiated under a JICA project.[8] Also remarkable has been the recent development in rice milling technology (Chap. 12 of this volume). Therefore, the Mwea Scheme, as of 2017, could be considered a top-class irrigation scheme in SSA.

10.3.2 Estimation of the Project Costs for Constructing the Mwea Scheme

Expenditure data for constructing the Mwea Scheme at its initial phase from 1954 to 1968 are reported by Sandford (1973). Although the current study's primary purpose is to estimate the total cost of constructing the Mwea Scheme as operating in 2017, we check the investment costs and its economic performance at the initial stage. We, therefore, estimate the costs of two 'new construction' projects, i.e., the Mwea 'As of 1968' Project with an irrigable area of 3,128 ha and the Mwea 'As of 2017' Project with 8,500 ha.[9]

Table 10.4 summarizes the estimated costs for constructing the Mwea Scheme and compares them with those of twentieth-century irrigation projects. The unit project cost at the initial phase of 1954–1968 is estimated to be US$ 10,071 /ha in 2016 prices, which is higher than the average unit cost of twentieth-century 'successful' new construction projects in SSA, probably because the Mwea Scheme is smaller than the average system size of successful projects, at 13,000 ha (Tables 7 and 12 of Inocencio et al. 2007).[10]

The total project cost to construct the Mwea Scheme as of 2017 is estimated in two steps: we first estimate the 'Civil-work' cost and then add the overhead costs by assuming three levels for the ratio of the total project cost to 'Civil-work' cost ('TPC/CWC' ratio).[11] For the low estimate of 1.5, the unit project cost is estimated to be US$13,706/ha, substantially higher than in the initial construction phase. This cost increase is expected because the Mwea Project 1990 both rehabilitated the existing

---

[8] 31 May 2022 interview of Mr. Masato Tamura, who was a JICA training specialist in the early 90 s.

[9] For the details of the cost estimation, see SP-M Section I-1 of Kikuchi et al. (2021) for 'As of 1968' Project and SP-M Section I-2 for 'As of 2017' Project.

[10] In this study, as in Inocencio et al. (2007), a large-scale irrigation project is considered 'successful' if its ex-post IRR ≥ 10%.

[11] For details in the estimation of the 'Civil-work' cost and the overhead costs, see SP-M Sub-section I-2–1 and Sub-section I-2–2 of Kikuchi et al. (2021), respectively.

Table 10.4 Estimated project cost to construct the Mwea Scheme at the initial construction phase and as of 2017[a]

				Remarks
I	Initial construction phase (as of 1968)[b]			
1	Project cost[c]	US $ '000	3,925	In 1960 prices
2	Project area	Ha	3,129	Irrigated area developed by 1968
3	Unit cost per ha	US $ / ha	1,255	In 1960 prices
		US $ / ha	10,071	In 2016 prices
II	After modernization (as of 2017)[d]			
1	Project costs[e]:			In 2016 prices
	Civil-work cost	US $ million	77.69	
	Low estimate	US $ million	116.53	'TPC/CWC' ratio = 1.5
	Middle estimate	US $ million	132.06	'TPC/CWC' ratio = 1.7
	High estimate	US $ million	155.37	'TPC/CWC' ratio = 2.0
2	Project area	ha	8,502	Irrigated area in 2017
3	Unit cost per ha:			In 2016 prices
	Civil-work cost	US $ /ha	9,137	
	Low estimate	US $ /ha	13,706	
	Middle estimate	US $ /ha	15,533	
	High estimate	US $ /ha	18,275	
III	Unit cost of twentieth-century projects in SSA[f]			In 2016 prices
1	New construction	US $ /ha	8,347	
2	Rehabilitation	US $ /ha	5,085	

[a]The deflator used is constructed by linking Word Bank's world GDP implicit deflator (1960–2017) with IMF's world export price index (1945–1960). See Chapter I of the Supplementary Material (SP-M) (p.2). For the years concerned, the deflator takes the following values: 2016 = 1.0000, 2000 = 0.6860, and 1960 = 0.1246

[b]Actual capital and construction-related recurrent expenditures for 1954–1968. For data sources and data compilation, see Section I-1 of the SP-M in Kikuchi et al. (2021)

[c]Consists of civil-work costs, management costs, ag-support costs, and other overhead costs

[d]Project costs, which are the costs if the Scheme is constructed now as a brand-new scheme with irrigation infrastructure in place in 2017. For estimation details, see Section I-2 of the SP-M in Kikuchi et al. (2021)

[e]Three levels of 'Project cost/Civil-work cost' ('TPC/CWC') ratio are assumed. For estimation details, see Sub-section I-2–2 of the SP-M in Kikuchi et al. (2021)

[f]For 'success' projects. Data are from Table 7 of Inocencio et al. (2007), converted from 2000 to 2016 prices

irrigation infrastructure and constructed new ones. If the high estimate of 2.0 is applied, the unit project cost would be US$18,275/ha, which is more than twice as high as that of the twentieth-century 'successful' new construction projects in SSA.

10.4   Economic Viability of Mwea Scheme Construction

We examine the economic profitability of constructing the Mwea Scheme as of 2017 as a new scheme by estimating the IRR of the investment based on the estimated project costs. The primary purpose of this study is to examine whether it is economically worth investing in large-scale irrigation projects financed and implemented by public institutions to enhance food security in SSA. The IRR we estimate is the 'economic' IRR, not the 'financial' IRR that measures private profitability. Although the IRR is the most used method to assess both ex-ante and ex-post economic performance of large-scale irrigation projects, it has often been criticized for its many defects (Tiffen 1987; World Bank 2010). The most serious weakness is its inability to assess the sustainability of projects: benefits may decline over time, as was the case for many twentieth-century irrigation projects (Plusquellec 2019). The static nature of IRR also makes it difficult to cope with the risk and uncertainty associated with the estimation of costs and benefits. Tiffen (1987) states that given this uncertainty, an IRR of 8% or less should be ruled out as within the margin of error that could include a negative outcome. Although all these arguments remain valid, we use the IRR because' cost–benefit analysis can be a powerful tool when appropriately applied' (World Bank 2010, 50).

10.4.1   The Internal Rate of Return (IRR)

In this study, the IRR is defined as r that equates the costs and the benefits of the irrigation project, evaluated at the time of the project completion:

$$(1+r)^m K = \sum_{j=1}^{J} [j(R-c)/(J+1)](1+r)^{J-j} + \sum_{n=1}^{N} (R-c)/(1+r)^n \quad (10.1)$$

where K = project investment (US\$/ha), R = returns from the investment (US\$/ha/year), c = O&M cost (US\$/ha/year), m = average gestation period of investment in years, J = the number of years during which only partial return accrues, N = lifespan of the scheme in years, and r = internal rate of return. It is assumed that the partial return, $[j(R-c)/(J+1)]$, increasing linearly from the year which is $(J-1)$ years before the project completion, eventually reaches the full return, $(R-c)$, in the year following project completion. The returns from investment (R) received by farmers, local traders, and millers in Mwea and urban traders and retailers in Nairobi are measured as the increase in value-added (income) adjusted for the opportunity cost of the non-land inputs used in rice production:

$$R = P\alpha\beta Y = P\alpha\beta\gamma y \quad (10.2)$$

where P = shadow price of rice in Nairobi retail market (US$/ton of milled rice), α = rice milling rate, β = value-added ratio, Y = increase in paddy production (t/ha/year) due to the project, y = paddy yield per season, and γ = cropping intensity (no. of crops/year). Since the area where Mwea Scheme was constructed had been vacant except for extensive stock grazing (Moris 1973), we assume no output in the area before the project.

Note that the costs and benefits in Eq. (10.1) are confined to those directly related to the project. There are indirect costs, such as adverse environmental effects, as well as indirect benefits, such as positive linkage and multiplier effects of increased agricultural production, both brought about by the project. These indirect costs and benefits are not included in this study, as is the case for irrigation project reports in general, because of the difficulty in obtaining necessary data.

Substituting Eq. (10.2), Eq. (10.1) can be re-written as:

$$(1+r)^m K - \sum_{j=1}^{J}\left[j(P\,\alpha\beta\gamma y - c)/(J+1)\right](1+r)^{J-j}$$
$$- (P\,\alpha\beta\gamma y - c)[(1+r)^{(N-1)}/r(1+r)^N] = 0 \qquad (10.3)$$

The 'r' that satisfies Eq. (10.3) can be obtained using the Goal Seek function of Microsoft EXCEL.

Because we aim to obtain general policy implications to overcome the food security problem in sub-Saharan Africa, we assume Thai 5% broken as representative of ordinary quality rice in the world market. Thai 5% broken price is comparable to that of Pakistani long grain, which accounted for 59% of imported rice to Kenya in 2017 (International Trade Center 2022). We used the Thai 5% broken price in Tanzania and the Mombasa CIF price as a parameter value of shadow price P for the 'As of 1968' project as in Kikuchi et al. (2021). In this chapter, we adjust these border prices to the Nairobi retail market prices to take into account the value added generated by post-harvest marketing activities for the 'As of 2017' project.[12]

We compare these hypothetical cases, applicable to SSA in general, with the case using the price of improved Basmati, which accounts for most rice cultivated in Mwea (Njeru et al. 2016; Mano et al. 2022; Chap. 12 of this volume), to reflect the current situations in Kenya.[13] The other variables and parameters assumed in the IRR estimation are given in Kikuchi et al. (2021) SP-M Section II, together with their data sources (SP-M Table II-1 for the 'As of 1968' project and 'As of 2017' project). Note that paddy yield parameter 9.3 t/ha/year for the 'As of 2017' project

---

[12] To obtain the Nairobi retail price, we added the importers' handling charge and margin (assuming 10% based on Kikuchi et al. 2016b), transport costs from Mombasa to Nairobi (550 km) in a 20 t container at the assumed rate of US$ 0.11/t/km (Rashid and Minot 2010), and the retailer margin at the rate of 1.08 (Kikuchi et al. 2016a) to the Mombasa CIF price (Kikuchi et al. 2021).

[13] The price of improved Basmati used is its Nairobi retail market price adjusted down for the rice tariff and its related costs in the same manner adopted for Thai 5% broken.

is obtained from our primary survey on farmers cultivating the improved Basmati, which likely underestimates yield for the high-yielding variety.

10.4.2 Results of Estimation

The results of the IRR estimation are summarized in Table 10.5. In the late twentieth century, the World Bank and other international donor agencies used interest rates of 10–12% as the threshold levels of the IRR, below which projects were considered unacceptable (Belli et al. 1998; Inocencio et al. 2007). The interest rate for lending has declined this century.[14] However, considering the argument that an IRR less than 8% could be within the margin of error, it is desirable for a project to have an IRR of higher than 8%, or even more preferably, than 10%.

Table 10.5 Internal rates of return (IRR; %) to the investment for newly constructing Mwea Irrigation Scheme in 1968 and in 2017

I. As of 1968 (in 1960 prices) [b]					
1. $\beta = 0.8$					
Rice price = US$ 89 /t		19.1			
Rice price = US$ 105 /t		22.1			
2. $\beta = 0.5$					
Rice price = US$ 89 /t		12.1			
Rice price = US$ 105 /t		14.3			
II. As of 2017 (in 2016 prices) [c]	Low cost	Medium cost		High cost	
1. $\beta = 0.8$					
Rice price = US$ 481 /t (Thai 5% broken; low)	11.9	10.7		9.3	
Rice price = US$ 600 /t (Thai 5% broken; middle)	14.2	12.9		11.3	
Rice price = US$ 719 /t (Thai 5% broken; high)	16.3	14.9		13.1	
Rice price = US$ 1,393 /t (Improved Basmati)	25.7	23.7		21.2	
2. $\beta = 0.5$					
Rice price = US$ 481 /t (Thai 5% broken; low)	7.8	6.8		5.6	

(continued)

---

[14] The Bank's interest rate in the 2010s is lower than these averages, ranging about 1% in 2014 to about 5% in 2018 (World Bank 2020b).

Table 10.5 (continued)

I. As of 1968 (in 1960 prices) [b]					
Rice price = US$ 600 /t (Thai 5% broken; middle)	9.6		8.6		7.2
Rice price = US$ 719 /t (Thai 5% broken; high)	11.2		10.1		8.7
Rice price = US$ 1,393 /t (Improved Basmati)	18.7		17.1		15.2

[a]The IRR is r that satisfies Eq. (10.3). For both 'projects,' the unit project costs (K) are from Table 4. Common parameters and variables assumed are: a (rice milling rate) = 0.65, β = 0.8 (value-added ratio), β = 0.5 (value-added ratio adjusted for opportunity costs of non-land production factors by applying respective market prices), N (project lifespan) = 30 years, For details including data sources, see SP-M Table II-1 in Section II of the SP-M in Kikuchi et al. (2021)

[b]Parameters and variables assumed are: m (average gestation period) = 1 year. c (O&M cost) = US$ 9.1 /ha/year. γ (cropping intensity) = 1.0, y (paddy yield per ha) = 6.4 t/ha/season, J (the year a partial benefit starts accruing before the full benefit is attained) = 0, P = US$ 89 /t (the rice price in Tanzania or P = US$ 105 /t (Mombasa CIF price of Thai A1). For details including data sources, see SP-M Table II-1 in Kikuchi et al. (2021)

[c]Parameters and variables assumed are: m = 4.5 years. c = US$ 189 /ha/year. γ = 2.0, Y (paddy production per year) = 9.3 t/ha/year, J = 3 years, and P = Nairobi market price of Thai 5% broken, estimated based on the Mombasa CIF (low price = 1986–2004 average, medium price = 2014–2018 average, high price = 2008–2013 average; see SP-M Table II-1 in Kikuchi et al. 2021), adopting the transportation costs and traders' margins from Kikuchi et al. (2021), and Nairobi market price of Improved Basmati in 2018 adjusted similarly for the tariff and import-related costs

The 'As of 1968' project is a 'successful' project, even for the lower rice price and the lower value-added ratio. Sandford (1973) estimated the project's net present value (NPV) with the lower rice price for three discount rates, 5%, 10%, and 15%. The NPV declines as the rate increases but remains positive at 15%, and its declining trend indicates that it would reach nil at the discount rate of about 18%, consistent with our estimation for value-added ratio β = 0.8 and the rice price of US$89/t.

The 'As of 2017' results show that the IRR is sensitive to the value-added ratio. If β = 0.8, that is, if there is no opportunity cost for non-land factors, the IRR is higher than 8% regardless of the rice price and project cost level. For β = 0.5— that is, if the opportunity costs of non-land factors are fully accounted for at market prices—the case that is closest to the contemporary Mwea situation, the IRR is lower than 8% for the low rice price of US$481 /t with any project cost level and for the medium price of US$600 /t with high project cost. The new construction of the Mwea Scheme with the irrigation infrastructure as of 2017 is not economically viable for the rice price as low as the level prevailing during the late twentieth century (Fig. 10.1). This result is consistent with the fact that costly large-scale irrigation projects mostly disappeared then (Inocencio et al. 2007). At the middle-level rice price prevailing over the last decade, the economic viability of medium- to high-cost irrigation projects could be insufficient if the value-added ratio were low. The rice price at the level prevailing during the mini rice price crisis of 2008–2013 brings up the IRR to more than 8% for the high-cost project even for β = 0.5. Using the rice price and yield that improved Basmati commands, the new construction of the Mwea Scheme is definitely economically viable. This Basmati price, if adjusted to an international price (the FOB Bangkok), is about US$1,000/t in 2016 prices, which is nearly 70% higher than the highest peak price of Thai 5% broken (US$595/t) in 2008.

10.4.3   Other Large-Scale Irrigation Projects in the Twenty-First Century

As noted, the Mwea Scheme is one of the most successful irrigation schemes in SSA in terms of water availability, rice yield, and cropping intensity. It is a simple river-diversion type surface irrigation scheme with no water storage capacity, making its construction cheaper than those requiring the construction of dams or the use of pumps. The investment project to construct it is successful. However, the Mwea Scheme might be an exception because of the adoption of proper rice farming intensification practices, in contrast to other large-scale unviable irrigation construction projects in SSA. Here we compare the performance of our Mwea 'As of 2017' project with some large-scale projects under implementation, appraisal, or study in this century. The analyses conducted for these additional large-scale irrigation projects and studies are given in SP-M Chapter IV in Kikuchi et al. (2021). We have assumed P = US$600 /t for all the cases to reflect the medium Nairobi retail market price of Thai 5% broken, rather than the high Mombasa CIF price US$550 used in the original paper.[15]

Like many other large-scale irrigation schemes in SSA, the Mwea Scheme is a rice scheme in which rice is the best (i.e., most profitable) crop to be planted. Our observations clearly indicate that rice is the farmers' most preferred crop in the scheme—not only for farmers in the main scheme, who are requested by the scheme's central management to plant rice, but also for farmers in out-grower areas, who can cultivate any crop at their discretion, yet still plant rice without exception. Some projects to be compared in this section may include non-rice projects. In this comparison, we assess the project performance of these projects by converting their project benefits in value to rice yield ('paddy equivalent'). It is reasonable to use rice as the standard of comparison as long as the economic performance of rice is better than, or comparable to, other staple food crops, which are important for the food security in SSA.[16]

---

[15] The analyses conducted for these additional large-scale irrigation projects and studies are given in SP-M Section IV of Kikuchi et al. (2021). Common assumptions on variables and parameters to estimate IRR, or the crop performance required to attain the planned IRR of these projects and studies, are presented in SP-M Table IV-1.

[16] See the SP-M Sub-section IV-3 in Kikuchi et al. (2021) for further justifications for using rice as the standard of comparison.

The first example is the Shire Valley Transformation Program in Malawi. The planned unit project costs of this project are US$15,000/ha, the same level as the medium estimate for the Mwea 'As-of-2017' project. This project is not a pure new construction project as a large part of the net irrigated area is already irrigated by pumps (World Bank 2017). To achieve an IRR of 11%, as targeted in the project appraisal report, the required performance of irrigated agriculture in the project area must be better than that of the Mwea Scheme at present: in terms of 'paddy equivalent,' a yield of 10.1 t/ha/year is required.[17]

The second example is from several irrigation developments envisaged by the Millennium Challenge Corporation in Burkina Faso, Mali, Senegal, and Ghana, the indicative unit project costs of which are US$34,300/ha, US$17,200/ha, US$14,800/ha, and US$5,600/ha, respectively (Merrey and Sally 2017). If the target IRR of these projects is 10%, the required levels of crop performance in 'paddy equivalent' are 20.2, 10.1, 8.7, and 3.3 t/ha/year, respectively.[18] The projects in Burkina Faso and Mali require crop production performance beyond that of the Mwea Scheme, while the other projects are attainable with the crop production performance level of the Mwea Scheme. In particular, the performance level needed for the project in Burkina Faso is far higher than the level actually attained in the Mwea Scheme.

The third example is the ongoing Mwea Irrigation Development Project supported by JICA (the 'Mwea Project 2022'). This is a modernization/rehabilitation project with a unit project cost of US$20,783/ha (JICA 2010), higher than our high estimate for new construction and an ex-ante IRR of 10.8%. The crop performance required to satisfy this IRR is 13.3 t/ha/year (8.7 t/ha/season) in 'paddy equivalent,' which exceeds the actual yield of 9.3 t/ha/year.[19]

It should be noted that the unit project costs of these recent large-scale irrigation projects tend to be higher than those of the twentieth-century 'success' projects (Table 10.4), and yet these projects tend to aim at an IRR of more than 10%. The crop performance that satisfies the target IRR is accordingly higher than that of the last century. The Mwea case suggests the importance of adopting modern inputs and improved rice cultivation practices, which are facilitated by proper land preparation using tractors and oxen, to realize such high benefits.

A fourth example is You et al. (2010), who examine how much irrigation potential Africa would have for large-scale irrigation development if the water stored by the existing dams were diverted for irrigation.[20] They show a potential area of 15 million ha in SSA if the unit investment cost (K) is US$3,000/ha and the project selection

---

[17] See SP-M Table IV-2 in Kikuchi et al. (2021) for the results of our analyses of the Shire Valley Project, variables and parameters assumed, analyses conducted, and data sources. Two-thirds of the target area is newly developed fields, while the rest is improvement of fields that used to have one cycle of 2t/ha/crop.

[18] See SP-M Table IV-3 in Kikuchi et al. (2021) for Millennium Challenge Corporation's projects.

[19] For JICA Mwea Rehabilitation Project, see SP-M Table IV-4 in Kikuchi et al. (2021).

[20] For details of our follow-up examination of You et al. (2010) data, see SP-M Sub-section IV-2–1 and SP-M Table IV-5 in Kikuchi et al. (2021).

criterion is IRR > 0%. The potential shrinks to 1.35 million ha if the selection criterion is raised to IRR > 12%. The mean IRR for the projects in SSA with IRR > 12% is 15.2%. Our follow-up calculation of IRR for K = US$8,000/ha, using their assumptions, gives IRR = 6.4% for this mean, which indicates that the economically viable potential of large-scale irrigation development in SSA would shrink from 1.35 million ha to one-half or even less, using a more plausible level of the unit investment cost. There could be some economically viable large-scale irrigation projects in SSA, but the extent of such potential projects would be tiny compared to their maximum estimate of 15 million ha.

The fifth example is You et al. (2014), who apply the same framework to Kenya. This paper states, 'We showed that there is considerable scope for the expansion of […] dam-based […] irrigation in Kenya' (34). They estimate the potential for large-scale irrigation development in Kenya to be 1.0 million ha if K = US$5,000/ha and IRR > 0% and 460,000 ha if K = US$8,000/ha and IRR > 12%.[21] Our follow-up calculation of the net crop return ([R–c] in Eq. 10.1), using their assumptions, reveals a possible overestimation of irrigation benefit. For example, an IRR of 74%, the highest IRR reported for a dam project for K = US$8,000/ha, requires (R–c) = US$26,000/ha/year, which is 200 t/ha/year in 'paddy equivalent,' a level of crop performance which, no doubt, is impossible to attain. Allowing for a possible benefit overestimation, the 'paddy equivalent' crop performance required for their cut-off IRR of 12% is reproduced as 11 t/ha/year, a level higher than the crop performance of the Mwea Scheme, the best irrigation scheme in Kenya, attained in 2017. All this suggests that the claimed large-scale irrigation potential should be adjusted downward accordingly.

10.5 Conclusions

The historical trend of the world rice price in Fig. 10.1 reminds us that the boom in irrigation investment in the last quarter of the twentieth century was induced and enhanced by repeated food crises in the 1960s and 1970s (Hayami and Kikuchi 1978). The pause in large-scale irrigation investment from the late 1990s until recently may have been due as much to the low-price regime in the world rice market— making it difficult to justify costly irrigation projects—as to concerns about the poor performance of large-scale irrigation projects.

Our exercise in evaluating the economic viability of large-scale irrigation development by estimating the costs of constructing the Mwea Scheme, one of the best irrigation schemes in SSA, as a brand-new scheme, shows that the investment performance of such a project exceeds the acceptable IRR level of 8–10% when a high-price regime of 2008–2013 prevails in the world rice market. The results imply that medium or high rice prices, coupled with the high performance of irrigated agriculture (i.e.,

---

[21] For details of our additional examination of You et al. (2014) data, see SP-M Sub-section IV-2-2 and SP-M Table IV-6.

more than 9 t/ha/year in terms of rice yield), could justify large-scale irrigation development if the project costs are less than US$15,000/ha. Though rare, some irrigation schemes in SSA attain such high levels of crop performance. We doubt that there is significant untapped potential in SSA for large-scale irrigation developments that are economically justifiable. Should we pursue the untapped potential, however small it is?

For the answer to this question to be 'Yes,' many conditions must be satisfied. In the case of Mwea, proper rice cultivation training was provided since the establishment of the Mwea scheme, while the rice market has been well developed since the liberalization in 1999. Farmers have adopted modern inputs and improved rice cultivation practices facilitated by thorough land preparation using tractors and oxen since the early period of irrigation development (Veen 1973), making Kenya one of the top five SSA countries in terms of rice productivity (Chap. 1 of this volume.). These observations are consistent with the importance of proper rice cultivation training and complementary technologies, as emphasized throughout this volume (see Chaps. 2 and 7 for overviews and other chapters for specific cases).[22]

Furthermore, we would ask, contra the 'Hiding Hand' thesis of Hirschman (1967), whether we have invented a way to overcome the 'malevolent hiding hand' that often works in large-scale irrigation projects. Have we found ways to prevent cost under-estimation and benefit overestimation? How to break the vicious cycle of the 'build-neglect-rebuild' syndrome to prevent moral hazard in scheme maintenance? What about appropriate institutional frameworks for effective O&M for scheme sustainability? The rising project overhead costs, including planning, preparing, and training for O&M, are one reason for the escalation of project costs in recent years. However, expenditures for these purposes may be insufficient to realize a good return on the investment unless sufficient resources are allocated every year after project completion to operate and maintain the scheme and to build institutional capacity for O&M. This has rarely been fulfilled.

Various other types of irrigation development have been identified and documented in SSA, which are likely to be more profitable than large-scale projects and to deliver benefits sooner (Woodhouse et al. 2017). When planning a large-scale irrigation project, a serious assessment must be made of these irrigation alternatives based on detailed grass-root studies of the area planned for the project. Unless the problems and defects inherent in large-scale irrigation development are overcome, we conclude that the promotion of such projects results in a substantial waste of resources.

---

[22] Chapters 3–6 of this volume provide new evidence on the impact of training by agricultural extension in the intensification of rice production in Tanzania, Cote d'Ivoire, Mozambique, and Uganda. We confirm that positive impacts of training were realized even without any improvement in irrigation, marketing, or credit programs.

References

Adams WM (1990) How beautiful is small? Scale, control and success in Kenyan irrigation. World Dev 18(10):1309–1323

Balasubramanian V, Sie M, Hijimans RJ, Otsuka K (2007) Increasing rice production in sub-Saharan Africa: challenges and opportunities. Adv Agron 94:55–133

Baldwin E, Washington-Ottombre C, Dell'Angelo J, Cole D, Evans T (2015) Polycentric governance and irrigation reform in Kenya. Governance September 2015. https://doi.org/10.1111/gove.12160

Barker R, Herdt RW, Rose B (1985) The rice economy of Asia. Resources for the Future and IRRI, Washington, D.C. and Manila

Bartier B, Jamin JY, Ouedraogo H, Diarra A, Barry B (2014) Irrigation investment trends and economic performance in the Sahelian countries in West Africa. In: Namara R, Sally H (eds) Irrigation in West Africa: current status and a view to the future. IWMI, Colombo, p 21–35

Belli P, Anderson J, Barnum H, Dixon J, Tan JP (1998) Handbook on economic analysis of investment operations. UNDP. https://www.adaptation-undp.org/sites/default/files/downloads/handbo okea.pdf

Biswas AK (1986) Irrigation in Africa. Land Use Policy 3(4):269–285

Borgia C, García-Bolañosa M, Li T, Gómez-Macpherson H, Comas J, Connor D, Mateos L (2013) Benchmarking for performance assessment of small and large irrigation schemes along the Senegal Valley in Mauritania. Agric Water Manag 121(C):19–26

Burney JA, Naylor RL, Postel SL (2013) The case for distributed irrigation as a development priority in sub-Saharan Africa. PNAS 110(31):12513–12517

Chambers R (1969) Settlement schemes in tropical Africa. Routledge and Kegan Paul, London

Chambers R (1973) The history of the scheme. In: Chambers R, Moris J (eds) Mwea: An irrigated rice settlement in Kenya. Weltforum Verlag, Munich, pp 64–78

Crow-Miller B, Webber M, Molle F (2017) The (re)turn to infrastructure for water management? Water Altern 10(2):195–207

de Fraiture C, Giordano M (2014) Small private irrigation: a thriving but overlooked sector. Agric Water Manag 131(C):167–74 https://doi.org/10.1016/j.agwat.2013.07.005

Diao X, Headey D, Johnson M (2008) Toward a green revolution in Africa: What would it achieve, and what would it require? Agric Econ 39(S1):539–550

Ejeta G (2010) African Green Revolution needn't be mirage. Science 327:831–832

Estudillo JP, Otsuka K (2012) Lessons from the Asian Green Revolution in rice. In: Otsuka K, Larson DF (eds) An African Green Revolution: finding ways to boost productivity on small farms. Springer, New York and London, pp 17–42

FAO and WFP (2022) Hunger hotspots. FAO-WFP early warnings on acute food insecurity: June to September 2022 Outlook. Rome. https://docs.wfp.org/api/documents/WFP-0000139904/dow nload/?_ga=2.13574004.2134344741.1654890341-67065628.1637678351

Fujiie H, Maruyama A, Fujiie M, Takagaki M, Merrey DJ, Kikuchi M (2011) Why invest in minor projects in sub-Saharan Africa? an exploration of the scale economy and diseconomy of irrigation projects. Irrig Drain 25(1):39–60

Harrison E (2018) Engineering change? the idea of 'the scheme' in African irrigation. World Dev 111:246–255

Hayami Y, Kikuchi M (1978) Investment inducements to public infrastructure: irrigation in the Philippines. Rev Econ Stat 60:70–77

Higginbottom TP, Adhikari R, Dimova R, Redicker S, Foster T (2021) Performance of large-scale irrigation projects in sub-Saharan Africa. Nat Sustain 4(6):501–508. https://doi.org/10.1038/s41 893-020-00670-7

Hirschman AO (1967) Development projects observed. The Brookings Institution, Washington, DC

Huppert W, Svendsen M, Vermillion DL (2003) Maintenance in irrigation: multiple actors, multiple contexts, multiple strategies. Irrig Drain 17(1–2):5–22

Inocencio A, Kikuchi M, Tonosaki M, Maruyama A, Merrey D, Sally H, de Jong I (2007) Costs and performance of irrigation projects: a comparison of sub-Saharan Africa and other developing regions. IWMI Research Report 109. International Water Management Institute

International Trade Center (2022) Trade map: trade statistics for international business development. https://www.trademap.org/Bilateral_TS.aspx?nvpm=1%7c404%7c%7c586%7c%7c1006%7c%7c%7c4%7c1%7c1%7c1%7c2%7c1%7c1%7c1%7c1%7c1. Accessed 13 Aug 2022

IRRI (2000) World rice statistics. International Rice Research Institute, Manila

JICA (1988) Kenya kyowakoku Mwea-chiku kangai keikaku jizen chosa hokokusho [Preliminary-survey report on irrigation plan of Mwea area in the Republic of Kenya]. Japan International Cooperation Agency, Tokyo. http://open_jicareport.jica.go.jp/pdf/10729622.pdf. Accessed 13 Jan 2020

JICA (1989) Basic design study report on the project for Mwea Irrigation Settlement Scheme Development in the Republic of Kenya. Japan International Cooperation Agency, Tokyo. http://open_jicareport.jica.go.jp/833/833/833_407_10759017.html. Accessed 13 Jan 2020

JICA (1997) Kenya Mwea kangai nogyo kaihatsu keikaku gaiyo [Mwea Irrigation Agricultural Development Project: Outline]. The Japan International Cooperation Agency, Tokyo. http://open_jicareport.jica.go.jp/pdf/11440146_01.pdf. Accessed 13 Jan 2020

JICA (2010) Mwea kangai kaihatsu jigyou Jizen hyouka hyo [Project appraisal list: Mwea Irrigation Development Project]. https://www2.jica.go.jp/ja/evaluation/pdf/2010_KE-P27_1_s.pdf. Accessed 17 Jan 2020

JICA Research Institute & Nippon Koei (2018) Estimation of construction cost for Mwea Irrigation Development Project: an Empirical Analysis of Expanding Rice Production in Sub-Sahara Africa Phase 2 (Final report). JICA Research Institute, Tokyo

Jones WJ (1995) The World Bank and irrigation. The World Bank, Washington, DC

Kabutha C, Mutero C (2002) From government to farmer-managed smallholder rice schemes: the unresolved case of the Mwea Irrigation Scheme. In: Blank HG, Mutero CM, Murray-Rust H (eds) The changing face of irrigation in Kenya. IWMI, Colombo. http://publications.iwmi.org/pdf/h030840.pdf

Kikuchi M, Haneishi Y, Tokida K, Maruyama A, Asea G, Tsuboi T (2016a) The structure of indigenous food crop markets in sub-Saharan Africa: the rice market in Uganda. J Dev Stud 52(5):646–664

Kikuchi M, Haneishi Y, Maruyama A, Tokida K, Asea G, Tsuboi T (2016b) The competitiveness of domestic rice production in East Africa: a domestic resource cost approach in Uganda. J Agric Rural Dev Trops Subtrop 117(1):57–72

Kikuchi M, Mano Y, Njagi TN, Merrey D, Otsuka K (2021) Economic viability of large-scale irrigation construction in Sub-Saharan Africa: What if Mwea Irrigation scheme Were constructed as a brand-new scheme? J Dev Stud 57(5):772–789. https://www.tandfonline.com/doi/full/10.1080/00220388.2020.1826443

KNA (2018, Nov 14) Hope for farmers as Sh7.35B irrigation project set to be revived. Business Today. https://businesstoday.co.ke/hope-farmers-sh7-35-billion-gravity-irrigation-project-set-revived/

Lankford B, Makin I, Matthews N, McCornick PG, Noble A, Shah T (2016) A compact to revitalise large-scale irrigation systems using a leadership-partnership-ownership "Theory of Change." Water Altern 9:1–32

Lefore N, Giordano M, Ringler C, Barron J (2019) Sustainable and equitable growth in farmer-led irrigation in sub-Saharan Africa: What will it take? Water Altern 12(1):156–168

Mano Y, Njagi TN, Otsuka K (2022) An inquiry into the process of upgrading rice milling services: the case of Mwea Irrigation Scheme in Kenya. Food Policy 106(C)

Merrey DJ, Sally H (2017) Another well-intentioned bad investment in irrigation: the millennium challenge corporation's "compact" with the Republic of Niger. Water Altern 10:195–203

Moigne GL, Barghouti S (1990) How risky is irrigation development in sub-Saharan Africa? In: Barghouti S, Moigne GL (eds) Irrigation in sub-Saharan Africa: the development of public and private systems. World Bank Technical Paper 123, p 45–59

Moris J (1973) The Mwea environment. In: Chambers R, Moris J (eds) Mwea: an irrigated rice settlement in Kenya. Weltforum Verlag, Munich, pp 16–63

Moris J (1986) Irrigation as a privileged solution in African development. Dev Pol Rev 5:99–123

Nakano Y, Bamba I, Diagne A, Otsuka K, Kajisa K (2012) The possibility of a rice Green Revolution in large-scale irrigation schemes in sub-Saharan Africa. In: Otsuka K, Larson DF (eds) An African Green Revolution: finding ways to boost productivity on small farms. Springer, New York and London, pp 43–70

NEPAD (2003) Comprehensive Africa agricultural development programme (CAADP). New Partnership for Africa's Development, Midrand, South Africa. https://www.nepad.org/caadp/public ation/au-2003-maputo-declaration-agriculture-and-food-security. Accessed 20 July 2019

NIB (2010) Final design report: consultancy for detailed design, tender documents preparation and construction supervision. Mwea Irrigation Scheme Water Management Improvement Project, vol 1. GIBB Africa Ltd., Nairobi

NIB (2018) Bura irrigation rehabilitation project. National Irrigation Board (Kenya). https://nib.or. ke/projects/flagship-projects/bura-gravity-project

Njeru TN, Mano Y, Otsuka K (2016) Role of access to credit in rice production in sub-Saharan Africa: the case of Mwea Irrigation Scheme in Kenya. J Afr Econ 25(2):300–321

Olivares J (1989) The agricultural development of sub-Saharan Africa: the role and potential of irrigation. Nat Resour Forum 13(4):268–274. https://doi.org/10.1111/j.1477-8947.1989.tb0 0349.x

Ostrom E (1992) Crafting institutions for self-governing irrigation systems. ICS Press and Institute for Contemporary Studies, San Francisco, California

Otsuka K, Larson DF (2012) Towards a green revolution in sub-Saharan Africa. In: Otsuka K, Larson DF (eds) An African Green Revolution: finding ways to boost productivity on small farms. Springer, New York and London, pp 281–300

Plusquellec H (2019) Overestimation of benefits of canal irrigation projects: decline of performance over time caused by deterioration of concrete canal lining. Irrig Drain 68(3):383–388

Rashid S, Minot N (2010) Are staple food markets in Africa efficient? Spatial price analyses and beyond. Paper presented at the COMESA policy seminar "Food price variability: causes, consequences, and policy options" on January 25–26 2010 in Maputo, Mozambique under the COMESA-MSU-IFPRI African Agricultural Markets Project (AAMP). https://www.researchgate.net/publication/46470919_Are_Staple_Food_Markets_in_ Africa_Efficient_Spatial_Price_Analyses_and_Beyond

Reliefweb (2007) Kenya: Kuwait finances the Bura Irrigation and Settlement Scheme Rehabilitation Project. Report from Government of Kenya on Dec 13 2007. https://reliefweb.int/report/kenya/ kenya-kuwait-finances-bura-irrigation-and-settlement-scheme-rehabilitation-project. Accessed July 20 2019

Rockström J, Lannerstad M, Falkenmark M (2007) Assessing the water challenge of a new green revolution in developing countries. PNAS 104:6253–6260

Sanchez P, Denning G, Nziguheba G (2009) The African Green Revolution moves forward. Food Secur 1:33–44

Sandford S (1973) An economic evaluation of the scheme. In: Chambers R, Moris J (eds) Mwea: an irrigated rice settlement in Kenya. Weltforum Verlag, Munich, pp 393–438

Suhardiman D, Giordano M (2014) Is there an alternative for irrigation reform? World Dev 57:91–100

Tiffen M (1987) Dethroning the internal rate of return: the evidence from irrigation project. Dev Policy Rev 5:361–377

Veen JJ (1973) The production system. In: Chambers R, Moris J (eds) Mwea: an irrigated rice settlement in Kenya. Weltforum Verlag, Munich, pp 99–131

Veldwisch G, Bolding A, Wester P (2009) Sand in the engine: the travails of an irrigated rice scheme in Bwanje valley, Malawi. J Dev Stud 45:197–226

WFP (2022) 2022 Global report on food crises: Joint Analysis for Better Decisions. https://docs.wfp.org/api/documents/WFP-0000138913/download/?_ga=2.226221786.137 9254128.1660037865-461457561.1660037865. Accessed 9 Aug 2022

WCD (2000) Dams and development. The World Commission on Dams. Earthscan, Webber, London. https://www.internationalrivers.org/sites/default/files/attached-files/world_commission_on_dams_final_report.pdf. Accessed 20 July 2019

Woodhouse P, Veldwish GJ, Venot JP, Brockington D, Komakech H, Manjichi A (2017) African farmer-led irrigation development: reframing agricultural policy and investment? Peasant Stud 44:213–233

World Bank (1990) Bura irrigation settlement project; Project completion report. http://documents.worldbank.org/curated/en/503041468046825515/pdf/multi-page.pdf. Accessed 20 July 2019

World Bank (2005) Shaping the future of water for agriculture: a sourcebook for investment in agricultural water management. http://siteresources.worldbank.org/INTARD/Resources/Shaping_the_Future_of_Water_for_Agriculture.pdf. Accessed 20 July 2019

World Bank (2007) Investment in agricultural water for poverty reduction and economic growth in sub-Saharan Africa. http://siteresources.worldbank.org/RPDLPROGRAM/Resources/459596-1170984095733/synthesisreport.pdf. Accessed 20 July 2019

World Bank (2010) Cost-benefit analysis in World Bank projects. http://documents.worldbank.org/curated/en/253101468340140595/pdf/624700PUB0Cost00Box0361484B0PUBLIC0.pdf. Accessed 25 Jan 2020

World Bank (2017) Malawi–Shire valley transformation program. http://documents.worldbank.org/curated/en/379081508551260039/pdf/Malawi-Project-Appraisal-Document-PAD-09282017.pdf. Accessed 20 July 2019

World Bank (2019) World Bank commodity price data (Updated on January 04, 2019). pubdocs.worldbank.org/en/226371486076391711/CMO-Historical-Data-Annual.xlsx. Accessed 20 July 2019

World Bank (2020a) World development indicators. https://data.worldbank.org/indicator/NY.GDP.MKTP.CD https://data.worldbank.org/indicator/NY.GDP.MKTP.KD https://data.worldbank.org/indicator/PA.NUS.FCRF?locations=KE-JP&view=chart. Accessed 23 Jan 2020a

World Bank (2020b) Lending rates & fees. https://treasury.worldbank.org/en/about/unit/treasury/ibrd-financial-products/lending-rates-and-fees#a. Accessed 20 Oct 2020b

You L, Ringler C, Nelson G, Wood-Sichra U, Robertson R, Wood S, Guo Z, Sun Y (2010) What is the irrigation potential for Africa? A combined biophysical and socioeconomic approach. IFPRI Discussion Paper 00993

You L, Xie H, Wood-Sichra U, Guo Z, Wang L (2014) Irrigation potential and investment return in Kenya. Food Policy 47:34–45

Zwart SJ (2013) Assessing and improving water productivity of irrigated rice systems in Africa. In: Wopereis MCS et al (eds) Realizing Africa's rice promise. CABI International, Boston, USA, p 265–275

Masao Kikuchi is a professor emeritus of Chiba University since 2010, after teaching agricultural development for 18 years. He received Ph.D. in agricultural economics from Hokkaido University in 1976.

Yukichi Mano is a professor at Hitotsubashi University, Japan, and is a fellow at Tokyo Center for Economic Research (TCER). He received Ph.D. in Economics from the University of Chicago in 2007. His scholarly interests include agricultural technology adoption, horticulture and high-value crop production, business, and management training (KAIZEN), human capital investment, migration and remittance, and universal health coverage in Asia and sub-Saharan Africa.

Timothy N. Njagi is a Senior Research Fellow at Tegemeo Institute of Agricultural Policy and Development, Egerton University. He received a Ph.D. in Development Economics from the National Graduate Institute for Policy Studies (GRIPS), Japan in 2012. His current research focus is on technology adoption, irrigation, credit, governance, land issues, and resilience.

Douglas J. Merrey is currently an independent consultant. His former positions include Director of Research at Food, Agricullture, and Natural Resources Policy Analysis Network, Director for Africa at International Water Management Intitute (IWMI), and Director General for Programs at IWMI. He received his Ph.D. in 1983 from the University of Pennsylvania in Anthropology.

Keijiro Otsuka is a professor of development economics at the Graduate School of Economics, Kobe University and a chief senior researcher at the Institute of Developing Economies in Chiba. Japan since 2016. He received Ph.D. in economics from the University of Chicago in 1979. He majors in Green Revolution, land tenure and land tenancy, natural resource management, poverty reduction, and industrial development in Asia and sub-Saharan Africa.

Chapter 11
Irrigation Scheme Size and Its Relationship to Investment Return: The Case of Senegal River Valley

Takeshi Sakurai

Abstract Despite the boost in rice production over the last decade in sub-Saharan Africa (SSA), increased production is required to satisfy the demand from rapidly growing urban populations. To enhance rice production in SSA, it is critically important to increase investment in irrigation, as it played a major role in advancing the rice Green Revolution in Asia. However, the question of whether large-scale irrigation is more efficient than small-scale projects has remained contentious. The objective of this chapter is to make a positive contribution to the debate by providing empirical evidence from the Senegal River Valley (SRV), where many irrigation schemes coexist. Based on a survey of 173 farmers' groups that use irrigation schemes of different sizes, OLS regression analyses are used to examine the association between the size of the irrigation scheme and investment performance, which is defined as annual rice income per hectare minus the annual depreciation cost of investment in the irrigation scheme per hectare. After controlling for factors that may influence investment performance, it is found that irrigation scheme size is positively associated with investment performance due to the economy of scale involved in the unit cost of investment. However, the positive association is non-linear and becomes negative beyond 1600 ha. The analyses also show that government-financed irrigation schemes perform worst. Therefore, even if investment in large-scale irrigation is justified, the questions of who will invest and how it will be managed are also important factors affecting the performance of large-scale irrigation schemes.

11.1 Introduction

Despite the boost in rice production over the last few decades in sub-Saharan Africa (SSA), increased production is needed to satisfy the demand from rapidly growing urban populations (Arouna et al. 2021). During the five-year period after the food crisis (2007–2012), rice production in SSA grew at 8.4% per year—much higher

T. Sakurai (✉)
Department of Agricultural and Resource Economics, University of Tokyo, Bunkyo-ku, Tokyo 113-8657, Japan
e-mail: takeshi-sakurai@g.ecc.u-tokyo.ac.jp

© JICA Ogata Sadako Research Institute for Peace and Development 2023
K. Otsuka et al. (eds.), *Rice Green Revolution in Sub-Saharan Africa*, Natural Resource Management and Policy 56, https://doi.org/10.1007/978-981-19-8046-6_11

than the 3.2% from 2000 to 2007—with 71% of this growth attributed to yield increase (Saito et al. 2015). Cross-country regression analyses showed that the share of irrigated rice area among the total rice area was one of the factors contributing to the recent increase in yield (Saito et al. 2015). However, the growth rate of rice yield declined between 2012 and 2018 following the cessation of the emergency response to the food crisis. This can be seen in the decrease in the growth rate of investment in agriculture per hectare, which fell from 3.28% between 2008 and 2012 to 0.91% between 2012 and 2018 (Arouna et al. 2021). Thus, to increase rice production in SSA, it is necessary to increase investment in agriculture again. The key question, however, is where to invest. In Asia, irrigation was almost a prerequisite for the rice Green Revolution (Estudillo and Otsuka 2012). However, it has remained underdeveloped in SSA. Therefore, a high priority should be given to considering investment in irrigation in SSA.

With respect to irrigation investments, a fundamental question concerns what kind of irrigation should be promoted, as many large-scale irrigation projects implemented in the latter half of the twentieth century have performed poorly—particularly in SSA (Adams 1992; Inocencio et al. 2007). Following this poor performance and increasing concern about the negative environmental impact from the construction of large-scale irrigation facilities, small-scale irrigation schemes (usually managed by farmers) seem to have been encouraged instead (World Bank 2005, 2007). However, large-scale irrigation projects have been revived because of the food crisis in 2008 and also encouraged by recent advances in yield-increasing technologies that require irrigated conditions.

Despite the resurgence in interest, the prospect of a revival in larger scale projects has evoked negative reactions (Kikuchi et al. 2021). This debate is relevant to considering ways of enhancing rice production through investment in agriculture. Therefore, the objective of this chapter is to contribute to this debate by providing relevant empirical evidence from the Senegal River Valley (SRV). The SRV provides an ideal study site to test whether large-scale irrigation schemes are better targets for investment than small-scale irrigation schemes, as there are many coexisting irrigation schemes of different sizes in this region that produce rice using similar technologies.

To assess investment performance, the approach typically used in the literature is internal rate of returns (IRRs). Based on IRRs from 314 large-scale irrigation projects, Inocencio et al. (2007) showed a significant positive association between project size and IRR. The positive association was due to a strong scale economy of project size in the unit cost of irrigation projects. Their analyses seemed to support larger scale projects, but since their data did not include small-scale irrigation projects, they could not conclude that large scale is more advantageous. To answer this remaining question, Fujiie et al. (2011) conducted a similar analysis, including small- and micro-scale irrigation projects in SSA, confirming that a strong scale economy exists within each scale category, i.e., large (>100 ha), small (5–100 ha), and micro (<5 ha), respectively. However, Fujiie et al. (2011) also found a positive association between project size and unit cost of the project and a consequent negative association between project size and IRR if they combine all the scale categories. Based on this finding, Fujiie et al. (2011) suggested a need to promote small- or micro-scale irrigation projects in

SSA. However, since many factors, such as production technologies and irrigation management, differ between large-scale irrigation projects and small/micro-scale irrigation projects, the observed negative association between project size and IRR may be caused by factors other than size.

In addition, since IRRs are usually reported at the time of project completion, both Inocencio et al. (2007) and Fujiie et al. (2011) implicitly assumed that the product price and production technologies do not change over time—an unrealistic assumption. In this regard, Kikuchi et al. (2021) considered the economic viability of a large-irrigation scheme (Mwea Irrigation Scheme) in Kenya if it were newly constructed now (i.e., in 2017).[1] This means that they used rice price and rice productivity observed in 2017 while investment cost was converted to a 2017 price to calculate the current IRR rather than the IRR at the time of project completion. According to the authors, rice production intensity increased from one crop/year in 1960 to two crops/year, including ratoon harvesting in 2017, while the rice yield per season did not change significantly during this period. Kikuchi et al. (2021) concluded that, with this high rice productivity, if rice price increases to the level reached during the period of the 2008 food crisis, the IRR of the Mwea project will become high enough to justify the investment compared with the opportunity cost of the investment fund. However, since Kikuchi et al. (2021) did not compare the estimated IRR of the Mwea Irrigation Scheme with other irrigation projects, we cannot know if smaller scale irrigation projects perform better under the same assumptions.

Thus, this chapter, adopting Kikuchi et al. (2021)'s approach of incorporating the change in production technologies, compares economic returns among irrigation schemes of different sizes, like that undertaken by Fujiie et al. (2011). However, unlike Fujiie et al. (2011), this chapter tries to control for factors that may be correlated with size and economic performance through regression analyses. Since such comparisons are not found in the literature, it will be a novelty of this study. As mentioned above, such analyses are possible because there are many irrigation schemes with different sizes in the SRV.

To examine the relationship between size and investment performance, both investment cost and output must be considered. With respect to investment cost, by controlling for other factors, unlike Fujiie et al. (2011), it is expected that scale economies in the unit cost of investment will not disappear even if small-scale irrigation schemes are included. Thus, if output does not depend on investment size, investment in large-scale irrigation should perform better than small-scale irrigation due to the economies of scale. However, regarding output, it is not known which kind of irrigation scheme—large-scale or small-scale—generates more income per hectare. Consequently, the relationship between irrigation scheme size and investment performance is an empirical question to which this study seeks an answer.

The structure of this chapter is as follows. Sect. 11.2 describes the study site. Data and methodology are presented in Sect. 11.3, and regression results follow in

---

[1] See also Chap. 10, which is a revised version of Kikuchi et al. (2021).

Sect. 11.4. Based on the findings that irrigation scheme size is positively associated with investment performance in the previous section, Sect. 11.5 offers some concluding comments and discusses policy implications.

11.2 Study Site

11.2.1 Irrigation Schemes in the SRV

The study site is located in the SRV (Fig. 11.1). The Senegal River, originating in the highlands in Guinea, forms an 800 km-long boundary between Mauritania to the north and Senegal to the south. While irrigation schemes for rice production exist on both sides of the river, this study focuses only on the Senegalese side, where the total area supported by irrigation schemes reached about 110,000 ha in 2012 (Manikowski and Strapasson 2016).

The construction of large-scale irrigation schemes started in 1960 after independence. In particular, the construction of two dams (the Diama and Manantali Dams shown in Fig. 11.1) in 1988 made it possible for this country to develop large irrigated rice fields along the river (Manikowski and Strapasson 2016). A governmental agency called SAED (Société Nationale d'Exploitation des Terres du Delta du Fleuve Sénégal et des Vallées Sénégal et de la Falémé) was established in 1965 and has been responsible for the development of irrigation schemes in the SRV.

Fig. 11.1 The Senegal River valley

Currently, irrigation schemes in the SRV can be classified into three types based on the investors involved in construction, and hence the three scheme types are considered to be an investment category to be explained below. The first one comprises the schemes that the government has invested in. In these cases, the scheme size is relatively large and is usually equipped with multiple irrigation and drainage pumps as well as canal networks. They were formerly managed directly by SAED (i.e., publicly owned and managed), but since 1987 the management has been transferred to farmers' organizations (Diouf et al. 2015). Each scheme is divided into many sections with feeder canals, and a group of 20–30 farmers is responsible for the water distribution and feeder canal maintenance within a section. The second type consists of village-based irrigation schemes managed and operated by village-level management committees. They are smaller scale irrigation schemes constructed by SAED in collaboration with villagers.

The third type comprises privately funded schemes. In response to the liberalization of the agricultural market, private investment in irrigation increased during the 1980s in the SRV, reaching 42,600 ha in total in 1993. It then declined due to the devaluation of the CFA Franc in 1994, which led to decreased incentives for continuing rice production due to the increased prices of imported inputs, such as fertilizer and fuel (Dia 2001). Stimulated by the food crisis in 2008 and encouraged by a new government policy (GOANA, Grande Offensive in Agriculture for Food and Abundance) initiated in April 2008, private investment in irrigation schemes has been growing again. In general, privately funded schemes are the smallest among the three.

11.2.2 Rice Production and Scheme Size

Irrigated rice production in the SRV is known for its high productivity in SSA (Nakano et al. 2012; Sakurai 2016). In fact Tanaka et al (2015) reported that the mean yield in the wet season over the nine-year period from 2002 to 2010 was 5.0–5.6 t/ha depending on the location. However, the yield had stagnated (i.e., had not increased during the nine-year period) and remained short of reaching agronomically attainable yields by 2.2–3 t/ha (Tanaka et al. 2015). Tanaka et al. (2015) showed that delayed sowing was the primary factor leading to yield reduction and that the major reasons for delayed sowing were related to the availability of credit, machinery, and irrigation water. By contrast, rice yield in the dry season was higher than the wet season, and increased from 5.9 to 6.8 t/ha during the same period as the wet season data discussed above (from 2002–2006 to 2008–2011).[2] Brosseau et al. (2021), using data obtained in 2017, showed that some farmers are shifting rice single cropping in the wet season to the hot dry season, whereas other farmers are adopting two cropping—namely

---

[2] There are three differentiated seasons in the SRV: humid and hot (wet season, about 200 mm rainfall) from July to October, dry and cool (cool dry season) from November to February, and dry and hot (hot dry season) from March to June (Haefele et al. 2002).

rice in the hot dry season and vegetables in the cold dry season. They also pointed out that the rice double cropping area would not increase or even decrease even if rice double cropping was strongly promoted by the Senegalese government.

Borgia et al. (2013) compared small-scale and large-scale irrigation schemes on the Mauritanian side of the SRV, finding that the mean yield did not differ much (3.50 and 3.77 t/ha).[3] The mean yield on the Mauritanian side is much lower than on the Senegal side, although they likely grow rice in similar production conditions. These comparisons may help when considering the kinds of irrigation schemes that could be promoted—large scale or small scale. However, unlike the analyses in this chapter, they do not take account of investment costs.

11.3 Data and Methods

11.3.1 Data

One hundred and eighty farmers' groups were randomly sampled from the list of 3304 farmers' groups provided by SAED. They are located along the Senegal River in Dagana and Podor departments (Fig. 11.1). From each group, five rice producers were randomly selected from the member list. The interviews commenced in March 2021 and continued until December 2021. Because the enumerators could not identify some of the sampled groups and because some of the sampled groups rejected the interview, data were collected from 174 groups out of the initially sampled 180 groups. In addition, one group was dropped from the analysis due to missing values. Thus, this study used data from 173 farmers' groups to investigate the performance of investments in irrigation schemes. Since the SAED's list of farmers' organizations did not include any information on the investment category explained in the previous section, the random sampling did not consider the distribution of the investment category in the sample. As shown in Table 11.1, out of the 173 sample farmers' groups, 63 were under government-funded schemes, 69 managed village-based schemes, and 41 managed privately funded schemes.

It is important to explain the relationship between farmers' groups and irrigation schemes. In this chapter, "scheme" refers to the unit of irrigation investment or the whole structure of irrigation. Thus, in village-based and privately funded irrigation schemes, each group has a corresponding irrigation scheme that the group is managing. On the other hand, in the case of government-funded irrigation schemes, the farmers' group is not the unit of investment since the investment is made at the scheme level.

Table 11.1 compares key characteristics of farmers' groups by investment category. This categorization is based on the information obtained in group interviews of each farmers' group, i.e., farmers' own perceptions. Note that village-based and

---

[3] Borgia et al. (2013) do not specify the cropping season, but it can be assumed that production data were collected in the wet season.

Table 11.1 Characteristics of farmers' groups by investment category

	Government funded	Village-based	Privately funded
Main investor for initial construction			
Government	1	0.42	0.07
Villagers[a]	0	0.48	0.22
Private funds[a]	0	0.10	0.71
Size of the irrigation scheme (ha)[b]	588 (664)	44.8 (41.8)	28.7 (24.5)
Number of the groups in the scheme[c]	20.0 (24.7)	1	1
Command area managed by the group (ha)	38.8 (37.4)	44.8 (41.8)	28.7 (24.4)
Canal length managed by the group (m)	573 (279)	856 (905)	598 (421)
Years since the creation of the group	26.5 (9.07)	28.1 (9.08)	27.5 (9.23)
Number of members in the group	42.7 (43.0)	54.3 (88.2)	29.3 (53.0)
Male members	38.9 (40.2)	43.8 (72.4)	18.6 (33.1)
Female members	3.81 (6.63)	10.5 (20.7)	10.7 (26.2)
Group was formed based on a family (1 = Yes, 0 = No)	0.03	0.07	0.12
Number of farmers' groups	63	69	41

Standard deviations are in parentheses
[a] Including loans from financial institutions and government subsidies
[b] The scheme is the unit of irrigation construction and hence is considered as the unit of investment.
[c]Even if a group is physically attached to a large irrigation scheme, if the construction was done independently, the group does not belong to the large scheme. In such cases, the number of the groups in the scheme is 1

privately funded irrigation schemes did not exclusively depend on villagers' contributions or private funds for the initial construction because the government and donors subsidized the scheme construction. As for the scheme size, the size of government-funded schemes is more than ten times larger than the other type of schemes, and the average number of farmers' groups that belong to a government-funded scheme is about 20. Otherwise, at the group level, the three categories of farmers' groups are not very different, although privately funded ones tend to be smaller in terms of command area and the number of members. In addition, for unknown reasons, the number of female members is significantly smaller in farmers' groups from government-funded irrigation schemes.

11.3.2 Methodology

As mentioned above, the objective of this chapter is to examine the relationship between the size of the irrigation project and the performance of investments in project. Specifically, the size of the irrigation scheme described above is used as the project size in this chapter. As for investment performance, this study adopts the idea of Inocencio et al. (2007) and Fujiie et al. (2011), who used IRR as an indicator of the investment performance of irrigation projects. However, this study uses annual return—to be defined below—instead of IRR. Taking account of the initial investment cost by using annual depreciation cost, annual return per hectare of farmers' group i is given as follows:

$$R_i = \sum_s (\omega_{is} p_s Y_{is} - C_{is}) - D_i \tag{11.1}$$

where R_i is annual return per hectare of the command area of the farmers' group i, Y_{is} is rice (paddy) yield of farmers' group i in season s, p_s is the market price of paddy in season s,[4] ω_{is} is the exploitation rate of farmers' group i in season s (to be defined shortly), C_{is} is the production cost of farmers' group i in season s including membership fee, and D_i is the annual depreciation cost per hectare, to be defined by Eq. (11.2), which is implicitly incurred by the farmers' group i. Rice yield (Y_{is}) was obtained as the mean of farmers in group i who actually grew rice in season s. Although rice yield was generally high compared with the standard in SSA as mentioned above, most farmers' groups did not fully use their land for rice production. To capture this inefficiency, the exploitation rate (ω_{is}) is introduced in Eq. (11.1). The exploitation rate of farmers' group i is defined as the proportion of rice harvested area divided by total irrigation command area of farmers' group i in season s. Thus, $\omega_{is} Y_{is}$ is exploitation rate adjusted yield.

Depreciation cost (D_i) consists of three kinds of investment in this study:

$$D_i = \frac{1}{N} K_i + \frac{1}{M} B_i + \frac{1}{M} E_i \tag{11.2}$$

where K_i is initial investment cost per hectare to construct the irrigation scheme to which farmers' group i belongs.[5] Since investment is made at the scheme level, total investment cost was divided by the size of irrigation scheme. Then, the value was converted to the 2021 price by using the Senegalese consumer price index (CPI). B_i is rehabilitation investment per hectare. It is also converted to the 2021 price. The last term, E_i, is the total value of pumps used by farmers' group i. The values were estimated by respondents to answer the question "How much would it be if you bought it now?" for each pump they were using. By adopting a straight-line depreciation method with no salvage value, N and M are the number of usable years of the

---

[4] The market prices are constructed by averaging sample farmers' sale prices of paddy in each season.

[5] Investment in common infrastructure is not considered in this initial investment. The most important common infrastructure in the SRV is comprised of the Diama and the Manantali dams (Fig. 11.1). The former dam was designed to stop saline water intrusion, and the latter dam was designed to maintain river water levels throughout the year. The construction of the two dams at a total cost of UD$830 million brought a potential of 240,000 ha of irrigated agriculture to the SRV and electricity to Senegal (Manikowski and Strapasson 2016).

investment. For rehabilitation (B_i) and pumps (E_i), lifetime was uniformly assumed to be ten years, and hence $M = 10$. On the other hand, for the initial investment (K_i), two different lifetimes were assumed. The first consists of "high depreciation" (or, short lifetime), where the lifetime of the initial investment is assumed to be 30 years for hard (compacted) structures and ten years for soft (non-compacted or partially compacted) structures. Hence, $N = 10$ or 30 depending on the structure. The second refers to "low depreciation" (or long lifetime), where the lifetime of the initial investment is assumed to be 30 years regardless of the structure. Hence, $N = 30$. In the low depreciation case, particularly for soft structure, D_i tend to be smaller than in the high depreciation case, and consequently R_i becomes larger.

Then, to investigate the association between irrigation scheme size and investment performance, annual return defined in Eq. (11.1) and other performance indicators such as exploitation rate, rice yield, etc., were regressed on the size of the irrigation scheme. Then regression model is given below.

$$R_i = \alpha + \beta_1 S_i + \beta_2 S_i^2 + \beta_3 V_i + \beta_4 P_i + X_i \gamma + \varepsilon_i \qquad (11.3)$$

where S_i is scheme size and S_i^2 is its square. Since the relationship between scheme size and return is expected to be non-linear, the squared term is included. V_i and P_i are binary dummy variables for village-based and privately funded irrigation schemes, respectively. As discussed above, they are correlated with scheme size but may have different influences on the economic return, so they are used as control variables. X_i is the vector of other control variables, which are a binary dummy for construction quality (a dummy for hard structure), the number of years since the creation of the farmers' group, how the groups were formed (a dummy variable taking 1 if the group was based on a family), the number of male members, and the number of female members. α is constant and ε_i is an error term. Equation (11.3) is estimated by OLS. In addition, to examine if the relationship between scheme size and return is observed in each investment category, interaction terms between size variables and investment category dummies are incorporated in the following specification

$$\begin{aligned} R_i = {}& \alpha + \beta_1 S_i + \beta_2 S_i^2 + \beta_3 V_i + \beta_{31} V_i \cdot S_i + \beta_{32} V_i \cdot S_i^2 \\ & + \beta_{41} P_i + \beta_{41} P_i \cdot S_i + \beta_4 P_i \cdot S_i^2 + X_i \gamma + \varepsilon_i \end{aligned} \qquad (11.4)$$

11.4 Results

11.4.1 Descriptive Statistics of Investment

Table 11.2 compares the three categories of irrigation schemes in terms of investment. The share of hard construction and initial investment cost per hectare are higher in government-funded irrigation schemes than in other irrigation schemes. However,

Table 11.2 Investment in irrigation facilities by investment category

	Government funded	Village-based	Privately funded
Construction type (share in each category)			
Hard (compacted)	0.254	0.029	0.073
Soft (not compacted/partially compacted)	0.746	0.971	0.927
Initial investment for the construction (million FCFA/ha)[a]	2.3 (5.4)	2.1 (3.3)	0.8 (0.7)
Farmers' contribution (share in each category)			
Monetary contribution	0.032	0.217	0.341
Labor contribution	0.016	0.174	0.073
No contribution	0.952	0.609	0.585
Farmers' monetary contribution toward construction (10^3 FCFA/ha)[a]	0.003 (0.02)	0.020 (0.11)	0.18 (0.58)
Total monetary investment (million FCFA/ha)[a]	2.3 (5.4)	2.1 (3.3)	1.0 (1.1)
In USD/ha[b]	4138	3802	1825
Rehabilitation, monetary investment (million FCFA/ha)	0.026 (0.13)	0.035 (0.15)	0.026 (0.11)
Rehabilitation, farmers' contribution (million FCFA/ha)[a]	0.006 (0.027)	0.005 (0.016)	0.004 (0.012)
Total value of pumps (10^3 FCFA/ha)[c]	5.1 (14.2)	1.2 (7.3)	0.3 (0.6)
Number of farmers' groups	63	69	41

Standard deviations are in parentheses
[a] Values are converted to 2021 prices using the Consumer Price Index
[b] The exchange rate of 1 FCFA = 0.0018 USD in 2021 is applied
[c] The values were estimated by farmers who were asked how much the price of each pump would be if they purchased it now

the difference in initial investment between government-funded irrigation schemes and village-based ones is not as large, even without farmers' monetary contributions (2.3 million versus 2.1 million FCFA/ha). This may be because SAED designed and constructed both government-funded and village-based irrigation schemes. In the case of privately funded irrigation schemes, the cost is about half of the other two categories. Since farmers' monetary contributions are relatively small compared with total investment cost, the inclusion of their contributions does not change the tendency. The Senegalese government used 3 million FCFA/ha as the cost of creating a new irrigation scheme for the SRV in its rice development plan (Ministère de l'Agriculture 2009). As it would be about 3.4 million FCFA/ha in 2021 prices, the estimated investment cost in this study is much lower. As for rehabilitation, the investment cost is much lower than the initial investment, not only because it comprises rehabilitation but also because many irrigation schemes have not implemented any rehabilitation activities since construction, i.e., the actual rehabilitation cost is zero. If only positive observations are considered, the average rehabilitation investment (the sum of monetary investment and farmers' contributions) will increase from 35,000 to 122,000 FCFA/ha. However, these are still much lower than the planned government rate of 600,000 FCFA/ha (Ministère de l'Agriculture 2009). Finally, it is important to note the huge difference in the value of pumps. In the case of government-funded irrigation schemes, since the scheme size is large, high-capacity pumps are required not only for irrigation but also for drainage. Thus, even if the value is divided by the scheme size, the unit cost is much higher than the other types of irrigation schemes. On the other hand, privately funded irrigation schemes mainly seem to use small, inexpensive pumps.

11.4.2 Descriptive Statistics of Rice Production and Its Investment Return

The first part of Table 11.3 compares several indicators of the intensity of rice production at the farmers' group level. The first one is the number of times rice was produced during the last seven seasons, beginning in the hot dry season of 2019 and extending to the hot dry season of 2021—a total of three hot dry seasons, the most favorable season for rice production. Thus, if the value of this indicator is 3, the group is considered to produce rice once a year. The value exceeds 3 only in the case of farmers' groups in government-funded irrigation schemes, while the values of other groups are just above 2, indicating farmers' groups in village-based and privately funded irrigation schemes use their fields less frequently.[6] The second aspect is the average exploitation rate over the last seven seasons, where the exploitation rate is defined as the ratio of area under rice production to total command area given to the farmers' group. The average exploitation rate of farmers' groups in government-funded irrigation schemes is 0.32, which is much higher than that of the other two categories. The exploitation rates by season show that the exploitation rates are more than 0.4 in all the investment categories in a hot dry season. However, in other seasons, which are less favorable for rice production, most of the farmers' groups in village-based or privately funded irrigation schemes do not produce rice at all. It is not because of the shift from rice to vegetables because vegetable production in cold dry season is not so popular among farmers' groups in village-based or privately funded irrigation schemes either, as shown in the table.

The second part of Table 11.3 provides rice yield. Rice yield (kg/ha/season) at the farmers' group level was calculated from household survey data that also covers seven seasons from the hot dry season 2019 to hot dry season 2021.[7] Thus, the rice

---

[6] This indicator is for group-level intensity and does not capture plot-level intensity—namely, how many times a year the same plot is used for rice production. In fact, even if a farmer produces rice two times a year, the farmer may use different plots in each season.

[7] In some cases, no one among the five sample farmers in a farmers' group produced rice in a particular season, even though the group reported that it was a positive area for producing rice in

Table 11.3 Performance of rice production

	Government funded	Village based	Privately funded
Number of seasons in which rice was produced in the last 7 seasons[a]	3.35 (1.54)	2.23 (1.29)	2.07 (1.46)
Average share of rice area in total area[b]	0.32 (0.17)	0.21 (0.15)	0.22 (0.18)
In hot dry season (Feb/Mar–June/July)	0.50 (0.39)	0.42 (0.32)	0.40 (0.36)
In rain season (July/Aug–Nov/Dec)	0.24 (0.36)	0.08 (0.19)	0.13 (0.24)
In cold dry season (Oct/Nov–Mar/April)	0.14 (0.27)	0.03 (0.12)	0.05 (0.14)
Average share of other crop areas in total area[c]	0.06 (0.11)	0.01 (0.03)	0.01 (0.06)
Rice yield (kg/ha/season) in planted area[d]	5374 (1107)	5380 (1342)	5821 (1087)
Rice production cost (10^3 FCFA/ha/season)[d]	267 (412)	333 (394)	349 (347)
Rice income (10^3 FCFA/ha/year)[e]	551 (544)	322 (813)	265 (309)
Membership fee (10^3 FCFA/ha/season)	4.48 (16.3)	67.7 (259)	41.3 (137)
Annual depreciation cost, high rate (10^3 FCFA/ha/year)[f]	713 (1550)	312 (760)	128 (128)
Annual depreciation cost, low rate (10^3 FCFA/ha/year)[g]	585 (1437)	190 (732)	64.2 (74.8)
Annual return, high depreciation (10^3 FCFA/ha/year)	−167 (1604)	−59.8 (1160)	90.6 (360)
Annual return, low depreciation (10^3 FCFA/ha/year)	−38.9 (1533)	62.2 (1115)	154 (333)
Number of farmers' groups	63	69	41

Standard deviations are in parentheses

[a] There are three seasons in a year at the study site: hot dry season, rain season, and cold dry season. Using irrigation water, rice can be grown in any season, but the hot dry season is the best because of the high temperature and sunlight. The 7 continuous seasons include 3 hot dry seasons, two rain seasons, and two cold dry seasons. Although rice can be produced three times a year in one irrigation scheme, it does not necessarily mean that the same plots are used three times a year due to the overlapping of seasons

[b] The share is land area planted to rice over the total land area managed by a farmers' group

[c] The share is land area planted to other crops over the total land area managed by a farmers' group.

[d] Rice yield and rice production cost are the average from plots where rice was planted. Note that most farmers' groups use only some part of the land they are managing, as shown in the table

[e] Rice income is the average for the group. That is, it takes account of the exploitation rate

[f] For the initial investment (construction), the lifetime of the structure is assumed to be 30 years for a hard (compacted) structure and 10 years for a soft (non-compacted/partially compacted). The lifetime of rehabilitation investment and pumps is assumed to be 10 years. Annual depreciation is the value divided by the number of lifetime years

[g] In the case of a low depreciation rate, the lifetime of the initial investment is assumed to be 30 years regardless of the structure. The lifetime of other investments is the same as the high depreciation, i.e., 10 years

yield shown in the table is that of farmers who actually produced rice.[8] Rice yield is generally high, consistent with previous reports such as Nakano et al. (2012), Sakurai (2016), and Tanaka et al. (2015). Farmers' groups in privately funded irrigation schemes show the highest average yield. This is because farmers in such groups tend to produce rice only during the most favorable, hot dry season.

Next, using the rice production data at the farmers' group level, adjusted by the intensity and the exploitation rates as explained above, group-level annual rice income per hectare is estimated. As shown in Table 11.3, annual income is the highest in the farmers' groups in government-funded irrigation schemes, reflecting their higher intensity and exploitation rates, particularly in unfavorable seasons. The group-level rice income does not include income from vegetable production, although it has become an important part of household income (Brosseau et al. 2021; Manikowski and Strapasson 2016). The exclusion of vegetable income can be justified by the fact that vegetable production was not so popular among the sample farmers collected from a wide range of SRV areas. Vegetable production was mainly done by farmers' groups in government-funded irrigation schemes, as shown in Table 11.3. Moreover, the objective of the government's investment in irrigation schemes in the SRV was to enhance rice production to make the country self-sufficient in rice.

The last part of Table 11.3 shows the annual depreciation cost and return per hectare. There are two important observations. First, the depreciation cost is the highest in the farmers' groups in government-funded irrigation schemes and the lowest in privately funded irrigation schemes, with those in between comprising village-based irrigation schemes, regardless of depreciation rate. Second, the annual returns are in the opposite order among the three investment categories. The first observation may imply that there are no economies of scale in the unit costs of irrigation investment. The second observation may suggest that investment performance is higher in smaller scale irrigation projects. That is, despite the highest income of farmers' groups in government-funded irrigation schemes, the return to investment is the lowest due to their higher depreciation cost. These observations seem to be consistent with the findings of Fujiie et al. (2011). However, although scheme size is correlated with the three investment categories, from these simple mean comparisons, it is unclear whether the significant differences in depreciation costs and the return to investment in irrigation schemes are due to scheme size or investment category. Therefore, regression analyses will be conducted in the next section.

---

that season (i.e., other farmers produced rice). Such cases are not included in the calculation of rice yield in the table. However, for the regression analysis, an expected rice yield obtained from similar farmers' groups in the same season is used.

[8] Since most of the farmers' groups do not use the command area fully for rice production, as discussed above, the rice yield obtained from farmers who actually produced rice must be adjusted by the exploitation rates in each season in order to calculate the rice production per hectare for each farmers' group.

11.4.3 Regression Analyses

Table 11.4 presents the estimation results of Eq. (11.3). Columns (1)–(5) show that irrigation scheme size has no significant association with current rice production, i.e., rice production intensity, exploitation rate of irrigation command area, rice yield, rice production cost per hectare, and rice income per hectare. However, as shown in columns (6) and (7), irrigation scheme size has a significant, negative correlation with depreciation cost per hectare regardless of depreciation rate, suggesting economies of scale in the unit cost of investment. Consequently, there is a positive relationship between scheme size and annual return, as shown in Columns (8) and (9), which favors larger sized irrigation schemes. However, because of the negative coefficient for the squared term of quadratic function, the positive association is diminishing and becomes negative after 1730 ha in the case of a high depreciation rate or 1560 ha in the case of a low depreciation rate.

Regression results provided in Table 11.5 examine if the positive association between scheme size and annual return is also found in each investment category. As shown in Columns from (6) to (9), coefficients for irrigation scheme size and its square are similar to those in Table 11.4, and their interaction terms with investment category dummies do not have any significant association on depreciation cost or annual return to investment. Thus, the results suggest that the positive association between scheme size and annual return does not differ among the investment categories.

In addition to the relationship between scheme size and investment performance, the regression analyses provide some interesting findings about investment category. The most important one is that the depreciation cost is significantly higher in the government-funded investment in irrigation schemes regardless of depreciation rate, as shown in Columns (6) and (7) of Table 11.4. Farmers' groups in government-funded irrigation schemes generate more income per hectare due to more intensified rice production and a higher rate of exploitation (Columns (1), (2), and (5)). However, at the same time, the investment performance of government-funded schemes is lower than other investment categories, as shown in Columns (8) and (9).

11.5 Conclusions

The objective of this chapter is to investigate the relationship between the size of irrigation scheme and the return to investment using farmers' group data collected in the SRV. The SRV is suitable for this purpose because, unlike other places, there are many irrigation schemes of differing sizes coexisting in similar environments. This question is relevant to considering whether investment in large-scale irrigation projects can be justified in circumstances that require the enhancement of agricultural production. The regression analyses of this study, after controlling for several factors, including investment categories, found an economy of scale in the unit cost of investment in irrigation schemes and a consequent positive association between

Table 11.4 Determinants of investment return

	(1) Number of rice productions in 7 seasons (number)	(2) Share of rice planted area in the scheme (share)	(3) Rice yield, group level (kg/ha/season)	(4) Rice production cost, group level (10³ FCFA/ha/season)	(5) Rice income, group level (10³ FCFA/ha/year)	(6) Annual depreciation, high rate (10³ FCFA/ha/year)	(7) Annual depreciation, low rate (10³ FCFA/ha/year)	(8) Annual return, high rate of depreciation (10³ FCFA/ha/year)	(9) Annual return, low rate of depreciation (10³ FCFA/ha/year)
Irrigation scheme size (ha)	$9.22*10^{-4}$ $(7.52*10^{-4})$	$0.30*10^{-4}$ $(0.71*10^{-4})$	-0.26 (0.98)	0.032 (0.056)	-0.18 (0.22)	-1.74^{***} (0.56)	-1.55^{***} (0.27)	1.59^{***} (0.55)	1.38^{**} (0.53)
Scheme size squared (ha²)	$-3.69*10^{-7*}$ $(2.06*10^{-7})$	$-2.35*10^{-8}$ $(1.90*10^{-8})$	$-1.01*10^{-4}$ $(2.72*10^{-4})$	$-0.10*10^{-4}$ $(0.14*10^{-4})$	$0.09*10^{-4}$ $(0.58*10^{-4})$	$0.46*10^{-3***}$ $(0.16*10^{-3})$	$0.41*10^{-3}$ $(0.15*10^{-3***})$	$-0.46*10^{-3}$ $(0.16*10^{-3***})$	$-0.41*10^{-3}$ $(0.15*10^{-3***})$
Investment category (dummies)									
Village-based	-0.84^{**} (0.37)	-0.11^{**} (0.04)	-916 (772)	-0.067 (32.5)	-316^{**} (153)	-1227^{***} (435)	-1119^{***} (412)	850^{*} (442)	741^{*} (430)
Privately funded	-1.02^{**} (0.42)	-0.10^{**} (0.05)	-802 (988)	-9.97 (34.6)	-327^{**} (141)	-1506^{***} (468)	-1345^{***} (449)	1146^{**} (471)	985^{**} (463)
Construction quality (dummies)									
Hard structure	0.99 (0.34)	-0.02 (0.04)	-1291^{**} (612)	-47.2^{**} (25.2)	-72.1 (96.4)	-651^{**} (249)	-506^{**} (240)	603^{**} (245)	458^{*} (241)
Years since the creation of the group	0.01 (0.01)	0.00 (0.00)	27.0 (35.3)	3.22^{***} (1.22)	-1.98 (6.60)	18.1 (12.7)	17.5 (12.7)	-20.8 (14.6)	-20.2 (14.6)
Group based on a family (dummy)	-0.53^{*} (0.30)	-0.68 (0.42)	-141 (728)	-31.6 (22.8)	-73.0 (86.8)	1100 (757)	1174 (768)	-1235 (767)	-1309 (807)
Number of male members	-0.00 (0.00)	-0.00 (0.00)	12.9 (12.1)	-0.31 (0.22)	1.72 (1.28)	-2.02^{*} (1.16)	-1.98^{*} (1.11)	3.85^{**} (1.97)	3.82^{**} (1.86)

(continued)

Table 11.4 (continued)

	(1) Number of rice productions in 7 seasons (number)	(2) Share of rice planted area in the scheme (share)	(3) Rice yield, group level (kg/ha/season)	(4) Rice production cost, group level (10^3 FCFA/ha/season)	(5) Rice income, group level (10^3 FCFA/ha/year)	(6) Annual depreciation, high rate (10^3 FCFA/ha/year)	(7) Annual depreciation, low rate (10^3 FCFA/ha/year)	(8) Annual return, high rate of depreciation (10^3 FCFA/ha/year)	(9) Annual return, low rate of depreciation (10^3 FCFA/ha/year)
Number of female members	−0.00 (0.01)	−0.00 (0.00)	−27.0 (35.3)	−0.37 (0.47)	−3.55 (2.09)	2.22 (2.14)	2.66 (2.06)	−5.30 (3.00)*	−5.74 (2.88)**
Constant	2.83 (0.48)***	0.32 (0.06)***	3965 (921)***	28.4 (37.9)	670 (193)***	1114 (283)***	874 (250)***	−438 (308)	−198 (288)
R^2	0.19	0.13	0.05	0.08	0.06	0.26	0.26	0.18	0.17
Number of observations	173	173	173	173	173	173	173	173	173

Robust standard errors are in parentheses. * $p < 0.1$, ** $p < 0.05$, *** $p < 0.01$

Table 11.5 Effect of irrigation scheme size

	(1) Number of rice productions in 7 seasons (number)	(2) Share of rice planted area in the scheme (share)	(3) Rice yield, group level (kg/ha/season)	(4) Rice production cost, group level (10^3 FCFA/ha/season)	(5) Rice income, group level (10^3 FCFA/ha/year)	(6) Annual depreciation, high rate of depreciation (10^3 FCFA/ha/year)	(7) Annual depreciation, low rate of depreciation (10^3 FCFA/ha/year)	(8) Annual return, high rate of depreciation (10^3 FCFA/ha/year)	(9) Annual return, low rate of depreciation (10^3 FCFA/ha/year)
Irrigation scheme size (ha)	$7.84{*}10^{-4}$ $(7.77{*}10^{-4})$	$0.41{*}10^{-4}$ $(0.73{*}10^{-4})$	0.11 (0.98)	0.032 (0.058)	−0.13 (0.23)	−1.76 $(0.57)^{***}$	−1.54 $(0.54)^{***}$	1.64 $(0.57)^{***}$	1.42 $(0.55)^{**}$
Scheme size squared (ha^2)	$-3.24{*}10^{-7}$ $(2.09{*}10^{-7})$	$-2.67{*}10^{-8}$ $(1.94{*}10^{-8})$	$-2.16{*}10^{-4}$ $(2.73{*}10^{-4})$	$-0.10{*}10^{-4}$ $(0.15{*}10^{-4})$	$0.08{*}10^{-4}$ $(0.60{*}10^{-4})$	$0.46{*}10^{-3}$ $(0.16{*}10^{-3})^{***}$	$0.40{*}10^{-3}$ $(0.15{*}10^{-3})^{***}$	$-0.47{*}10^{-3}$ $(0.16{*}10^{-3})^{***}$	$-0.42{*}10^{-3}$ $(0.16{*}10^{-3})^{***}$
Investment category (dummies)									
Village-based	−1.63 $(0.53)^{***}$	−0.15 $(0.06)^{**}$	−325 (1448)	−31.8 (48.4)	−151 (275)	−1001 $(490)^{**}$	−845 $(468)^{*}$	854 (536)	698 (524)
Village-based x scheme size	0.03 $(0.01)^{**}$	0.02 (0.01)	−5.46 (33.2)	1.49 (1.29)	−4.44 (5.72)	−9.0 (8.5)	−10.8 (8.4)	1.67 (10.7)	3.46 (10.4)
Village-based x scheme size sq	−0.00 $(0.00)^{**}$	−0.00 $(0.00)^{**}$	−0.07 (0.14)	−0.01 (0.01)	0.01 (0.02)	0.05 (0.04)	0.06 (0.04)	−0.02 (0.05)	−0.03 (0.05)
Privately funded	−1.36 $(0.63)^{**}$	−0.12 (0.08)	−1203 (984)	−79.2 $(41.5)^{*}$	−193 (176)	−1516 $(541)^{***}$	−1408 $(523)^{***}$	1194 $(572)^{**}$	1087 $(563)^{*}$
Privately funded x scheme size	0.02 (0.02)	0.03 (0.03)	45.3 (40.3)	4.28 $(2.10)^{**}$	−4.92 (5.10)	1.15 (6.49)	2.41 (5.84)	−1.81 (9.99)	−2.13 (9.43)
Privately funded x scheme size sq	−0.00 (0.00)	−0.00 (0.00)	−0.54 $(0.31)^{*}$	0.04 $(0.02)^{**}$	0.01 (0.04)	0.04 (0.05)	−0.00 (0.04)	−0.06 (0.07)	−0.02 (0.07)
Other control variables	Yes	Yes	Yes	Yes	Yes	Yes	Yes	Yes	Yes

(continued)

Table 11.5 (continued)

	(1)	(2)	(3)	(4)	(5)	(6)	(7)	(8)	(9)
	Number of rice productions in 7 seasons (number)	Share of rice planted area in the scheme (share)	Rice yield, group level (kg/ha/season)	Rice production cost, group level (10^3 FCFA/ha/season)	Rice income, group level (10^3 FCFA/ha/year)	Annual depreciation, high rate (10^3 FCFA/ha/year)	Annual depreciation, low rate (10^3 FCFA/ha/year)	Annual return, high rate of depreciation (10^3 FCFA/ha/year)	Annual return, low rate of depreciation (10^3 FCFA/ha/year)
Constant	2.96 (0.48)***	0.31 (0.06)***	3875 (1019)***	30.3 (40.2)	621 (217)***	−644 (248)***	846 (267)***	−469 (336)	−221 (315)
R^2	0.21	0.15	0.07	0.11	0.08	0.26	0.26	0.18	0.18
Number of observations	173	173	173	173	173	173	173	173	173

Robust standard errors are in parentheses. * $p < 0.1$, ** $p < 0.05$, *** $p < 0.01$

scheme size and return to investment. This finding is consistent with Inocencio et al. (2007) and Kikuchi et al. (2021). In addition, this study made two contributions to the debate on the relationship between irrigation size and investment performance. First, this study found that the positive relationship between irrigation size and investment performance is non-linear and disappears once the irrigation scheme becomes more than a certain size. In our empirical case, the peak exists around 1600–1700 ha. Second, this study found that the investment performance of government-funded irrigation schemes is poorer than that of village-based and privately funded irrigation schemes because of the significantly higher depreciation costs of government-funded schemes.

Thus, this study suggests that the promotion of investment in large-scale irrigation schemes is not unconditionally recommended. We should pay attention to the negative association between government-funded irrigation schemes and return to investment. Government-funded irrigation schemes have better facilities (more compacted structure), higher pump capacity, and as a consequence, can be intensified (i.e., a larger amount of rice production in a year). They also have a higher exploitation rate, are more diversified (increased vegetable production), and produce more income per hectare. Nevertheless, due to the high investment costs—particularly in pumps—its investment performance is worse than village-based and privately funded irrigation schemes. In other words, the rice productivity of government-funded irrigation schemes is not sufficiently high enough to cover their high investment costs. Thus, although large-scale (but not too large) irrigation should be promoted, excessive investment per hectare must be avoided.

References

Adams WM (1992) Wasting the rain: river, people and planning in Africa. Routledge, New York

Arouna A, Fatognon IA, Saito K, Futakuchi K (2021) Moving toward rice self-sufficiency in sub-Saharan Africa by 2030: lessons learned from 10 years of the coalition for African rice development. World Dev Perspect 21:100291

Borgia C, García-Bolaños M, Li T, Gómez-Macpherson H, Comas J, Connor D, Mateos L (2013) Benchmarking for performance assessment of small and large irrigation schemes along the Senegal valley in Mauritania. Agric Water Manag 121:19–26. https://doi.org/10.1016/j.agwat.2013.01.002

Brosseau A, Saito K, van Oort PAJ, Diagne M, Valbuena D, Groot JCJ (2021) Exploring opportunities for diversification of smallholders' rice-based farming systems in the Senegal river valley. Agric Syst 193:103211

Dia I (2001) Private irrigation in the Senegal river delta: evolution and prospects. In: Sally H, Abernethy CL (eds) Private irrigation in sub-Saharan Africa. International Water Management Institute, Colombo, pp 121–126

Diouf A, Elbow K, Seck OK (2015) Large-scale investments in irrigated agricultural production in the Senegal river valley: making the most of opportunities to expand decentralized and participatory land governance. Paper presented at the 2015 World Bank conference on land and poverty. The World Bank, Washington DC, 23–27 March

Estudillo JP, Otsuka K (2012) Lessons from the Asian green revolution in rice. In: Otsuka K, Larson DF (eds) An African green revolution: finding ways to boost productivity on small farms. Springer, New York, pp 17–42

Fujiie H, Maruyama A, Fujiie M, Takagaki M, Merrey DJ, Kikuchi M (2011) Why invest in minor projects in sub-Saharan Africa? An exploration of the scale economy and diseconomy of irrigation projects. Irrig Drain 25(1):39. https://doi.org/10.1007/s10795-011-9111-4

Haefele SM, Wopereis MCS, Wiechmann H (2002) Long-term fertility experiments for irrigated rice in the West African Sahel: agronomic results. Field Crops Res 78(2):119–131. https://doi.org/10.1016/S0378-4290(02)00117-X

Inocencio A, Kikuchi M, Tonosaki M, Maruyama A, Merrey D, Sally H, de Jong I (2007) Costs and performance of irrigation projects: a comparison of sub-Saharan Africa and other developing regions (IWMI Research Report No 109). International Water Management Institute, Colombo

Kikuchi M, Mano Y, Njagi TN, Merrey D, Otsuka K (2021) Economic viability of large-scale irrigation construction in sub-Saharan Africa: what if Mwea irrigation scheme were constructed as a brand-new scheme? J Dev Studies 57(5):772–789

Manikowski S, Strapasson A (2016) Sustainability assessment of large irrigation dams in Senegal: a cost-benefit analysis for the Senegal river valley. Front Environ Sci 4. https://doi.org/10.3389/fenvs.2016.00018

Ministère de l'Agriculture (2009) Programme national d'autosuffisance en riz stratégie nationalede développement de la riziculture. http://www.jica.go.jp/english/our_work/thematic_issues/agricultural/pdf/senegal_en.pdf

Nakano Y, Bamba I, Diagne A, Otsuka K, Kajisa K (2012) The possibility of a rice green revolution in large-scale irrigation schemes in sub-Saharan Africa. In: Otsuka K, Larson DF (eds) An African green revolution: finding ways to boost productivity on small farms. Springer, New York, pp 43–70

Otsuka K, Larson DF (2012) Towards a green revolution in sub-Saharan Africa. In: Otsuka K, Larson DF (eds) An African green revolution: finding ways to boost productivity on small farms. Springer, New York, pp 281–300

Saito K, Dieng I, Toure AA, Somado EA, Wopereis MCS (2015) Rice yield growth analysis for 24 African countries over 1960–2012. Glob Food Sec 5:62–69

Sakurai T (2016) On the determinants of high productivity rice farming in irrigated areas in Senegal: the efficiency of large compared with small-scale irrigation schemes. In: Otsuka K, Larson DF (eds) Pursuit of an African green revolution: views from rice and maize farmers' fields. Springer, Tokyo, pp 119–143

Tanaka A, Diagne M, Saito K (2015) Causes of yield stagnation in irrigated lowland rice systems in the Senegal river valley: application of dichotomous decision tree analysis. Field Crops Res 176(May):99–107

World Bank (2005) Shaping the future of water for agriculture: a sourcebook for investment in agricultural water management. http://siteresources.worldbank.org/INTARD/Resources/Shaping_the_Future_of_Water_for_Agriculture.pdf. Accessed 21 Sept 2021

World Bank (2007) Investment in agricultural water for poverty reduction and economic growth in sub-Saharan Africa. http://siteresources.worldbank.org/RPDLPROGRAM/Resources/459596 1170984095733/synthesisreport.pdf. Accessed 21 Sept 2021

Takeshi Sakurai is a professor of the University of Tokyo since 2014. He obtained a Ph.D. in agricultural economics from Michigan State University in 1995. His expertise is in agriultural economics and development economics.

Chapter 12
Rice Milling in Kenya: An Inquiry into the Process of Upgrading Rice Milling Services

Yukichi Mano, Timothy N. Njagi, and Keijiro Otsuka

Abstract Countries in sub-Saharan Africa (SSA) rely heavily on imported rice from Asia that is of superior quality compared to local rice. The objective of this study is to assess the impacts of the adoption of improved milling technologies and the associated structural transformation of the rice value chain from 2011 to 2019 using the original census of the rice millers in the Mwea Irrigation Scheme in Kenya. Our empirical analysis uses the doubly robust method and the endogenous switching model, which collectively reveal that the adopters of improved milling machines successfully improved the quality of locally milled rice. This allows it to compete with imported rice, thereby increasing the amount of domestic rice sold to supermarkets as well as urban traders and consumers. Through probit regression, it is also found that a few educated, entrepreneurial rice millers operating in rice milling clusters began using large-scale improved milling machines that included destoning capacity, importing them from China around 2010. Later, smaller improved machines were also introduced and these were more widely adopted into the rice milling clusters. In contrast, the many other millers that did not adopt improved machines were forced to downsize their business or exit the industry. These results suggest that adopting improved milling technologies—including destoners—is critical to improving the competitiveness of SSA's domestic rice against imported rice.

This chapter draws heavily on Mano et al. (2022).

Y. Mano (✉)
Graduate School of Economics, Hitotsubashi University, 2-1 Naka, Isono Building Room 324, Kunitachi-shi, Tokyo 186-8601, Japan
e-mail: yukichi.mano@r.hit-u.ac.jp

T. N. Njagi
Tegemeo Institute of Agricultural Policy and Development, Tegemeo Institute, Egerton University, Kindaruma lane, orr Ngong road, P.O. Box 20498, Nairobi 00200, Kenya
e-mail: tnjagi@tegemeo.org

K. Otsuka
Graduate School of Economics, Kobe University, 2-1 Rokkodai-Cho, Nada-ku, Fourth Academic Building, 5th floor, Room 504, Kobe 657-8501, Hyogo, Japan
e-mail: otsuka@econ.kobe-u.ac.jp

© JICA Ogata Sadako Research Institute for Peace and Development 2023
K. Otsuka et al. (eds.), *Rice Green Revolution in Sub-Saharan Africa*, Natural Resource Management and Policy 56, https://doi.org/10.1007/978-981-19-8046-6_12

12.1 Introduction

Countries in sub-Saharan Africa (SSA) have made serious efforts to increase domestic staple food production to improve food security and reduce poverty, especially after the 2007–08 food crisis. Meanwhile, rice is an up-and-coming crop underpinning the increase in food production in SSA (Chap. 1 of this volume; Otsuka and Larson 2013, 2016; CARD 2019; Otsuka 2019). The annual growth rate of rice production was as high as 6.8% between 2009 and 2019 (Soullier et al. 2020; Chap. 2 of this volume). However, rice imports are estimated to be increasing more rapidly, at 7.8% per annum, and around one-third of the rice consumed in SSA is still imported from Asia in varying degrees ranging from 10 to 93% across the countries of the region (Mendez del Villar and Lançon 2015; Saito et al. 2015; FAO 2018).

Population growth, urbanization, and increasing incomes raise rice consumption. Urban consumers in coastal countries typically prefer imported Asian rice of superior quality relative to domestic rice (e.g., Mauritania, the northern part of Senegal, Liberia, Cote d'Ivoire, Ghana, Togo, Benin and Cameroon) (Diako et al. 2010; Demont 2013; Futakuchi et al. 2013; Demont et al. 2017; Ibrahim et al. 2020). By contrast, in landlocked countries or coastal countries that host a center of rice domestication, consumers traditionally prefer domestic rice (e.g., the southern Casamance region of Senegal, Guinea, Guinea-Bissau, Sierra Leone) (Britwum and Demont 2021; Demont 2013; Demont and Ndour 2015; Demont et al. 2017; Soullier et al. 2020). In Eastern Africa, Indian immigrants brought fragrant rice to Tanzania, which acted as a center of rice cultural heritage.[1] Nevertheless, massive rice imports may change consumer preference for imported rice, and it is crucial to improve the quality of African rice to strengthen its competitiveness. Among the many factors affecting grain quality, including varieties, cultivation, harvesting, threshing, and storage technologies (Futakuchi et al. 2013), the use of appropriate milling machines is critical if domestic milled rice is to compete with imported rice from Asia (Fiamohe et al. 2018; Ragasa et al. 2020). Research based on framed field experiments demonstrates that urban consumers are more willing to purchase rice with high cleanliness levels and low breakage rates of grains (Demont and Ndour 2015; Demont et al. 2017), indicating the importance of appropriate rice milling technology. Importantly, case studies by Tokida et al. (2014) and Kapalata and Sakurai (2020) reveal that adopting destoners and other improved milling machines also allows rice millers to charge higher milling fees and increases their profitability in Uganda and Tanzania.[2] However, the evidence remains limited on whether improved milling technologies enable domestic rice to compete with imported rice.

This paper analyzes the development of a rice milling cluster in Kenya's most prominent rice production area, the Mwea irrigation scheme, situated 90 km northeast

---

[1] Due to the influence of neighboring Tanzania, consumers in Kenya may also traditionally prefer domestic rice. But as we discuss below, supermarkets started selling domestic milled rice only after the millers adopted improved milling technologies.

[2] The adoption of destoners also played a crucial role in improving the quality of milled rice when Japan imported rice from Taiwan in the 1930s (Irukayama 1939; Arimoto and Sakane 2017).

of Nairobi. This cluster has 8,500 hectares of irrigated paddy area, grows primarily improved Basmati rice[3] and has achieved a rice yield of 6.2 tons per hectare with two crop cycles (Chap. 10 of this volume; Njeru et al. 2016; Kikuchi et al. 2021). Until the early 2010s, millers and traders would hire casual workers to remove stones from the milled rice, which slowed the domestic rice supply to consumers. Some rice millers have gradually adopted improved milling machines over the past ten years, and the sale of Mwea rice in supermarkets began after the millers installed destoners. The rice milled by these millers is of higher quality and successfully competes with imported rice from Asia in urban markets, including in Nairobi, the capital of Kenya. On 11 December 2018, we observed supermarkets in Nairobi selling improved Basmati rice from Mwea at 140–200 Kshs per kg, compared with Pakistani long grain at 100–120 Kshs per kg.[4] Interviews with local experts suggest that high-quality rice, primarily consumed in the urban areas, comprises about 15–20% of total consumption. These observations indicate that African rice can compete with Asian rice if improved milling machines are introduced to the SSA (Chaps. 7 and 13 of this volume).

Data were collected from the commercial millers operating in Mwea in 2011, 2016, 2018, and 2019 to analyze the adoption process of destoners and other improved milling technologies and their effects on the quality of the milling service, their business performance, and survival rates in the market. Because destoners and other improved devices used to be mainly provided as part of large-scale multi-stage milling machines, only 3 out of the 82 millers sampled in 2011 and 15 out of 103 millers in 2016 were using destoners. Later, small- and medium-scale multi-stage milling machines with destoners became locally available, and 8 and 11 additional millers adopted destoners in 2018 and 2019, respectively. By contrast, many millers who did not adopt improved technologies were forced to shrink their businesses or exit the market. Using the doubly robust method and endogenous switching regression model to address the potential endogeneity of destoner adoption, we find that the adoption of improved rice milling technologies is associated with higher milling fees, a greater amount of milled rice being produced, and higher profitability. Moreover, the millers using destoners are more likely to survive in the market. These findings confirm the critical importance of improved rice milling technology in enhancing the quality of African rice.

The rest of this paper is organized as follows. Section 12.2 explains the data and describes the characteristics of sample rice millers in Mwea. Section 12.3 sets out the hypotheses tested in this study and explains the empirical strategy, and Sect. 12.4 describes the results. Section 12.5 discusses the results and draws policy implications.

---

[3] Improved Basmati is a cross-breed between Basmati and high-yielding modern varieties and is widely grown in India and Pakistan. It is of lower quality but is higher yielding than original Basmati rice. A small amount of other rice varieties is produced solely for farmers' domestic consumption.

[4] In 2019, Pakistan accounted for 67% of imported rice to Kenya, followed by Thailand with 25%, Republic of Korea with 3%, and India with 2% (KNBS 2021). The Pakistani rice is not Basmati but a type of long grain nonaromatic rice. According to our informal interviews with local rice traders, some sellers blend Mwea rice with imported rice from Pakistan and sell it as "Mwea rice." However, note that consumers prefer high-quality Asian rice. For example, Jasmine rice from Thailand retailed at 350 Kshs per kg while Mwea rice was sold at 140–200 Kshs per kg.

12.2   Millers in Mwea

Figure 12.1 presents rice consumption, production, imports, and paddy price in Kenya from 2010 to 2019. After the 2007–2008 food crisis, the paddy price declined over time, consistent with the decline in the global price of rice (Chap. 10 of this volume; Kikuchi et al. 2021). While consumption and imports almost doubled, rice production was stagnant during this period in Kenya. A severe drought in 2016 and 2017 in the Horn of Africa region depressed rice production in the country over this period.

The Mwea Irrigation Scheme produces 80–88% of the domestic rice in Kenya (Samejima et al. 2020). We studied the rice millers in Mwea to inquire into the development process of the rice milling sector and found that learning new technologies from overseas plays a crucial role in the development of rice milling. In what follows, we discuss the history of the development of the rice milling industry and explain our primary data collection in the context of Mwea.

12.2.1   Brief History

In the Mwea Irrigation Scheme, the history of modern rice milling dates back to 1967. A public rice miller, Mwea Rice Mill Ltd. (MRM), had monopoly rights over rice milling services and began operating four lines of a large multi-stage milling

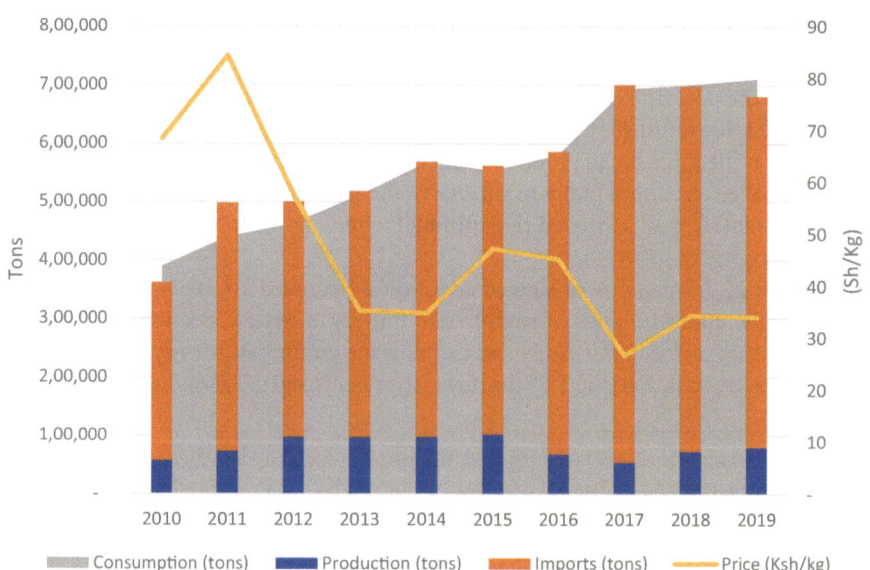

Fig. 12.1 Rice consumption, production, imports, and paddy price in Kenya. *Source* Kenya National Bureau Statistics (2022)

machine imported from Germany. In 1999 the rice sector was liberalized, and MRM started competing with other millers to provide milling services. In the early period of this process, the operators of multi-stage milling machines received on-the-job training at MRM, and some moved to other mills later.

In 2000, the Mwea Rice Growers Multipurpose Cooperative Society Ltd. (MRGM) purchased a large Chinese multi-stage milling machine without a destoner because they did not realize the importance of this function as they bought the machine from a broker in Nairobi without visiting China. In 2007, a private miller, Mwega, purchased a second-hand small-sized multi-stage milling machine with destoners. But this installation did not influence other millers, probably because Mwega's mill was small at the time.

A significant event occurred in 2010 when the chairman and two workers of MRGM went to China to learn new milling technologies and purchase a new multi-stage milling machine. Notably, the new machine this time was equipped with a destoner, which improved the quality of milled rice by removing small stones and other impurities. In the following period, some owners of milling services who had expertise in rice marketing or had accumulated wealth through other business activities also started inquiring into new rice milling technologies by visiting machine suppliers in China. As a result, a little over ten private millers and the MRM installed large multi-stage milling machines in the first half of the 2010s. Aside from visiting China, these owners employed experienced operators, especially from MRM and MRGM.

Because small- or medium-sized multi-stage milling machines were introduced from China in the late 2010s to the local area, the importance of their destoner capability became widely recognized in rice milling clusters. Consequently, other millers, mainly those who were highly educated, also adopted destoners. By contrast, the remaining millers either reduced their business while using the traditional types of milling machines without destoners (so-called "Jets" and "fridges") or exited the market.

12.2.2 Data

The authors visited the Mwea Irrigation Scheme in 2011, 2017, and 2020 to conduct a census of commercial rice millers. As with the other rice millers in SSA, the task of rice millers is primarily to provide a milling service for farmers and traders rather than purchasing paddy or selling milled rice to the market. Because December and January are the primary months for rice harvesting and milling in Mwea, our survey focused on rice millers' characteristics and performance in January 2011, December 2016, December 2018, and December 2019.

We aimed to interview all the commercial rice millers in the region. However, there was no official list of rice millers during our survey in 2011 and 2017, so the coverage of the survey gradually expanded as we discovered additional millers operating in this area. In 2011, we visited the millers in Mwea town (locally known as Wanguru or Ngurubani), the primary rice milling cluster, stretching over 5 km along the Embu-Nairobi highway that connects Mwea with Nairobi in about 1.5 h. We then visited the millers in Mutithi, the secondary cluster located about 7 km south of Wanguru that extends about a kilometer along the highway. We also visited Kandongu town, which is about 4 km off the highway. In 2017, we maintained this strategy but also visited additional mills deemed commercial following a local rice extension officer's advice. The 2020 survey is therefore the most comprehensive. We compared our list of rice millers with the list obtained from the Kirinyaga County government, which had also tried to construct a census of all rice millers in the county, and again found additional millers to contact.

12.2.3 Descriptive Statistics

Table 12.1 presents our sample millers' basic characteristics by destoner adoption status, representing improved rice milling technologies, including pre-cleaners and graders (see Appendix Table 12.9), as of December 2019. The survey coverage expanded over the research period, and 62 and 45 millers were added in the 2017 and the 2019 surveys, respectively. The early adopters are those millers who adopted improved machines by 2016, and three did so in 2011. The adoption rate was low in 2011 but increased to nearly 15% in 2016. The late adopters are those millers that adopted destoners and other improved devices after 2016. Because of their adoption, the adoption rate increased to roughly 25% in 2018 and to 33% in 2019. At this time, almost all millers exiting the market were non-adopters.

The mid-point year of the establishment was 2009, with early and late adopters establishing themselves in 2005 and 2011 on average, respectively. Only one-quarter of non-adopters but most adopters of improved machines are located in the primary and secondary clusters, Mwea town and Mutithi, along the Embu-Nairobi highway. Although rice millers in the clusters are not tied by ethnicity or other social groups, they know each other very well and regularly exchange production and marketing knowledge. This frequent communication and information exchange facilitates the diffusion of new technology and marketing methods within each cluster. The average age of the main decision-makers was 50 years old, and they were mostly male. Only 45% of the non-adopters were high school graduates, whereas almost all the decision-makers among the adopters had completed high school education. As for their previous occupations, many decision-makers in the group of non-adopters were previously farmers, whereas the majority of the early and the late adopters had gained work experience in other businesses or formal jobs.

Table 12.2 presents the total milling capacity, the value of milling machines, and the number of employees. Almost all the respondent rice millers had only one milling machine.[5] In 2011, when only three early adopters used destoners, the average milling

---

[5] Only nine millers had two machines, and only one had three machines in 2019.

Table 12.1 Basic characteristics of sample millers in 2019 by destoner adoption[a]

	Total	Nonadopters	Early adopters[b]	Late adopters[c]
No. of millers (% destoner adoption)				
2011	82 (3.7)	72 (0)	6 (50)	4 (0)
2016	103 (14.6)	82 (0)	15 (100)	6 (0)
2018	84 (26.1)	57 (0)	14 (100)	13 (61.5)
2019	95 (34.7)	62 (0)	14 (100)	19 (100)
Exit				
2012–2016	41	40	1	0
2017–2019	53	52	1	0
Establishment year	2008.9 (8.8)	2008.9 (6.3)	2005.4 (12.6)	2011.4 (11.6)
Location				
Mwea town (=1)	0.401	0.226	0.786	0.684
Mutithi (=1)	0.084	0.048	0.143	0.158
Other (=1)	0.515	0.726	0.071	0.158
Characteristics of decision-maker				
Age	49.8 (10.4)	50.4 (10.9)	49.7 (10.8)	47.8 (8.6)
Female (=1)	0.31	0.33	0.21	0.31
High school/above (=1)	0.61	0.45	1	0.84
Farmer (=1)	0.38	0.43	0.21	0.32
Rice milling & trading (=1)	0.17	0.22	0.14	0.05
Business & formal jobs (=1)	0.35	0.22	0.64	0.58
Mechanic (=1)	0.10	0.13	0	0.05

Notes
[a]The standard deviations of the continuous variables are in parentheses. The survey coverage expanded over time, and 62 and 45 millers were added in 2016 and 2019 surveys, respectively
[b]The early adopters are those millers that had adopted destoners by 2016. Three of them did so in 2011
[c]The late adopters are those millers that adopted destoners only after the 2016 survey. Eight of them did so in 2018

capacity was similar among the three groups of rice millers.[6] From 2011 to 2016, the average capacity of early adopters increased significantly. In 2019, the non-adopters continued to use the small traditional milling machines. While the early adopters

---

[6] Although we visited MRM and MRGM in 2011, we did not formally interview them because their technologies and business size seemed to be too distinct from other millers. We interviewed them in 2017 and 2019.

Table 12.2 Total milling capacity, the current market value of milling machines, and the number of employees of millers by destoner adoption

	Total	Nonadopter	Early adopter	Late adopter
Capacity (kg/h)				
2011	468.8	472.6	445.0	435.0
	(227.9)	(237.5)	(89.1)	(225.7)
2016	1139.0	538.0	4453.3	966.6
	(3032.2)	(385.3)	(7175.7)	(581.9)
2018	2088.8	819.8	6196.4	2838.4
	(4377.6)	(575.2)	(8328.0)	(4710.2)
2019	2076.7	815.9	6196.4	3478.9
	(4298.7)	(574.3)	(8327.9)	(5113.0)
Value (million Kshs)				
2016	2.15	0.27	13.0	0.44
	(6.44)	(0.15)	(12.3)	(0.27)
2018	3.19	0.19	13.5	4.4
	(8.11)	(0.09)	(14.1)	(7.1)
2019	3.02	0.19	13.0	4.9
	(7.34)	(0.10)	(13.5)	(6.2)
Employees				
2011	1.5	1.4	3.0	1.0
	(1.4)	(1.1)	(3.3)	(0)
2016	1.6	0.90	5.4	1.3
	(2.9)	(0.63)	(6.1)	(0.5)
2018	2.6	0.61	10.4	3.1
	(10.3)	(1.19)	(25.3)	(5.3)
2019	2.6	0.54	10.0	3.8
	(10.4)	(0.73)	(25.3)	(5.7)

Notes Standard deviations are in parentheses. The total capacity of the machines and the total current market value of all the milling machines are reported

had large-sized multi-stage milling machines, the late adopters installed small- or medium-sized multi-stage milling machines.[7]

We also collected data on the total current market value of the milling machines assessed by the decision-makers in 2016, 2018, and 2019 (Table 12.2). The current market value of the early adopters' milling machines was substantially higher than that of the late adopters' milling machines, and the non-adopters' traditional milling machines were the least valuable.[8] Furthermore, the non-adopters reduced their labor

---

[7] The Appendix Table 12.9 shows that the areas of workshop, storage, and yard were also largest for the early adopters, followed by the late adopters, and the non-adopters had the smallest space. The early adopters attracted traders by offering their large storage space for free.

[8] We asked the decision-makers the total current market value of all the milling machines in the present state. In this paper, all the monetary values are deflated using the GDP deflator and presented in real 2019 values. The average exchange rate in December 2019 was 101.5 Kshs to 1 USD.

Table 12.3 Monthly performance of milling service by destoner adoption

	Total	Nonadopter	Early adopter	Late adopter
Milling fee (Kshs/kg)				
2011	1.90 (0.28)	1.89 (0.27)	2.10 (0.42)	1.87 (0.19)
2016	2.12 (1.18)	2.03 (1.27)	2.83 (0.54)	1.61 (0.42)
2018	2.06 (0.62)	1.81 (0.52)	2.78 (0.26)	2.29 (0.59)
2019	2.01 (0.57)	1.76 (0.46)	2.64 (0.39)	2.37 (0.42)
Rice milled for customers (tons)				
2011	58.6 (50.2)	51.8 (48.9)	108.7 (30.7)	99.7 (39.4)
2016	65.4 (163.3)	26.3 (38.8)	270.3 (351.0)	55.6 (11.6)
2018	130.0 (293.5)	32.3 (65.5)	422.3 (539.7)	220.3 (287.7)
2019	198.4 (791.2)	21.6 (38.3)	882.9 (1839.9)	255.1 (507.2)

Notes Standard deviations are in parentheses. The milling fee is deflated with the GDP deflator (in 2019 values)

force over time, whereas the early adopters and the late adopters increased theirs as they installed milling machines with greater capacity (Table 12.2).

Table 12.3 presents the monthly milling performance of the sample millers for January 2011, December 2016, December 2018, and December 2019.[9] The average milling fees—which reflect the milling quality—of the early adopters and the late adopters were substantially higher than the milling fee of the non-adopters. Over time, the non-adopters' milling fees increased between 2011 and 2016, after which they declined, likely reflecting decreased demand for their milling service. The early adopters' milling fees increased substantially between 2011 and 2016 when all of them adopted destoners. Similarly, the late adopters' milling fees increased between 2016 and 2019. We do not know why the late adopters' milling fees were lower than that of non-adopters. It may well be that their strategy was to attract customers by reducing milling fees in the face of competition with the early adopters. As may be expected, in 2019, their average milling fee was slightly lower than that of the early adopters.

The amount of rice milled by non-adopters for customers declined between 2011 and 2019 (Table 12.3), whereas the early adopters' milled rice increased during the same period. The late adopters' amount of milled rice also increased substantially

---

[9] Appendix Table presents the annual amount of rice sold in 2019. We believe the monthly data we use is more accurate than the annual data because of the shorter recall period.

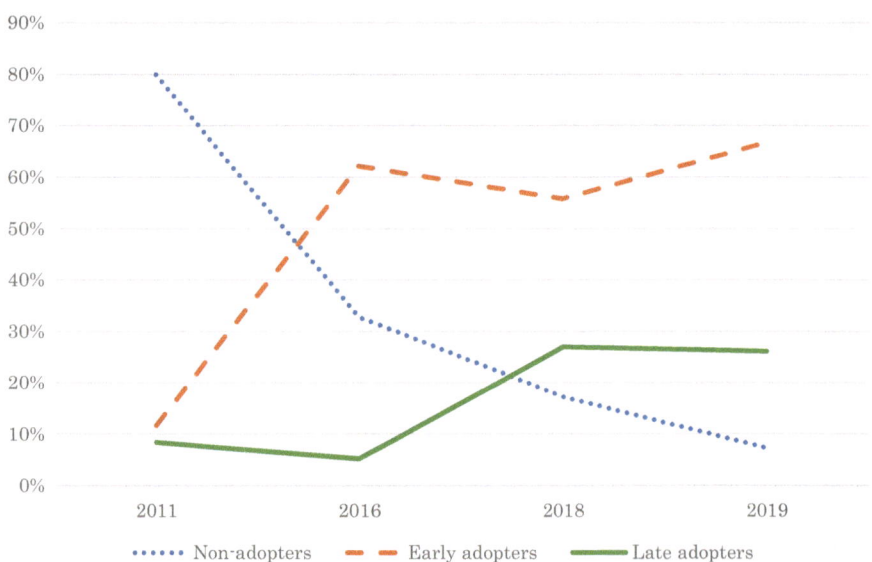

Fig. 12.2 Market share of rice milled for customers by destoner adoption. *Notes* The early adopters are the millers that adopted destoners by 2016, and three of them did so in 2011. The late adopters are the millers that adopted destoners only after the 2016 survey

between 2016 and 2019, as they increased destoner adoption. Figure 12.2 documents the changes in the market share of the amount of milled rice by these three groups. In 2011, the non-adopter group dominated the milling service market with a market share of 80%, whereas the early adopter group rapidly increased its market share to 62% in 2016. The late adopters had a market share of 8% in 2011, but their market share has more than tripled in recent years. The non-adopters lost significant market share to the improved technology adopters and had only 7% of the market share by 2019. The near demise of non-adopters strongly indicates that milling machines with improved devices are profitable.

About 40% of sample rice millers purchased paddy[10] and sold milled rice to consumers and traders (Table 12.4). The proportion of millers that sold milled rice was similar across the three groups and did not change over time. The early adopters sold a substantially greater amount of milled rice than the late adopters. The amount of rice sold by the late adopters increased from 2016 to 2018 but remained small compared to the amount of rice milled for customers. These observations suggest that some early adopters shifted their business focus to selling rice, while the late adopters increased their share in the milling service. The non-adopters' and the early adopters' prices for sold rice declined between 2016 and 2019. The late adopters'

---

[10] There is no contract farming system for rice, and these millers do not provide farmers with credit.

Table 12.4 Monthly performance of rice sales by destoner adoption

	Total	Nonadopter	Early adopter	Late adopter
Sell milled rice (=1)				
2016	0.38	0.37	0.40	0.50
2018	0.41	0.44	0.36	0.37
2019	0.42	0.41	0.35	0.47
Sold rice (tons)				
2016	16.4 (142.7)	5.2 (6.6)	248.6 (588.5)	12.1 (9.5)
2018	21.9 (126.5)	7.0 (7.6)	301.8 (515.6)	54.1 (35.9)
2019	16.8 (83.8)	6.8 (8.8)	216.8 (323.2)	38.0 (49.4)
Price of sold rice (Kshs/kg)				
2016	127.5 (22.9)	124.4 (24.7)	141.0 (6.9)	132.8 (14.0)
2018	126.4 (16.3)	120.5 (10.5)	138.2 (17.7)	140.6 (22.2)
2019	120.8 (15.4)	115.4 (9.1)	131.4 (18.3)	130.4 (5.1)

Notes Standard deviations are in parentheses. The rice price is deflated with the GDP deflator (in 2019 values)

prices for sold rice substantially increased between 2016 and 2018 but fell in 2019, roughly comparable to the early adopters' price.[11]

Table 12.5 lists the paddy suppliers, the use of brand names in marketing, and the market channels of millers purchasing paddy and selling milled rice in December 2019 by destoner adoption. The non-adopters sourced paddy mainly from farmers, whereas the early and late adopters primarily purchased paddy from local traders. None of the non-adopters used brand names, but 50% of the early adopters and 25% of the late adopters used brand names. As for the marketing channels of millers who did not use brand names, the non-adopters sold milled rice to local consumers and traders. The early adopters and the late adopters without brand names sold milled rice to consumers and traders.[12] Notably, the adopters of brand names also sold milled

---

[11] Paddy quality is also crucial in determining overall rice quality, and farm-level agronomy affects the paddy quality, including leveling, fertilizer application, water usage during crop establishment, and pest and disease control. Mechanical harvesting has also helped reduce losses. Traders determine the paddy quality through observation and checking for moisture content. Many traders have moisture meters and usually check for the moisture content when purchasing. We do not observe a significant difference in the minimum required moisture content across the millers (Appendix Table 12.9). They also observe the paddy grain to ensure the pods are full, a sign of good quality paddy. But there is no grading system or pricing differential on paddy quality.

[12] According to local experts, these consumers and traders come from both local and urban areas.

Table 12.5 Paddy suppliers, brand names, and market channels of the millers purchasing paddy and selling milled rice in December 2019 by destoner adoption

	Total	Nonadopter	Early adopter	Late adopter
Paddy suppliers (%)				
Farmers	55.4	67.8	32.0	31.4
Traders	41.4	31.7	68.0	54.2
Other	3.1	0.4	0	14.2
Brand names (%)	25.0	0.0	50.0	25.0
Market channel (%) without brand name				
Consumers	35.2	35.9	21.0	39.7
Traders	61.9	64.1	79.0	43.6
Supermarkets	0	0	0	0
Other mills	2.9	0	0	16.7
Other	0	0	0	0
Market channel (%) with brand name				
Consumers	38.0	0	36.7	38.9
Traders	21.8	0	26.7	18.5
Supermarkets	15.6	0	25.0	9.3
Other mills	0	0	0	0
Other	24.7	0	11.7	33.3

rice directly to supermarkets, which require an ample supply of high-quality milled rice.

We calculated the monthly capacity utilization rate to examine milling productivity by dividing the actual amount of rice milled by the expected "full" amount of rice milled when all milling machines are operated for 200 h per month (Table 12.6).[13] The non-adopters reduced their capacity utilization rate over time between 2011 and 2019, showing decreased demand for the non-adopters' milling service. The early adopters reduced their capacity utilization rate between 2011 and 2016 when they initially adopted large-scale multi-stage milling machines but they were unable to attract many customers for their milling service. However, the early adopters increased capacity utilization between 2016 and 2019, when they substantially increased their milling service, as observed in Table 12.3. The late adopters initially had a high capacity utilization rate in 2011, which dropped in 2016 but they maintained their capacity utilization rate in 2018 and slightly increased it in 2019, unlike the declining performance of the non-adopters. This monthly capacity utilization rate reflects the milling performance in the main season. We also calculated the annual capacity utilization rate in 2019 (Appendix Table 12.9). We found that the non-adopters had 0.13 and the late adopters had 0.30, a slightly reduced total but

---

[13] The monthly capacity utilization rate is the total amount of milled rice and sold rice (tons per month) divided by the milling capacity of machines for 200 h of operation (tons per month).

Table 12.6 Monthly capacity utilization rate, gross profit, gross profit per capacity by destoner adoption

	Total	Nonadopter	Early adopter	Late adopter
Capacity utilization rate				
2011	0.70	0.60	1.33	1.37
	(0.57)	(0.50)	(0.67)	(0.65)
2016	0.32	0.30	0.40	0.38
	(0.35)	(0.37)	(0.26)	(0.14)
2018	0.34	0.25	0.64	0.38
	(0.45)	(0.33)	(0.73)	(0.35)
2019	0.52	0.21	1.36	0.45
	(1.65)	(1.36)	(3.03)	(0.58)
Profit (million Kshs)				
2016	1.71	−0.08	11.5	0.03
	(18.9)	(0.49)	(48.3)	(0.87)
2018	1.99	−0.05	10.3	1.04
	(18.1)	(0.23)	(42.7)	(2.90)
2019	0.92	−0.06	2.84	2.64
	(11.3)	(0.20)	(27.6)	(9.0)
Profit per capacity (million Kshs/kg)				
2016	−0.03	−0.04	0.001	−0.03
	(0.22)	(0.20)	(0.35)	(0.17)
2018	0.01	−0.01	−0.02	0.12
	(0.20)	(0.08)	(0.34)	(0.31)
2019	−0.01	−0.01	−0.20	0.12
	(0.36)	(0.09)	(0.75)	(0.45)

Notes Standard deviations are in parentheses. The profit is deflated with the GDP deflator (the 2019 value). The capacity utilization rate is the ratio between the total amount of milled rice and sold rice (tons per month) and the milling capacity of 200 h of operation (tons per month). Profit per capacity is the ratio between the profit (million Kshs per month) and the milling capacity of 200 h of operation (kg per month)

comparable to their performance in the main season. The early adopters, on the other hand, had 0.45, suggesting that the early adopters substantially reduced their milling performance in the offseason, leaving their large-scale milling machines idle.

The profit of the non-adopters was negative during the period between 2016 and 2019 (Table 12.6).[14] The early adopters' profit was far the largest but declined substantially. By contrast, the late adopters' profit increased between 2016 and 2019

---

[14] We calculated profit by subtracting the labor costs, the purchasing cost of paddy, electricity costs, maintenance costs and the depreciation costs of milling machines from the revenue of milling service and selling rice and bran. The depreciation cost is estimated by assuming the linear depreciation over the usable lifetime, $D = (V - S)/(12 \times L)$, where D = monthly depreciation, V = the value of machine, S = the salvage value of the machine after its useful life, assumed to be 10% of the machine value, L = useful life of the milling machine, assumed to be 10 years (Tokida and Barrett 2014; Norbu 2018). The estimation results are essentially robust to parameter values in reasonable

as they upgraded their milling machines. Since the absolute amount of profit depends on the amount of investment, we also calculated the ratio of profit to milling capacity per month as an alternative indicator of the milling business's profitability. The non-adopters had negative ratios between 2016 and 2019. The early adopters also had negative and declining ratios,[15] suggesting overcapacity for early adopters, particularly after the late adopters also adopted the improved milling machines. In contrast, although the late adopters initially had a negative ratio in 2016, they increased profit per capacity in 2018 and 2019.

12.3   Conceptual Framework and Empirical Strategy

12.3.1   Conceptual Framework

To understand Kenya's current rice value chain transformations, it may be helpful to learn about Asia's experience (Reardon et al. 2014), which can be summarized as follows: (1) medium and large rice mills upgrade milling equipment, which improves the quality of milled rice; (2) some of them introduce branding; (3) as a result, small mills are closed; and (4) vertical coordination takes place between farmers and medium and large mills, including contract farming, and between supermarkets/large urban retailers/wholesalers and medium and large mills. The descriptive analysis in the previous section suggests that processes (1)–(3) are taking place in Mwea, although we did not observe the existence of any contract farming.

Figure 12.3 summarizes the structural transformation of the rice value chain in Mwea. Traditionally, farmers and local traders bring paddy to rice millers who use traditional machines, primarily providing a milling service. Local traders and consumers purchase the rice milled by these traditional millers. By contrast, farmers sell paddy to local traders under the new system that brings paddy to rice millers adopting large-scale improved machines. Contract farming is not happening though, primarily because large-scale millers find it costly to trade with many small farmers directly.[16] Finally, some millers with improved machines have introduced brand names and sell milled rice to urban supermarkets directly. In the transition period, this type of rice miller also provides a simple milling service.

---

ranges. Because of the data limitation, we did not consider other expenses such as the rental costs for the workspace and storage, marketing costs, or tax payments.

[15] The early adopters' average profits are positive because of the large positive profits made by some millers using large milling machines despite the negative profits of other millers. But the average profit-capacity ratios reduced the weight of those successful large millers and turned out to be negative.

[16] Contract farming is designed to provide inputs on credit and production instruction (Otsuka et al. 2016). According to our previous study (Njeru et al. 2016), a credit market is working in Mwea. Furthermore, farmers understand production methods well due to the efficient extension system. Thus, the advantage of contract farming is nil. These observations may also explain the absence of vertical coordination between millers and farmers.

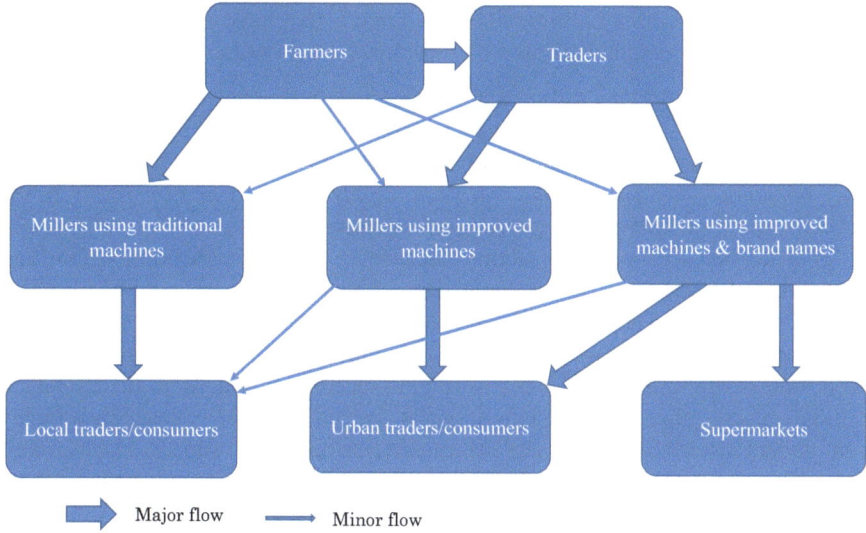

Fig. 12.3 A schematic view of structural transformation of rice marketing

One of the key questions is who plays the role of the innovator that introduces new, improved machines. Most adopters of improved milling technologies operated in the rice milling clusters (Table 12.1). Advancing Marshall's (1920) argument of agglomeration economies in industrial clusters, Sonobe and Otsuka (2006, 2011, 2014) provide a theory of cluster-based industrial development with ample supporting evidence from Asia and SSA. They argue that cluster-based industries typically expand through the entry of new firms imitating successful pioneers. However, the increasing supply of products without quality improvement reduces their prices, and hence, profitability—thereby inducing innovative firms to improve product quality to restore profitability. They also update marketing strategies to sell their new differentiated and improved products. Following these innovators, a small number of competent firms succeed in imitating such products and marketing innovations and expanding their businesses. In contrast, other firms fail to do so and eventually exit the market because of reduced profitability.

In the early 2010s, a few educated millers operating in the major rice milling clusters along the highway had learned the importance of the destoner component. The improved technologies were primarily available only with expensive large-sized multi-stage milling machines. Thus, the early adopters are likely to already understand the urban market and have sufficient wealth to make a significant initial investment. By the time small- and medium-sized multi-stage milling machines became available in the late 2010s, the importance of specialist knowledge about the marketing of high-quality rice and wealth had declined. However, the general human capital represented by education is expected to remain important in making the proper adoption decisions of milling machines. Moreover, because the millers operating in the rice milling clusters along the highway can attract more large urban traders keen on procuring high-quality milled rice,[17] they have a greater incentive to improve their milling quality. Based on these considerations, we hypothesize that:

Hypothesis 1: "Those educated millers who had acquired knowledge about the urban market or accumulated wealth through previous occupations or operations in rice milling clusters tended to adopt improved milling technologies early. Following the early adopters, other millers who were also highly educated and operating in the rice milling clusters adopted improved technologies later."

As we learned in the previous section, because the only available improved milling machines in the early 2010s were large-sized, the early adopters could be expected to be large millers. They would have provided high-quality milling services and improved market performance in those years. In the late 2010s, small- and medium-sized multi-stage milling machines were introduced to Mwea. With the expanded milling capacity, the early and late adopters were more likely to purchase paddy from traders instead of farmers and sell milled rice directly to supermarkets as well as urban traders and consumers (Soullier et al. 2020). Based on these observations, we hypothesize that:

Hypothesis 2: "Both the early and the late adopters improved their product quality and expanded their business with improved milling technologies."

During the 2010s, millers who adopted improved machines are expected to have improved their milling and overall business performance. In contrast, the remainder kept using traditional milling machines and performed poorly or exited the market because they lost the milling service business to the adopters. Thus, we also test the following hypothesis:

Hypothesis 3: "The non-adopters tend to exit the market because they lose milling service business to the adopters of improved machines, who were more likely to survive in the business."

12.3.2   Doubly Robust and Endogenous Switching Regression

We consider an estimation strategy to explicitly control the selection of observables and match millers with similar characteristics. Earlier studies on agricultural technologies have used such estimation methods, including propensity score matching, inverse probability weighting, and the doubly robust (DR) method, when no other plausible instruments are available (Takahashi and Barrett 2014; Bellemare and Novak 2016; Kahn et al. 2019; Mano et al. 2020; Chap. 8 of this volume). We apply the DR method, or more precisely, an inverse probability weighted regression adjustment, that combines the regression and propensity score weighting. It is more robust than the propensity score matching estimator and the inverse probability weighting estimator. It can provide a consistent estimator as long as either

---

[17] Those rice millers can more easily source paddy rice because the clusters attract local traders and farmers.

the propensity score for destoner use or the outcome regression in terms of miller characteristics is correctly specified (Wooldridge 2007, 2010, Sect. 21.3.4).[18] More specifically, we first estimate the binary response model of improved technology adoption:

$$D_{it}^* = x_{it}'\beta + \epsilon_{it} \text{ with } D_{it} = \begin{cases} 1 \text{ if } D_{it}^* > 0 \\ 0 \text{ if } D_{it}^* \leq 0 \end{cases} \tag{12.1}$$

where D_{it}^* is the latent variable of improved technology adoption, x_{it} is the vector of miller i's characteristics in a month of the main season in year $t = 2011, 2016, 2018, 2019$, β is the parameter to be estimated, ϵ_{it} is the error term, and the probability of improved technology adoption conditional on the miller's characteristics can be expressed as the probit model:

$$P(D_{it} = 1|x_{it}) = \Phi\left(x_{it}'\beta\right) \equiv p(x_{it})$$

where $p(x_{it})$ is the propensity score of improved technology adoption. The miller's characteristics are the decision-makers' age, a high school dummy that takes the value of 1 if the decision-maker graduated from high school, the former occupation dummies, and the rice mill's establishment year. Year dummies are also used. Using the estimated propensity $\hat{p}(x_{it})$, we estimate the regression parameters γ's by using the following set of inverse probability weighting linear least-squares problems:

$$\min_{\gamma_1} \sum_i \sum_t D_{it}\left(Y_{it} - x_{it}'\gamma_1\right)^2 / \hat{p}(x_{it})$$

$$\min_{\gamma_0} \sum_i \sum_t (1 - D_{it})\left(Y_{it} - x_{it}'\gamma_0\right)^2 / \left[1 - \hat{p}(x_{it})\right]$$

We estimate the average treatment effect on the treated (ATT) of the improved technology adoption on outcome Y as the average of the difference in predicted values of outcomes:

$$\hat{\tau}_{ATT,DR} = \overline{x}_1'\left(\hat{\gamma}_1 - \hat{\gamma}_0\right) \tag{12.2}$$

where $\overline{x}_1 = n_1^{-1} \sum_i \sum_t D_i x_{it}$ is the vector of average characteristics of millers (including the constant term) over the improved technology adopters, and n_1 is the number of improved technology adopters. Here, $\hat{\tau}_{ATT,DR}$ is the DR estimator of ATT. The outcome variables are the milling fee, the amount of rice milled for customers, the rice purchasing and selling dummy, which takes the value 1 if the miller purchased paddy and sold rice and 0 otherwise, the price of sold rice, the amount of sold rice,

---

[18] We used the STATA command *teffects ipwra* to implement the DR method.

the capacity utilization rate, the gross profit, the gross profit per capacity, and the survival dummy, which takes the value 1 if the miller continued operation until the next period of our observation and 0 otherwise.

The other empirical strategy that we use is the endogenous switching regression (ESR). This method is also often used in the studies of agricultural technology adoption to address the endogeneity bias due to unobserved characteristics, which is assumed away by the DR method (Di Falco et al. 2011; Khonje et al. 2018; Bairagi et al. 2020). The first step is to estimate the binary response model of improved technology adoption (12.1), and the second step specifies the relationship between the outcome variable Y_{it} and a vector of miller's characteristics x_{it} for improved technology adopters and non-adopters separately:

$$\text{Destoner adopters} \quad Y_{1it} = x'_{1it}\delta_1 + \mu_{1it} \text{ if } D_{it} = 1 \quad (12.3)$$

$$\text{Nonadopters} \quad Y_{0it} = x'_{0it}\delta_0 + \mu_{0it} \text{ if } D_{it} = 0 \quad (12.4)$$

where subscripts 1 and 0 represent adopters and non-adopters, respectively, δ's are the vectors of regression coefficients, and μ's are the random error terms. The error terms in Eqs. (12.1), (12.3), and (12.4) are assumed to be jointly and normally distributed with mean vector zero and the following covariance matrix:

$$\Omega = cov(\epsilon_{it}, \mu_{1it}, \mu_{0it}) = \begin{bmatrix} \sigma_\epsilon^2 & \sigma_{\epsilon 1} & \sigma_{\epsilon 0} \\ \sigma_{\epsilon 1} & \sigma_1^2 & \cdot \\ \sigma_{\epsilon 0} & \cdot & \sigma_0^2 \end{bmatrix}$$

where $\sigma_\epsilon^2 = var(\epsilon_{it}) = 1$, $\sigma_1^2 = var(\mu_{1it})$, $\sigma_0^2 = var(\mu_{0it})$, $\sigma_{\epsilon 1} = cov(\epsilon_{it}, \mu_{1it})$, $\sigma_{\epsilon 0} = cov(\epsilon_{it}, \mu_{0it})$, and the covariance between μ_{1it} and μ_{0it} are not defined because they are not observed simultaneously (Green 2012). The conditional expectation of the outcome of the destoner adopters in the actual case of adopting destoners can be expressed as:

$$E(Y_{1it}|D_{it} = 1, x_{1it}) = x'_{1it}\delta_1 + E(\mu_{1it}|D_{it} = 1, x_{1it}) = x'_{1it}\delta_1 + \sigma_{\epsilon 1}\lambda_{1it} \quad (12.5)$$

where $E(\mu_{1it}|D_{it} = 1, x_{1it}) = \sigma_{\epsilon 1}\dfrac{\phi\left(x'_{1it}\beta\right)}{\Phi\left(x'_{1it}\beta\right)} \equiv \sigma_{\epsilon 1}\lambda_{1it}$. Analogously, the destoner adopters' expected outcome in the counterfactual case of not adopting destoners can be expressed as:

$$E(Y_{0it}|D_{it} = 1, x_{1it}) = x'_{1it}\delta_0 + E(\mu_{0it}|D_{it} = 1, x_{1it}) = x'_{1it}\delta_0 + \sigma_{\epsilon 0}\lambda_{1it} \quad (12.6)$$

Following Heckman et al. (2001) and Di Falco et al. (2011), we calculate the covariate-specific effect of the treatment on the treated as the difference between Eqs. (12.5) and (12.6):

$$E(Y_{1it}|D=1, x_{1it}) - E(Y_{0it}|D=1, x_{1it}) = x'_{1it}(\delta_1 - \delta_0) + (\sigma_{\epsilon 1} - \sigma_{\epsilon 0})\lambda_{1it}$$

Taking the average of this value over the destoner adopters, we obtain the ATT estimator of the destoner adoption using the endogenous switching regression:

$$\hat{\tau}_{ATT,ESR} = \overline{x}'_1\left(\hat{\delta}_1 - \hat{\delta}_0\right) + \left(\hat{\sigma}_{\epsilon 1} - \hat{\sigma}_{\epsilon 0}\right)\overline{\lambda}_1$$

where $\overline{x_1} = n_1^{-1}\sum_i\sum_t D_i x_{it}$ is the vector of average characteristics of millers (including the constant term) over the destoner adopters, $\overline{\lambda}_1$ is the average inverse mill's ratio calculated using the regression parameter estimated in model (1), δ's and σ's are parameters of regression (12.5) and (12.6) to be estimated.[19] Because we do not have a decent instrument for destoner adoption, we rely on the nonlinearity of the inverse mill's ratio for identification in the parameter estimation (Wooldridge 2010). Analogously, we also estimate the average treatment effect on the untreated (ATUT), the hypothetical effect of adopting improved technologies for non-adopters:

$$\hat{\tau}_{ATUT,ESR} = \overline{x}'_0\left(\hat{\delta}_1 - \hat{\delta}_0\right) + \left(\hat{\sigma}_{\epsilon 1} - \hat{\sigma}_{\epsilon 0}\right)\overline{\lambda}_0$$

where $\overline{x_0} = n_0^{-1}\sum_i\sum_t(1 - D_i)x_{it}$ is the vector of average characteristics of millers (including the constant term) over the non-adopters, $\overline{\lambda}_0$ is the average inverse mill's ratio for non-adopters. We use ATUT estimates to examine whether non-adopters had lower expected profitability, which may explain why they did not adopt improved technologies.

12.4 Estimation Results

Table 12.7 presents the estimation results of the probit model of improved technology adoption of the early adopters with pooled data between 2011 and 2019 and of the late adopters between 2018 and 2019, when some or all of them had adopted improved machines. The decision-makers with a high school education or above and previous work experience in business and formal jobs tended to adopt large-scale improved milling technologies early. The main decision-makers of the late adopters were also found to have higher education but less previous work experience in rice marketing or mechanics. Both the early adopters and the late adopters were more likely to operate in the rice milling clusters. These results are consistent with Hypothesis 1.

Table 12.8 presents the estimation results of the DR method and the ESR method. We prefer the ESR method because it addresses potential selection on unobservables. Overall, however, the ATT estimates of improved technology adoption are similar between the two methods. We also estimated the ATT for early and late adopters

---

[19] We used the STATA command *movestay* to implement the ESR method.

Table 12.7 Determinants of improved technology adoption of early adopters and late adopters (probit)

	Early adopter	Late adopter
High school	0.095***	0.113*
	(0.031)	(0.067)
Rice miller & trading	0.054	−0.175**
	(0.046)	(0.062)
Business and formal jobs	0.137***	0.071
	(0.048)	(0.099)
Mechanics	−0.044	−0.181***
	(0.040)	(0.059)
Year of establishment	−0.002	0.005
	(0.002)	(0.003)
Mwea town	0.266***	0.306***
	(0.040)	(0.072)
Mutithi	0.118**	0.311***
	(0.044)	(0.100)
Log pseudo-likelihood	−79.731	−44.834
Obs.	332	147

Notes The marginal effects are reported. The numbers in parentheses are robust standard errors. ***, **, and *indicate statistical significance at the 1, 5, and 10% levels, respectively. The base outcome is non-adoption. We pooled data from 2011 to 2019 for the analysis of early adopters and data from 2018 and 2019 for the analysis of late adopters. Year dummies are also controlled

separately using the DR method and the ATUT of improved technology adoption using the ESR method. The adoption of improved milling technologies increased milling fees for the adopters in general, the early adopters, and the late adopters. At the same time, it would have also increased the milling fees of the non-adopters if they had adopted the improved machines. These findings are consistent with Hypothesis 2. The improved technology adoption also increased rice milled for customers for the adopters in general, the early adopters, and the late adopters. At the same time, it would have also increased the amount of milled rice for the non-adopters.

The DR estimate of ATT of improved technology adoption on the probability of selling rice was significantly positive for the late adopters only. According to the corresponding ESR estimate, the improved technology adoption increased the probability of selling rice for the adopters in general. In contrast, it would have reduced the non-adopters' probability of selling rice. We do not know why the adoption of improved milling machines would decrease the non-adopters' likelihood of selling rice. We suspect that because the decision-makers of the non-adopters lack former work experience in business or general education (Table 12.1), they were incapable of increasing the profitability of selling rice simply by adopting improved machines. The adoption of improved machines increased the price of sold rice for the adopters in general and the late adopters, while it would have also increased the price of sold

Table 12.8 Effects of improved milling technologies (doubly robust and endogenous switching regression)

	Milling fee (Kshs/kg)	Rice milled (ton)	Sell milled rice (=1)	Price of rice (Kshs/kg)	Rice sold (ton)	Capacity utilization	Profit (mill. Kshs)	Profit per capacity (mill. Kshs/kg)	Survival (=1)
Doubly robust									
ATT	0.807*** (0.098)	311.177*** (112.464)	0.033 (0.102)	13.340*** (4.235)	145.693*** (62.668)	0.291* (0.175)	5.907 (3.787)	0.010 (0.060)	0.393*** (0.152)
Early adopter's ATT	0.804*** (0.142)	414.242** (162.333)	−0.053 (0.114)	7.654 (5.948)	236.936** (111.311)	0.410 (0.259)	8.074 (5.947)	−0.069 (0.080)	–
Late adopter's ATT	0.755*** (0.119)	176.692* (96.319)	0.246* (0.141)	15.967*** (6.857)	39.080*** (11.645)	0.115 (0.101)	2.168 (1.460)	0.144* (0.077)	–
Endogenous switching									
ATT	0.432*** (0.019)	437.296*** (33.176)	0.265*** (0.026)	15.253*** (0.637)	144.576*** (29.305)	0.174*** (0.041)	5.763** (2.390)	0.018 (0.023)	0.739*** (0.020)
ATUT	0.536*** (0.013)	322.601*** (13.573)	−0.307*** (0.015)	30.649*** (0.541)	264.450*** (22.673)	0.269*** (0.017)	−2.281*** (0.564)	0.989*** (0.013)	0.491*** (0.009)

Notes ATT is the average treatment effect on the treated, and ATUT is the average treatment effect on the untreated. Robust standard errors are in parentheses. ***, **, and *indicate statistical significance at the 1, 5, and 10% levels, respectively. The miller's characteristics are also controlled in the analyses: the age of decision-makers; the high school dummy (which takes 1 if the decision-maker graduated from high school); the former occupation dummies; the establishment year of the rice miller; the cluster dummies; and the year dummies

rice for the non-adopters. These findings are consistent with Hypothesis 2. The adoption of improved technologies increased the amount of sold rice for the adopters in general, the early adopters, and the late adopters. Moreover, it would have increased the rice sold for the non-adopters.

The adoption of improved machines increased the capacity utilization rate for the adopters in general, and it would have also increased it for the non-adopters. The DR estimates of the effect of improved machine adoption on profit were not statistically significant. However, according to the ESR estimates, improved technology adoption increased the profit for the adopters and would have reduced the profit for the non-adopters. As was explained earlier, because the decision-makers of the non-adopters lack work experience in business or general education, we suspect that they were incapable of increasing their profits simply by adopting improved machines. The estimated effect of improved machine adoption on the early adopters' profit-capacity ratio was insignificant, whereas the adoption of improved machines significantly increased the late adopter's profit-capacity ratio. According to the ESR estimates, improved technologies would have increased the profit-capacity ratio for the non-adopters.

We also estimated the ATT and ATUT of destoner adoption on the millers' survival rates. Because we observed the late adopters' technology adoption only during the last survey in 2020, we cannot analyze its effect on survival. The adoption of improved technologies increased the probability of early adopters' survival, and it would have also increased the non-adopters' survival probability. This result is consistent with Hypothesis 3.

12.5 Discussion

SSA heavily relies on imports from Asia to meet its rapidly increasing rice consumption due to population growth and urbanization, despite the efforts of various governments to improve food security by increasing rice farming productivity (Chap. 2 of this volume). Although consumers in some countries traditionally prefer domestic rice, massive rice imports have induced or may induce consumers to prefer imported rice. Thus, improving Africa's milled rice quality is vital for competing with Asia's milled rice (Chaps. 7 and 13 of this volume). Mwea rice is of higher quality than imported Pakistani rice and, hence, is more expensive. This study explored why this is the case.

We found that learning improved milling technologies from abroad triggered the rice milling industry's transformational improvement in Kenya's largest rice production area, Mwea. After the 2007–2008 food crisis, a few educated and experienced decision-makers of rice milling companies located in the rice milling clusters visited China. They learned the importance of destoners and other improved rice milling machines in improving the quality of milled rice. They adopted large-scale multistage rice milling machines with destoners and other improved technologies in the first half of the 2010s. They successfully enhanced rice milling quality and sales of milled rice to urban supermarkets as well as consumers and traders. In the late 2010s, followers in the rice milling clusters introduced small- and medium-scale multi-stage rice milling machines to Mwea. Improved rice milling technologies were more widely adopted in the clusters, contributing to Mwea rice's competitiveness against imported rice in the market. These findings strongly indicate the critical importance of adopting advanced technologies from abroad to improve enterprises' performance in SSA.

12.5.1 Policy Implications

These findings suggest that the development process of the rice milling sector in Mwea is similar to Asia's rice value chain transformations (as described by Reardon 2014) and may have significant implications for other SSA countries struggling to improve the quality of milled rice (Soullier et al. 2020; Chap. 13 of this volume). The critical point is that the adoption of destoners and other improved milling technologies has significantly improved the quality of local milled rice and its competitiveness against imported Asian rice. Moreover, we observed the associated transformation of rice value chains beyond the millers' adoption of new technologies. The adopters of improved mills with greater milling capacity sourced paddy from local traders. Some adopters also branded their milled rice and developed a direct marketing channel to supermarkets as well as urban consumers and traders.

The previous National Rice Development Strategies (NRDS) of African countries under the Coalition for African Rice Development (CARD) initiative focused on increasing rice production (Demont 2013). However, in recognizing the importance of upgrading rice value chains to increase the quality-based competitiveness of domestic rice (Demont and Ndour 2015; Demont et al. 2017), African governments have added the installment of improved rice mills into the new set of NRDS. The current study provides supporting evidence that adopting destoners and other improved rice technologies increases the sales and price of domestic milled rice in urban markets.

We also observed that the introduction of small- and medium-scale multi-stage rice milling machines reduced the financial burden on potential adopters and assisted the widespread adoption of the new technology. In other words, the choice of appropriate technology is likely to be of crucial importance for SSA because the profitability of investment in improved machines varies and may not always be positive. To the extent that knowledge of various improved milling machines is a local public good, there is room for the government to provide appropriate information about the cost and benefit of various rice milling machines. Furthermore, we must note that, although the adoption of improved milling machines significantly improved the quality of rice and the performance of millers in Mwea, paddy produced in Mwea is a high-quality improved Basmati type that is uncommon in SSA. Whether the adoption of milling machines with destoners will enhance the quality of milled rice and millers'

Table 12.9 Other characteristics of millers in 2019 by destoner adoption

	Nonadopter	Early adopter	Late adopter
Functions of milling machines (=1)			
Pre-cleaner	0	0.93	1
De-husker	1	1	1
Polisher	0.98	1	1
Grader	0.03	0.93	0.68
Color sorter	0	0.57	0.11
Area of workshop & storage (m^2)	262.5 (336.0)	2595.2 (2470.3)	949.4 (898.6)
Area of the yard (m^2)	23.9 (69.9)	3801.5 (6076.3)	553.6 (1813.6)
Milled rice to paddy ratio (%)	62.0 (3.56)	65.7 (3.8)	64.1 (2.7)
The minimum moisture content (%)	12.6 (1.1)	12.0 (1.2)	12.9 (0.8)
Price of bran sold (Kshs/kg)	5.7 (3.1)	12.9 (2.7)	12.0 (3.3)
The annual amount of bran sold (tons)	37.1 (253.5)	196.9 (521.9)	280.2 (1097.2)
The annual amount of rice milled (tons)	136.2 (203.3)	3766.6 (5222.8)	4019.9 (11,760.4)
The annual amount of rice sold (tons)	17.0 (53.1)	1198.4 (4263.4)	57.5 (145.9)
Capacity utilization rate (annual)	0.13 (0.22)	0.45 (0.50)	0.30 (0.51)

Notes Standard deviations of continuous variables are in parentheses

performance in areas where more popular rice varieties are grown is a critical issue to be explored in future studies.

Appendix

References

Arimoto Y, Sakane Y (2017) Japanese agriculture and rural issues (Nihon nougyo to nouson mondai). In: Fukao K, Nakamura N, Nakabayashi M (eds) Iwanami lecture series: history of Japanese economy (Iwanami-koza: Nihon keizai no rekishi). Iwanami Shoten Pub, Tokyo, pp 139–181

Bairagi S, Mishra AK, Durand-Morat A (2020) Climate risk management strategies and food security: evidence from Cambodian rice farmers. Food Secur 95:101935. https://doi.org/10.1016/j.foodpol.2020.101935

Bellemare M, Novak L (2016) Contract farming and food security. Am J Agric Econ 99(2):357–378. https://doi.org/10.1093/ajae/aaw053

Britwum K, Demont M (2021) Tailoring rice varieties to consumer preferences induced by cultural and colonial heritage: lessons from New Rice for Africa (NERICA) in The Gambia. Outlook Agric 50(3):305–314. https://doi.org/10.1177/0030727021101975

Coalition for African Rice Development (CARD) (2019) Rice for Africa. https://riceforafrica.net. Accessed 21 February 2021

Demont M (2013) Reversing urban bias in African rice markets: a review of 19 national rice development strategies. Glob Food Sec 2(3):172–181. https://doi.org/10.1016/j.gfs.2013.07.001

Demont M, Fiamohe R, Kinkpe AT (2017) Comparative advantage in demand and the development of rice value chains in West Africa. World Dev 96:578–590. https://doi.org/10.1016/j.worlddev. 2017.04.004

Demont M, Ndour M (2015) Upgrading rice value chains: experimental evidence from 11 African markets. Glob Food Sec 5:70–76. https://doi.org/10.1016/j.gfs.2014.10.001

Di Falco S, Veronesi M, Yesuf M (2011) Does adaptation to climate change provide food security? A micro-perspective from Ethiopia. Am J Agric Econ 93(3):829–846. https://doi.org/10.1093/ ajae/aar006

Diako C, Sakyi-Dawson E, Bediako-Amoa B, Saalia FK, Manful JT (2010) Consumer perceptions, knowledge, and preferences for aromatic rice types in Ghana. Nat Sci 8(12):12–19. http://197. 255.68.203/handle/123456789/1233

Fiamohe R, Demont M, Saito K, Roy-Macauley H, Tollens E (2018) How can West African rice compete in urban markets? A Demand Perspective for Policymakers. Eurochoices 17(2):51–57. https://doi.org/10.1111/1746-692X.12177

Food Agricultural Organization (FAO) (2018) As rice import bills rise, African countries must sustain growth. http://www.fao.org/africa/news/detail-news/en/c/1154254/. Accessed 21 February 2021

Futakuchi K, Mnful J, Sakurai T (2013) Improving grain quality of locally produced rice in Africa. In: Wopereis MCS, Johnson D, Ahmadi N, Tollens E, Jalloh A (eds) Realizing Africa's rice promise. Africa Rice Center, Benin, pp 311–323

Green WH (2012) Econometric analysis, 7th edn. Pearson Education Limited, New York

Heckman JJ, Tobias JL, Vytlacil EJ (2001) Four parameters of interest in the evaluation of social programs. South Econ J 68(2):210–223. https://doi.org/10.2307/1061591

Ibrahim LA, Sakurai T, Tachibana T (2020) Local rice market development in Ghana: experimental sales of standardized premium quality rice to retailers. J Agric Econ 22:118–122. https://doi.org/ 10.18480/jjae.22.0_118

Irukayama S (1939) Regarding rice inspection in Taiwan (Taiwan ni okeru beikoku kensani tsuite (Chuu-no-ni)). Taiwan Agricultural Association Bulletin (Taiwan noukaihou), 1–6 June 1939

Kapalata D, Sakurai T (2020) Adoption of quality-improving rice milling technologies and its impacts on millers' performance in Morogoro region, Tanzania. J Agric Econ 22:101–105. https:// doi.org/10.18480/jjae.22.0_101

Kenya National Bureau of Statistics (KNBS) (2022) Economic Survey 2021

Khan MF, Nakano Y, Kurosaki T (2019) Impact of contract farming on land productivity and income of maize and potato growers in Pakistan. Food Policy 85:28–39. https://doi.org/10.1016/ j.foodpol.2019.04.004

Khonje MG, Manda J, Mkandwire P, Tufa AH, Alene AD (2018) Adoption and welfare impacts of multiple agricultural technologies: evidence from eastern Zambia. Agric Econ 49(5):599–609. https://doi.org/10.1111/agec.12445

Kikuchi M, Mano Y, Njagi TN, Merry D, Otsuka K (2021) Economic viability of large-scale irrigation construction in Sub-Saharan Africa: what if the Mwea irrigation scheme were constructed as a brand-new scheme? J Dev Stud 57(5):772–789. https://doi.org/10.1080/00220388.2020.182 6443

KNBS (2021) Trade Map. https://www.trademap.org/. Accessed 11 July, 2021

Mano Y, Njagi TN, Otsuka K (2022) An inquiry into the process of upgrading rice milling services: the case of the Mwea Irrigation Scheme in Kenya. Food Policy 106:102195. https://doi.org/10.1016/j.foodpol.2021.102195

Mano Y, Takahashi K, Otsuka K (2020) Mechanization in land preparation and agricultural intensification: the case of rice farming in Cote d'Ivoire. Agric Econ 51(6):899–908. https://doi.org/10.1111/agec.12599

Marshall A (1920) Principles of Economics. Macmillan, London, UK

Mendez del Villar P, Lançon F (2015) West African rice development: beyond protectionism versus liberalization? Glob Food Secur 5:56–61. https://doi.org/10.1016/j.gfs.2014.11.001

Njeru TN, Mano Y, Otsuka K (2016) Role of access to credit in rice production in Sub-Saharan Africa: the case of Mwea irrigation scheme in Kenya. J Afr Econ 25(2):300–321. https://doi.org/10.1093/jae/ejv024

Norbu K (2018) Cost analysis of operating a medium-sized rice processing mill in Bhutan. Bhutanese J Agricul 1(1):70–76

Otsuka K (2019) Evidence-based strategy for a rice green revolution in sub-Saharan Africa. Japan International Cooperation Agency Research Institute Policy Note No. 5. JICA-RI, Tokyo

Otsuka K, Larson D (eds) (2013) An African green revolution: finding ways to boost productivity on small farms. Springer, Dordrecht, Netherlands

Otsuka K, Larson D (eds) (2016) In pursuit of an African Green Revolution: views from rice and maize farmers' fields. Springer, Dordrecht, Netherlands

Otsuka K, Nakano Y, Takahashi K (2016) Contract farming in developed and developing countries. Ann Rev Resour Econ 8:353–376. https://doi.org/10.1146/annurev-resource-100815-095459

Ragasa C, Andam K, Asante S, Amewu S (2020) Can local products compete against imports in West Africa? Supply- and demand-side perspectives on chicken, rice, and tilapia in Accra Ghana. Glob Food Sec 26:100448. https://doi.org/10.1016/j.gfs.2020.100448

Reardon T, Chen KZ, Minten B, Adriano L, Dao TA, Wang J, Gupta SD (2014) The quiet revolution in Asia's rice value chains. Ann N Y Acad Sci 1331:106–118. https://doi.org/10.1111/nyas.12391

Saito K, Dieng I, Toure AA, Somado EA, Wopereis MCS (2015) Rice yield growth analysis for 24 African countries over 1960–2012. Glob Food Sec 5:62–69. https://doi.org/10.1016/j.gfs.2014.10.006

Samejima H, Katsura K, Kikuta M, Njinju SM, Kimani JM, Yamauchi A, Makihara D (2020) Analysis of rice yield response to various cropping seasons to develop optimal cropping calendars in Mwea. Kenya. Plant Prod Sci 23(3):297–305. https://doi.org/10.1080/1343943X.2020.1727752

Sonobe T, Otsuka K (2006) Cluster-based industrial development: an East Asian model. Palgrave Macmillan, Hampshire, UK

Sonobe T, Otsuka K (2011) Cluster-based industrial development: a comparative study of Asia and Africa. Palgrave Macmillan, Hampshire, UK

Sonobe T, Otsuka K (2014) Cluster-based industrial development: Kaizen management for MSE growth in developing countries. Palgrave Macmillan, Hampshire, UK

Soullier G, Demont M, Arouna A, Lançon F, Mendez del Villar P (2020) The state of rice value chain upgrading in West Africa. Glob Food Sec 25:100365. https://doi.org/10.1016/j.gfs.2020.100365

Takahashi K, Barrett CB (2014) The system of rice intensification and its impacts on household income and child schooling: evidence from rural Indonesia. Am J Agric Econ 96:269–289. https://doi.org/10.1093/ajae/aat086

Tokida K, Haneishi Y, Tsuboi T, Asea G, Kikuchi M (2014) Evolution and prospects of the rice mill industry in Uganda. Afri J Agric Res 9(33):2560–2573. https://doi.org/10.5897/AJAR2014.8837

Wooldridge JM (2007) Inverse probability weighted estimation for general missing data problems. J Econom 141(2):1281–1301. https://doi.org/10.1016/j.jeconom.2007.02.002

Wooldridge JM (2010) Econometric analysis of cross section and panel data, 2nd edn. MIT Press, Cambridge MA

Yukichi Mano is a professor at Hitotsubashi University, Japan, and is a fellow at Tokyo Center for Economic Research (TCER). He received Ph.D. in Economics from the University of Chicago in 2007. His scholarly interests include agricultural technology adoption, horticulture and high-value crop production, business, and management training (KAIZEN), human capital investment, migration and remittance, and universal health coverage in Asia and sub-Saharan Africa.

Timothy Njagi is a seasoned Development Economist and a Research Fellow at Tegemeo Institute of Agricultural Policy and Development of Egerton University. He holds a Ph.D. in Development Economics and a Master's Degree in International Development from the National Graduate Institute for Policy Studies (GRIPS), Japan. His current research focus is on agricultural productivity, technology adoption, governance and efficiency of agricultural systems, climate resilience, and evaluation within the agriculture sector.

Keijiro Otsuka is a professor of development economics at the Graduate School of Economics, Kobe University, and a chief senior researcher at the Institute of Developing Economies in Chiba, Japan since 2016. He received Ph.D. in economics from the University of Chicago in 1979. He majors in Green Revolution, land tenure and land tenancy, natural resource management, poverty reduction, and industrial development in Asia and sub-Saharan African countries.

Chapter 13
Toward Quality Upgrading of Rice Production in SSA: Experimental Evidence from Northern Ghana

Tatsuya Ogura, Joseph A. Awuni, and Takeshi Sakurai

Abstract Quality improvement of domestic milled rice is an urgent issue in sub-Saharan Africa (SSA) because domestic rice cannot compete with imports in the growing urban market. Past studies considered poor milling facilities to be a major factor leading to the inferior quality of domestic rice. However, even with the modern milling facilities recently established in SSA, the quality of milled rice is not yet necessarily competitive with imports. Thus, finding ways to improve the quality of paddy remains an important question in SSA. We hypothesize that the lack of knowledge of paddy quality and its relationship with price causes farmers to continue producing low-quality paddy. We conduct a field experiment in northern Ghana to verify this hypothesis. We randomly selected 108 villages and 10 rice producers from each of the villages. From this sample, we randomly chose 54 treatment villages and we provided the farmers with information about quality-enhancing technologies and quality parameters appreciated by the market. Utilizing data collected before and after the intervention, we found that the intervention significantly influenced farmers in the adoption of some quality-enhancing practices. Moreover, the intervention induced important behavioral changes among the treated farmers: they sold more aromatic varieties of paddy outside the village than the control farmers and received a higher sales price. Thus, we conclude that the provision of information about paddy quality and quality-based pricing improved farmers' paddy production management and market sales. It is noteworthy that even a low-cost intervention without any technical training was able to generate sufficiently desirable outcomes.

T. Ogura (✉) · T. Sakurai
Department of Agricultural and Resource Economics, Graduate School of Agricultural and Life Sciences, University of Tokyo, 1-1-1 Yayoi, Yayoi Campus, Bunkyo-ku, Tokyo 113-8657, Japan
e-mail: tatsuya.23.0125.ogu@gmail.com

T. Sakurai
e-mail: takeshi-sakurai@g.ecc.u-tokyo.ac.jp

J. A. Awuni
Department of Economics, School of Applied Economics and Management Sciences, University for Development Studies, Nyankpala Campus, P.O. Box TL1350, NT-0272-1946 Tamale, Northern Region, Ghana

© JICA Ogata Sadako Research Institute for Peace and Development 2023
K. Otsuka et al. (eds.), *Rice Green Revolution in Sub-Saharan Africa*, Natural Resource Management and Policy 56, https://doi.org/10.1007/978-981-19-8046-6_13

13.1 Introduction

Rice production in sub-Saharan Africa (SSA) has dramatically increased over the last decade (Africa Rice Center 2017; CARD 2018). However, it is insufficient to satisfy the increasing demand from urban consumers who prefer high-quality imported milled rice (Demont 2013; Demont and Neven 2013). The gap in quality between imported and domestic rice suggests that rice producers in SSA are not fully capturing the emerging economic opportunities, as urban consumers are willing to pay a premium price for domestically produced and milled rice if its quality is comparable to imports (Rutsaert et al. 2013; Demont and Ndour 2015; Fiamore et al. 2017). Therefore, quality improvement of domestic rice is an urgent issue in SSA that needs to be addressed to promote rice production further.

Consumers are concerned with the quality of the milled rice that they purchase and consume. In recent years, modern rice milling facilities have emerged in SSA to meet consumer demand (Soullier et al. 2020; Mano et al. 2022). Yet since the quality of paddy is not sufficiently high, such facilities alone cannot produce high-quality milled rice comparable to imports, and hence, rice millers are looking for high-quality paddy as inputs. Generally, paddy quality depends on farmers' rice production practices, including varietal choice and post-harvesting activities—such as drying and storing—as the types of seeds and physical characteristics of paddy are major determinants of milled rice quality[1] (Futakuchi et al. 2013; Rutsaert et al. 2013).

However, rice farmers are still producing low-quality paddy in SSA. Conceptually, both supply (producer) and demand (trader or consumer) sides constrain farmers from producing higher quality agricultural products (de Janvry and Sadoulet 2020; Abate et al. 2021). If paddy price does not depend on paddy quality due to the absence of demand for quality rice, it would be natural for farmers not to undertake any efforts to produce higher quality products. However, this is not necessarily the case. For example, if we ask farmers, they usually say that paddy quality affects the price in our study site[2] (Ogura et al. 2020). Instead, supply-side factors seem to matter. Indeed, farmers do not seem to know how to improve paddy quality to obtain higher prices. Firstly, this is because many rice production projects and extension activities, either foreign or domestic, have focused on enhancing the quantity of rice production rather than quality,[3] and secondly, farmers do not know which particular paddy quality is appreciated by buyers. The latter situation is typical of local staple food markets in rural SSA, where traders and millers are small-scale, and their transactions are in

---

[1] The types of seeds are based on genetic traits such as grain color, grain shape, aroma, and the extent of their mixture. Physical characteristics of harvested paddy are moisture content, degree of crack, maturity, color and damage to the grain, foreign materials such as stones and husks, and rice mixed in pre-milled paddy rice.

[2] For example, damaged paddy cannot be sold at the same average price as undamaged paddy.

[3] Some projects in the study area (e.g., AGRA project) of this paper have provided information on paddy quality and quality-enhancing technologies in recent years. They mainly conduct projects in irrigated rice production areas, which comprise only a relatively small part of this region. However, this study focuses on rainfed rice production since it dominates the supply of paddy in this region.

small lots and infrequent (Dillon and Dambro 2017; Bergquist and Dinerstein 2020; Bold et al. 2022). As it is very costly for farmers to investigate and compare the quality of paddy over space and time, a consistent relationship between price and quality cannot emerge. Thus, we hypothesize that the lack of adequate knowledge is a major constraint on upgrading the rice quality in SSA.

From these discussions, a question arises: if farmers understand that there is an opportunity to sell paddy to buyers that pay a premium for paddy quality, will they improve their paddy quality and sell it to such buyers at a better price? This chapter seeks to answer this question through a randomized controlled trial implemented in the northern region of Ghana. It shows that the provision of information about paddy quality and quality-based pricing improved farmers' paddy production management and market sales.

The remainder of this chapter is organized as follows. In Sect. 13.2, we review the literature. Following this, our data collection methods and intervention design are explained in Sect. 13.3. Data descriptions are given in Sect. 13.4, while Sect. 13.5 provides regression analyses and the results. Section 13.6 offers some concluding comments on this chapter.

13.2 Review of Related Literature

This chapter considers rice farmers' responses to a short training session, which provides them with information about paddy quality improvement and its pricing. We hypothesize that the impact of such new information will influence farmers' behavior in the following three step-wise aspects: first, farmers will update their knowledge of paddy quality and adopt quality-enhancing practices; second, such practices will improve paddy quality; and third, farmers will sell the improved paddy to buyers at a higher price.

With respect to the first step, this study belongs to the literature on agricultural technology adoption in developing countries. The vast number of studies provide various explanations of why potentially profitable agricultural technologies are not adopted or abandoned within a short period (e.g., see Magruder (2018) and Takahashi et al. (2020) for reviews). However, most existing studies have focused on quantity-increasing technologies with less focus on quality-enhancing technologies, although the number of studies on the latter issue has been gradually increasing recently.[4] Among these, the studies that are closely related to ours are Bernard et al. (2017) and Bold et al. (2022).

Bernard et al. (2017) analyzed the impact of quality certification (labeling) on the adoption of quality-enhancing technologies in the case of onions in Senegal.

---

[4] These studies are on onions in Senegal (Bernard et al. 2017), groundnuts in Ghana and Senegal (Magnan et al. 2021; Deutschmann et al. 2021), maize in Uganda (Bold et al. 2022), raw milk in Vietnam (Saenger et al. 2013), and dragon fruit in Vietnam (Park 2021; Park et al. 2021).

Their randomized intervention announced the upcoming introduction of a quality-certifying system, and they found that it induced farmers to adopt pre-harvest quality-enhancing farming practices (i.e., optimal compositions of chemical fertilizers). Bold et al. (2022) examined the impact of an extension service that aimed to enhance quality with and without an experimental quality premium in the case of maize in Uganda. They found that the extension service combined with quality premium induced farmers to adopt multiple dimensions of quality-enhancing technologies (especially post-harvest technologies), even though the extension service alone did not change farmers' behavior.[5]

Both studies were conducted in study sites where pricing with a quality premium did not exist, but quality-enhancing practices existed and were well known to farmers. These conditions are similar to our study in northern Ghana. However, our case differs from theirs in two ways. First, paddy quality is determined by several elements: different kinds of pre- and post-harvest technologies affect paddy quality, which is an important determinant of product (i.e., milled rice) quality. Therefore, the goal of improving paddy quality is much more complicated than with onions or maize. Second, the quality-based pricing we introduce to farmers is a real one that has been used by a large-scale private rice miller in our study site. On the other hand, quality certification in Senegal is a public regulation: although it had not been implemented at the time of their study, it was expected to become compulsory for all the onion producers in the study area. The quality premium in Uganda is just a hypothetical setting introduced for their study in cooperation with a private agrochemical company. Thus, this study, which deals with an existing pricing scheme adopted and implemented by a private firm, differs from the two previous studies.

As for the second step, it is important to examine whether the quality-enhancing technologies actually improve quality. This question has also been investigated by Bernard et al. (2017), Magnan et al. (2021), and Bold et al. (2022).[6] Bernard et al. (2017) assessed product quality as part of the experiment and found that their intervention increased the share of onions judged as "good." Bold et al. (2022) examined the impact of a quality-enhancing service package on product quality and found that their intervention improved the quality of maize. On the other hand, in Magnan et al. (2021) whose concern is food safety, finding ways to improve product quality (i.e., to reduce aflatoxin contamination in groundnut) was not well known to farmers at the time of the study. They found that their experimental intervention reduced aflatoxin contamination in groundnut in northern Ghana only when the technology information came with a drying sheet.

Our case should fall between the three studies regarding the prevalence of quality-enhancing technologies before the intervention: most rice farmers knew the technologies that may improve paddy quality to some extent, but they did not know

---

[5] These findings are similar to Magnan et al. (2021), but the main concern of Magnan et al. (2021) is food safety (aflatoxin contamination in groundnut).

[6] The emerging literature has examined the quality upgrading of food products in developing countries. In addition to the literature mentioned in Footnote 4, other papers examined raw milk quality in Indonesia and India (Treurniet 2021; Rao and Shenoy 2021).

precisely what quality parameters would be used to assess the quality of paddy in the market. As for the nature of quality-enhancing technologies, the technologies affect plant growth in the case of onions in Senegal, while in the case of maize in Uganda and the case of groundnut in Ghana, the technologies relate to the improvement of post-harvest treatment and storage. On the other hand, in the case of paddy in Ghana, the quality-enhancing technologies include a wide range of technologies from pre-planting to post-harvest/storage, and naturally the quality parameters that each technology improves differ. Because of the complicated relationship between the multiple quality-enhancing technologies and the multiple quality parameters, the impact of quality-enhancing practices on paddy quality may not be so clearly identified.

The third step concerns the relationship between paddy quality and its price. Namely, the question is to what extent paddy quality improvement translates into higher paddy prices. In the case of milled rice, several studies investigated consumer preferences toward rice in SSA by using experimental auctions (Demont et al. 2013, 2017; Demont and Ndour 2015; Diagne et al. 2017; Fiamore et al. 2017). They found that urban consumers prefer rice that is white, long-grain, aromatic, and high swelling, as well as less starchy and less broken. It also should have fewer contaminants, a soft texture, and a shorter cooking time. Such consumer preferences for milled rice should be reflected in the price when farmers sell paddy to traders or millers.[7] However, few studies support or deny this conjecture, although anecdotal evidence (either for or against) can be obtained from the field.[8] Since most paddy rice transactions between farmers and traders are carried out at the farmgate at the study site, collecting paddy samples and recording the paddy price for each transaction in the field is challenging. In addition, since adopting quality-enhancing practices is each farmer's choice, analyzing the impact of the adoption of such practices on paddy quality and price should lead to an endogeneity problem. Therefore, our approach experimentally introduces information about knowledge of paddy quality and a quality-based pricing scheme to treatment villages. It then assesses its impact on farmers' paddy quality and the sale price of paddy, regardless of their adoption of quality-enhancing practices, i.e., intention-to-treat impact.

Overall, our study differs from the three related studies—Bernard et al. (2017), Magnan et al. (2021), and Bold et al. (2022)—in three ways. First, the relationship between production practices and product quality is more complicated in our case than in the case of the three papers. Second, farmers in our study site can choose the

---

[7] Even in the retail market, it is unclear if milled rice price reflects the consumers' preferences. Regardless of product quality, uniform pricing is often observed in the agricultural retail market in SSA (Bergquist and Dinerstein 2020; Bold et al. 2022). In this regard, Ibrahim et al. (2020) showed that an experimental introduction of a wholesaler who adopted quality-based pricing in retail markets in Kumasi, the central region of Ghana, induced the establishment of quality-based pricing among retailers in the markets where the intervention occurred.

[8] In other staple food markets in SSA, Bergquist and Dinerstein (2020) and Bold et al. (2022) found no exact relationship between price and quality in the local maize market in the case of Kenya or Uganda. Do Nascimento Miguel (2022) found that there was a positive relationship between price and observable quality (not in unobservable quality) in local wheat markets in the case of Ethiopia.

type of buyer when they sell paddy, from either those who continue the traditional method of transactions by paying little attention to paddy quality or those who determine the purchase price of paddy depending on its quality. Third, all three papers analyze data from experiments in which they provided direct training or support for adopting quality-enhancing technologies. Our study only provided information on quality-enhancing technologies without providing any direct training or support for production. Thus, our contribution is to investigate the possibility of upgrading the multidimensional quality of staple foods under a natural market environment.

13.3 Data

13.3.1 Data Collection

Data sets used in this study were mainly collected as a part of the Coalition for African Rice Development (CARD) project, "An Empirical Analysis on Expanding Rice Production in Sub-Saharan Africa," funded by the JICA Research Institute.[9] The JICA research team, with researchers from the University for Development Studies (UDS) and the University of Tokyo, conducted data collection.

In March 2018, the research team conducted a stratified random sampling of rice farmers in the following way. First, we drew a 54 km by 54 km square on a 1/50,000 scale topographic map, at the center of which there is a large-scale rice milling plant,[10] and identified 435 villages within the square. Second, we divided the square into nine blocks of 18 km by 18 km square and randomly selected 12 villages from each block to obtain 108 sample villages.[11] Finally, we created a list of rice farmers in each sample village and randomly drew ten farmers as sample farmers.

---

[9] This project is also partially supported by JSPS scientific research grant (grant number: 16H02733).

[10] We chose this particular large-scale miller as the center of the study area because we adopted the pricing scheme used by this miller. When we designed our experimental study in March 2018, there was only one large-scale miller with modern milling facilities in this region, and its sole status continued until the end of our data collection in January 2021. As shown in Fig. 13.1, the miller is located close to Tamale, the capital city of the northern region where the central market is situated, and hence, the villages we selected for the study are around Tamale. The private milling facilities were established in this location in 2011 and, in 2014, the milling company introduced a new pricing scheme for purchasing paddy from farmers, in which the company determines the purchasing price based on their evaluation of paddy quality. For example, when the authors visited the company in January 2018, they observed a notice indicating that the company's purchase price is 1.25 GHS/kg for aromatic paddy and 1.10 GHS/kg for non-aromatic paddy in the case of grade A paddy, and 1.10 GHS/kg and 1.05 GHS/kg in the case of grade B paddy (The exchange rate was about 1 GHS = 0.22 USD in January 2018). Unlike in local markets, where seasonal price changes are significant, the paddy purchase price of the company is constant throughout the year. Such a pricing scheme attracts more paddy in the harvesting season and allows the rice milling plant to collect better-quality paddy.

[11] The number of sample villages in each block is determined by the data collection capacity of the research team.

This process generated 1,080 randomly selected sample rice farmers. Figure 13.1 shows the location of sample villages and the large-scale rice milling plant.

The baseline survey was conducted in August and September 2018 (see the timeline in Fig. 13.2). The survey gathered information about sample villages and farmers, including production and sales in the previous year, i.e., the 2017 season. Following the baseline survey, an experimental intervention (to be explained in the next section) was implemented through a group meeting in November and December 2018.

Rice production in the 2019 season was the first production after the intervention. Thus, we collected paddy samples from each participating farmer in November and

Fig. 13.1 Location of sample villages

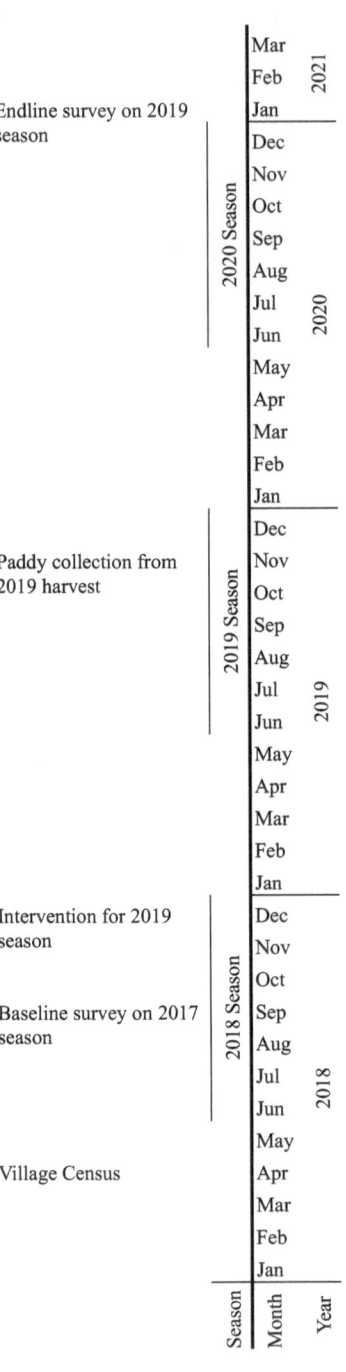

Fig. 13.2 Timeline

December 2019 and had their quality analyzed[12] by specialists at SARI (Savanna Agricultural Research Institute) near the large-scale milling plant. Finally, endline information regarding rice production and sales in the 2019 season was collected in December 2020 and January 2021.

13.3.2 Intervention

In order to implement an experimental intervention, we first randomly assigned treatment status to a half of the sample villages. The assignment was stratified by a block unit with nine blocks. Six villages from 12 villages were randomly selected for the treatment in each block. As a result, we had 54 treatment villages and 54 control villages.

In November and December 2018, we conducted a group meeting in each of the treatment villages to provide sample farmers with information on the existence of and details about the pricing system adopted by the large-scale rice milling plant.[13] The meeting was conducted in cooperation with the manager of the large-scale rice milling plant and agricultural extension agents of MoFA (Ministry of Food and Agriculture) assigned to treatment villages. All the sample households were encouraged to participate in the group meeting in the treatment villages.[14]

In the group meeting, the research team provided two kinds of information. The first one comprised business information, and the second one consisted of technical information about paddy quality.[15] The business information includes the concept of the grading and pricing schemes, the plant's location, the procedure to sell paddy

---

[12] The paddy quality analyses adopted the same eight parameters that the large-scale rice miller used, assessing aroma, moisture content, cracked rice, foreign materials, red grain, variety mixture, discolored grain, and immature grains.

[13] According to an official announcement from the large-scale rice milling company, the company will purchase paddy from any rice farmers or entities who bring paddy to the company. However, very few small-scale rainfed rice producers around the mill knew about the quality-based pricing in 2017 when we started casual interviews with them.

[14] Although most of the sample farmers participated in the group meeting, we provided the same handout used in the meeting to those who were absent from the meeting.

[15] It is important to note that "quality" used in this paper refers to something observable in paddy that potentially influences the sale price after milling. The important point about the quality grading system is that most parameters depend on farmers' production, harvesting, and post-harvest practices, although some parameters are additionally affected by weather conditions during production. Also note that the large-scale milling plant indicates the eight parameters, but local traders do not necessarily use all of the parameters in trading with farmers. Among the eight parameters, the aroma is mainly determined by rice varieties, and aromatic rice varieties have become popular in the Tamale area, as revealed in Ogura and Sakurai (2019) and He and Sakurai (2019). However, in 2012 when their data were collected, it was not yet clear whether local traders had paid a higher price for the paddy of aromatic rice than non-aromatic rice.

to the plant, and the contact phone number of the manager. The technical information includes the definition of eight quality parameters that the system uses to evaluate paddy quality (as explained in Footnote 12), the determinants of eight quality parameters, and recommended practices for attaining good paddy quality.[16] The two components are the same package the company uses in regular business promotions. The information was given to farmers not only orally but also through a handout written in the local Dagbani language.[17]

The exposure to the group meeting creates an exogenous variation in access to information about paddy quality and sales opportunities. Thus, we hypothesize that the intervention will induce farmers to update their knowledge of paddy quality, change their rice farming practices, improve paddy quality, and obtain higher paddy prices.

13.4 Data Descriptions

13.4.1 Descriptive Statistics of Baseline Characteristics

Table 13.1 shows the comparisons of baseline characteristics between the treatment and the control groups. Columns (1) and (2) are full samples interviewed during the baseline survey, and Columns (3) and (4) are subsamples that were available for the endline survey and used panel data analysis for this study.

Farmer (i.e., the household member responsible for rice cultivation) characteristics include gender, age, English literacy, household head status, birthplace, polygamous status, health condition, rice cultivation experience, ethnic group, and household asset holdings. Village characteristics are represented by the distance in km from the village to the central market of Tamale. As for the characteristics of the rice plots, plot ownership, slope, soil, water source, flood frequency, and plot size are compared. As shown in Columns (1) and (2) of Table 13.1, there is no significant difference between the two groups on average at the time of the baseline survey, except for the share of the Dagomba ethnic group and plots with a steep slope. Thus, apart from these aspects, the random samples are well balanced.[18]

---

[16] These are pre-planting practices (use of aromatic varieties, use of certified seed, seed selection in salt water, and uniform plant growth by leveling), pre-harvesting practices (timely harvesting by the use of machine harvester), and post-harvesting practices (threshing and quick drying on tarpaulin and winnowing). Most practices were already known to rice producers in the study site, but the intervention specifically told them about the relationship between those practices and the eight quality parameters.

[17] Dagbani language is spoken by the Dagomba ethnic group and is the dominant language in terms of the proportion of speakers in the study site.

[18] The total number of farmers selected from the list of rice farmers was 1080 (10 farmers × 108 villages). However, the survey team could not reach some, and hence, the total number of samples at the baseline survey is 1065.

Table 13.1 Household/plot baseline characteristics

	Randomly selected samples			Samples in the panel data		
	(1) Treatment	(2) Control	P-value	(3) Treatment	(4) Control	P-value
A: Household (respondent's) characteristics						
Gender (Male = 1, Female = 0)	0.95	0.96	0.38	0.95	0.96	0.37
Age (years)	41.0	41.4	0.49	40.9	41.5	0.41
English literacy (speak, read, and write) (1 or 0)	0.08	0.08	0.63	0.08	0.09	0.62
Household head (1 or 0)	0.67	0.67	0.95	0.67	0.67	0.86
Born in the current village (1 or 0)	0.77	0.78	0.70	0.78	0.79	0.71
Polygamously married (1 or 0)	0.44	0.45	0.92	0.44	0.44	0.91
Good health status (1 or 0)	0.98	0.98	0.84	0.98	0.98	0.64
Rice cultivation experience (years)	13.4	12.6	0.16	13.3	12.6	0.21
Value of asset holdings (1000 GHS)	13.3	17.9	0.34	13.6	15.6	0.65
Ethnic group (Dagomba = 1, others = 0)	0.89	0.84	0.01***	0.89	0.84	0.01***
Distance to the central market (km)	20.85	21.04	0.70	20.87	21.17	0.54
Number of observations[a]	534	531		514	509	
B: Plot characteristics (not used in regression analyses)						
Household's own plot (1 or 0)	0.81	0.78	0.19	0.83	0.80	0.35
Steep sloping (1 or 0)	0.05	0.07	0.09*	0.04	0.07	0.10
Fertile soil (1 or 0)	0.35	0.39	0.18	0.33	0.40	0.03**

(continued)

Table 13.1 (continued)

	Randomly selected samples			Samples in the panel data		
	(1) Treatment	(2) Control	P-value	(3) Treatment	(4) Control	P-value
Rainfed plot (1 or 0)	0.95	0.97	0.14	0.95	0.96	0.32
High frequent flood (1 or 0)	0.025	0.016	0.34	0.023	0.009	0.11
Rice plot size (ha)	1.41	1.26	0.27	1.40	1.24	0.23
Number of observations[b]	480	492		426	423	

$*p < 0.1, **p < 0.05, ***p < 0.01$

[a]We interviewed 1065 rice farmers during the baseline survey in 2018, but could only meet 1023 of them for the endline survey in 2020/21 because of death, sickness, long-term absence, rejection, etc. Thus, the 1023 farmers constitute a two-year panel for the analyses

[b]Some plot characteristics were intended to be obtained by actually visiting the plot in question with GPS equipment, and as a result we failed to obtain data from some sample plots. Because of the missing data, this paper does not use the plot characteristics for the analyses

Columns (3) and (4) compare the baseline characteristics of the sample farmers interviewed for the endline survey. Forty-two farmers were unavailable at the time of the endline survey due to being deceased, sick, absent long-term, or rejecting participation. Despite the attrition, it does not seem to have affected the balance between the treated and the control groups since the significance level of mean difference is almost the same for the baseline characteristics except for some plot characteristics.[19]

13.4.2 Descriptive Statistics: Outcome Variables

Table 13.2 shows the descriptive statistics of the outcome variables of interest that we will use in the regression analysis in the next section.

Knowledge

Panel A of Table 13.2 is for knowledge about paddy quality that we taught in the group meeting as an indicator of the effectiveness of the intervention. We use two knowledge variables: knowledge of good practices necessary for obtaining a good quality of paddy rice (i.e., quality-enhancing practices) and knowledge about the measurements of paddy quality (i.e., quality parameters). In the interview, we asked farmers to mention quality-enhancing practices and paddy quality parameters that they know.

[19] This does not mean that the attrition happened randomly. As shown in Table 13.6 in the Appendix, the baseline characteristics of the samples attrited are significantly different from the remaining samples in some aspects.

Table 13.2 Descriptive statistics of outcome variables

	Before intervention			After intervention			
	(1) Treatment	(2) Control	P-value	(3) Treatment	(4) Control	P-value	(5) DID
A: Knowledge of paddy quality							
Quality-enhancing practices (score: 0–7)	3.06	3.46	0.00***	3.48	3.64	0.12	0.25**
Paddy quality parameters (score: 0–9)	2.70	2.98	0.00***	3.33	3.23	0.35	0.36***
Number of observations	514	509		514	509		
B: Quality-enhancing practices							
Use of aromatic varieties (0 or 1)	0.35	0.31	0.11	0.38	0.34	0.09*	0.00
Use of certified seed (0 or 1)	0.08	0.12	0.04**	0.09	0.13	0.02**	-0.00
Seed selection by salt water (0 or 1)	0.002	0.002	0.50	0.004	0.004	0.50	-0.00
Plot leveling (0 or 1)	0.025	0.004	0.00***	0.023	0.003	0.00***	-0.00
Machine harvesting (0 or 1)	0.029	0.002	0.00***	0.039	0.002	0.00***	0.01*
Threshing on tarpaulin sheet (0 or 1)	0.48	0.44	0.15	0.51	0.44	0.02**	0.04***
Number of observations	514	509		514	509		
C: Paddy quality (laboratory examination)							
Foreign material (weight %)	NA	NA	NA	0.70	0.73	0.67	NA
Mixture of varieties (weight %)	NA	NA	NA	2.52	2.01	0.08*	NA
Aromatic (0 or 1)	NA	NA	NA	0.53	0.46	0.07*	NA
Damaged rice (weight %)	NA	NA	NA	68.5	68.7	0.93	NA
Red rice (weight %)	NA	NA	NA	16.2	16.3	0.97	NA
Immature rice (weight %)	NA	NA	NA	1.48	1.30	0.66	NA
Dehusked rice (weight %)	NA	NA	NA	0.68	0.69	0.83	NA

(continued)

Table 13.2 (continued)

	Before intervention			After intervention			
	(1) Treatment	(2) Control	P-value	(3) Treatment	(4) Control	P-value	(5) DID
Cracked rice (weight %)	NA	NA	NA	68.2	64.5	0.15	NA
Moisture (too wet) (% point)	NA	NA	NA	0.31	0.07	0.03**	NA
Moisture (too dry) (% point)	NA	NA	NA	3.83	4.30	0.06*	NA
Moisture (too wet/dry combined)	NA	NA	NA	4.14	4.37	0.22	NA
Number of observations				419	405		
D: Paddy sales							
Sales outside the village (0 or 1)	0.52	0.58	0.15	0.62	0.50	0.00***	0.18***
Sales of Aromatic varieties (0 or 1)	0.17	0.28	0.00***	0.44	0.42	0.60	0.12***
Paddy sale price (GHS/kg)	1.25	1.34	0.01**	1.17	1.10	0.07*	0.17***
Paddy price net costs (GHS/kg)	1.24	1.33	0.01**	1.17	1.10	0.06*	0.16***
Total paddy weight sold (kg)	1648	1397	0.14	1522	1221	0.10*	50.0
Paddy sales revenue (GHS)	2178	1928	0.38	1555	1331	0.10*	−24.7
Number of observations[a]	373	345		373	345		

Note DID is {(3)–(4)}–{(1)–(2)} for Panel A, B, and D; 1 GHS = 25 JPY in January 2018
*$p < 0.1$, **$p < 0.05$, ***$p < 0.01$
[a]We collected rice sales data from all the sample farmers who sold rice (either paddy or in any other form, such as milled rice or parboiled rice). However, for this analysis we use only farmers who sold paddy. Total number of farmers who sold rice is 926 at the baseline and 841 at the endline. Among these, 885 and 827 farmers sold paddy in the respective surveys. After eliminating some observations with outlier sale prices, a total of 718 constitute a balanced panel data for the analysis. We had to drop a significant number of observations because many farmers sold paddy in only one of the survey years

Then, we use the aggregated number of correct items as the measure of the two kinds of knowledge. As shown in Columns (1) and (2) of Panel A, for an unknown reason, farmers in the treatment villages had significantly less knowledge of both quality-enhancing practices and quality parameters before the intervention. However, after the intervention, the knowledge gaps became smaller, and the statistical significance of the gaps disappeared, as shown in Columns (3) and (4). Simple DID estimates confirm that the change in the treatment group is significantly larger than in the control group (Column (5)). Overall, the farmers who received the intervention tended to have increased knowledge, suggesting the intervention was effective.

Quality-Enhancing Practices

Quality-enhancing practices include pre-planting, harvesting, and post-harvesting practices, which are measured based on the use of each technology at the household level.

Pre-planting practices comprise "use of aromatic varieties," "use of certified seed" and "seed selection in salt water"—all of which are binary dummies. The dummy for aromatic varieties takes 1 if the farmer states that they planted AGRA or Jasmine 85, which are aromatic rice varieties that have been disseminated as improved, high-yielding varieties at the study site. As shown in Panel B of Table 13.1, more than 30% of farmers grew aromatic rice. The share of farmers who grew aromatic rice became significantly higher in treatment villages than in control villages after the intervention, but no DID impact is observed. On the other hand, the share of certified seed use was significantly higher in control villages than in treatment villages before as well as after the intervention. However, according to DID estimates, the differences were not significant following the intervention. As for seed selection, the adoption rate was very low and did not differ before or after the intervention. Plot leveling had been introduced in the study site as a yield-enhancing technology, but it also contributes to quality-enhancing thanks to its effect on the uniform growth of plants. Although the adoption rate was significantly higher in treatment villages before and after the intervention, the adoption rates were generally low and were not affected by the intervention. Thus, overall, the intervention seems to have had little impact on pre-planting quality-enhancing practices.

Harvesting and post-harvesting practices are represented by "machine harvesting" and "threshing on tarpaulin sheet." Machine harvesting (i.e., harvesting using combine harvesters) is not just for labor-saving but will improve paddy quality by enabling farmers to avoid excessively dry paddy due to harvest delays, while the use of a tarpaulin sheet is designed to avoid contamination from the soil. The adoption rates of these practices were different both before and after the intervention: they are generally low in the case of machine harvesting (i.e., the use of a combine harvester), while they sit at more than 40% in the case of using a tarpaulin sheet for threshing. However, the intervention had a positive and significant impact on both practices. Therefore, the intervention seems to have increased the adoption of harvesting/post-harvesting quality-enhancing practices.

Paddy Quality

We use paddy quality parameters as the outcomes for paddy quality. We obtained a paddy sample only after the intervention; hence, there is no baseline data. The quality parameters are foreign material (% by weight), a mixture of varieties (% by weight), detection of aroma (binary dummy), damaged rice (% by weight), immature rice (% by weight), dehusked rice (% by weight), and cracked rice (% by weight). Moisture (excessively wet) indicates to what extent the sample paddy is wetter than the optimal moisture content (12–14%) measured by the absolute value of the difference in moisture content (if the sample is drier than the optimal, the value is set to be zero). Similarly, moisture (excessively dry) indicates the extent of dryness by the absolute difference in moisture content on the dry side (i.e., the drier, the larger the value becomes). While "excessively wet" can be reduced by drying under the sun, "excessively dry" cannot be adjusted once the paddy becomes too dry, causing broken rice after milling. Therefore, we consider that they are different problems. However, we created a combined indicator of suboptimal moisture content, referring to the absolute value of the percentage point difference from the optimal level.

As shown in Panel C of Table 13.1, which offers a simple mean comparison after the intervention, only two parameters, i.e., the use of aromatic varieties and moisture content (excessively dry), are better for farmers in treatment villages than those in control villages. On the other hand, a mixture of varieties and moisture content (excessively wet) are better in control villages. Higher aroma detection and lower dryness in the treatment villages seem to be consistent with the quality-enhancing practices shown in Panel B—namely, farmers in treatment villages tend to use aromatic varieties and harvesting machines after the intervention. On the other hand, the higher mixture rate of different varieties in treatment villages also seems to reflect the lower use of certified seeds in treatment villages, as shown in Panel B. Although the use of tarpaulin for threshing is higher in the treatment villages after the intervention, as shown in Panel B, it does not significantly reduce the contamination of foreign material. Thus, although the overall quality of paddy is not as high in treatment villages compared to control villages, we can observe some improvement in important parameters, namely aroma and moisture (dryness side), which is consistent with the change in quality-enhancing practices.

Paddy Sales

For the analysis of the rice sales, the rice sales data are averaged at the household level. Since we use farmers who sold paddy (not in other forms like milled rice and parboiled rice) both in 2017 (recorded in the baseline data) and 2019 (recorded in the endline data), the number of observations in each year is 718. One of the outcomes is whether farmers always sold paddy outside the village, which includes cases where farmers brought paddy to a market or a miller outside the village and cases where farmers sold paddy to buyers coming from outside the village. This behavior is considered to involve seeking a better price compared to selling paddy to a local buyer who lives in the village. Other outcomes comprise whether they always sold aromatic varieties, the average paddy sale price (GHS/kg), the average paddy sale price net of transportation costs (GHS/kg), the total amount of paddy sales (kg), and the total amount of paddy sales revenue (GHS). Panel D of Table 13.2 shows that farmers in treatment villages increased paddy sales outside of the village and sales of aromatic varieties much more than those in control villages. For unknown reason the paddy price went down during the two surveys in the study site, but the price decline was much smaller among farmers in the treatment villages than the control villages.

13.5 Regression Analysis

13.5.1 Regression Specifications

The impact of the exposure to the exogenous information provision on the farmers' decisions is statistically assessed using an ANCOVA model (McKenzie 2012) as follows:

$$Y_{ikm1} = \alpha + \beta Y_{ikm0} + \delta T_{km} + \gamma X'_{ikm0} + \omega_m + \varepsilon_{ikm} \tag{13.1}$$

where subscript i is for household, k is for village, and m is for block. Time indicator takes 1 if the observation is of the endline or takes 0 if it is of the baseline. Y_{ikm1} and Y_{ikm0} are the outcomes in the 2019 season and in the 2017 season, respectively. T_{km} is the indicator variable that takes the value of 1 if the village is assigned to the treatment. X_{ikm0} is the set of control variables of characteristics of farmers at the baseline, including the distance (km) from the village to the central market in Tamale, which is a village-level variable. In addition, ω_m is block fixed effect, α is constant, and ε_{ikm} is the error term. In the case of paddy sales, each sale made by sample farmers was recorded. However, for the analysis, they are averaged at the household level for each year, with Eq. (13.1) also applied. Standard errors are clustered at the village level, which is the unit of the random assignment to the treatment, which accounts for the correlation in error terms within villages. Given the exogenous assignment of T_{km}, the coefficient δ is interpreted as the average treatment effect. As noted above, some treatment sample farmers did not attend the group meeting, so the effect is considered as an intention-to-treat effect.

Since we measured the quality of the paddy sample only after the intervention, we do not apply the ANCOVA model specified as Eq. (13.1). Instead, we simply conduct a mean comparison between the two groups as shown in Panel C of Table 13.2. An OLS regression can be applied, but since the results are qualitatively very similar, this study does not show regression results. As paddy samples were obtained from each plot of the sample households, the outcome variables are at the plot level.

13.5.2 Regression Results

Knowledge

Table 13.3 shows that the impact of our intervention on both kinds of knowledge, either quality-enhancing practices or paddy quality parameters, is not statistically significant in spite of the significant estimates of simple DID as shown in Panel A of Table 13.2. As shown in Fig. 13.2, the endline survey was conducted two years after the intervention. Thus, the insignificant impacts may imply that the knowledge tends to be lost after two years. However, it does not necessarily mean that the intervention

Table 13.3 Impact on knowledge about paddy quality

	(1) Knowledge: quality-enhancing practices	(2) Knowledge: paddy quality parameters
Intervention	−0.015 (0.240)	0.212 (0.200)
Household-level baseline variables	Yes	Yes
Block fixed effect	Yes	Yes
Number of observations	1023	1023

Source Author's own
Note Standard errors clustered at the village level are in parentheses

was ineffective for rice production or sales because rice production started only six months after the information provision, as shown in Fig. 13.2. In fact, as will be shown in the following sections, farmers in treatment villages adopted some of the quality-enhancing practices and obtained higher sales prices.

Technologies

Table 13.4 shows the adoption of the quality-enhancing practices. Significant DID estimates are obtained for "machine harvesting" and "threshing on tarpaulin sheet," as shown in Panel B of Table 13.2 and confirmed by ANCOVA regressions. The results indicate that the intervention had a positive impact on harvesting and post-harvesting technologies to improve paddy quality, but not pre-planting technologies. As discussed above, these significant changes in quality-enhancing practices seem to contribute to the reduction of excessive dryness, but not the reduction of foreign materials. As for the aroma, farmers may have increased the share of aromatic varieties and/or chosen better aromatic varieties. Such a change in practice is not reflected in the binary dummy for the use of aromatic varieties.

Sales Results

Finally, Table 13.5 shows the results of paddy sales. First, there is a statistically significant impact on paddy sales: farmers in treated villages sold paddy more outside their villages than the farmers in control villages (Column (1)). However, we cannot find any significant impact on the sales of aromatic varieties (Column (2)), which is consistent with Table 13.4, showing that the intervention did not increase the share of aromatic varieties significantly. As for the paddy sales price (net of transportation cost), the average price is higher in treatment villages than in control villages regardless of the sales outlets or rice varieties (Column (3)). Since sales prices should be affected by the sales outlet as well as rice varieties, we include those variables as explanatory variables. Because they are endogenous binary dummy variables, the intervention is used as an instrumental variable and an endogenous treatment model is applied. The results in Columns (4) and (5) indicate that both sales outside the village and aromatic varieties have a significantly positive effect on paddy sale prices. The coefficient for sales outside the village (0.662) seems to be too large compared to

Table 13.4 Impact on adoption of quality-enhancing practices at plot level

	(1) Plot leveling	(2) Certified seed	(3) Aromatic varieties	(4) Seed selection	(5) Machine harvesting	(6) Threshing on sheet
Intervention	0.002 (0.004)	−0.004 (0.004)	0.010 (0.022)	−0.000 (0.002)	0.016* (0.009)	0.037** (0.015)
Control variables	Yes	Yes	Yes	Yes	Yes	Yes
Block fixed effect	Yes	Yes	Yes	Yes	Yes	Yes
# of observations	1023	1023	1023	1023	1023	1023

Source Author's own
Note Standard errors clustered at the village level are in parentheses
*$p < 0.1$, **$p < 0.05$

the average paddy prices net of transportation cost (from 1.10 to 1.33) for unknown reasons. On the other hand, the coefficient for aromatic varieties (0.100) is reasonable compared with the aroma premium paid by the large-scale miller, as discussed in footnote 10. Thus, our intervention increases the paddy sales price through the change in sales location and the use of aromatic varieties. We also confirm that the intervention increases farmers' paddy sales volume and the revenue from the sales, as demonstrated in Columns (6) and (7).

13.6   Conclusion

This study examined the impact of the information provision regarding paddy quality and quality-based paddy pricing on farmers' rice production and marketing in the Northern region of Ghana through a randomized controlled trial. The effectiveness of the intervention is examined by assessing the change in farmers' knowledge about paddy quality-enhancing practices and paddy quality assessment parameters. Although simple DID estimations show significant impacts of the intervention on farmers' knowledge, regression results do not fully confirm it. Nevertheless, farmers who receive the paddy quality information seem to try to improve the quality, particularly through the adoption of harvesting/post-harvesting technologies. Although we cannot directly demonstrate the consequences of such technology adoption on paddy quality due to data limitations, there are positive associations between the intervention and improved quality parameters, such as more aroma and optimal moisture (less excessive dryness). Our analyses further found several important impacts on paddy marketing: farmers in treatment villages increased sales outside the village and aromatic varieties, and as a result, the average sale price of paddy is higher in treatment villages than in control villages. We consider that these changes in paddy

Table 13.5 Impact of intervention on paddy sales

	(1) Sales outside the village (dummy)	(2) Sales of aromatic varieties (dummy)	(3) Paddy sales price (GHS/kg)	(4) Paddy sales price (GHS/kg)	(5) Paddy sales price (GHS/kg)	(6) Total paddy sales in weight (kg)	(7) Total paddy sales in value (GHS)
Intervention	0.145** (0.056)	0.028 (0.045)	0.061* (0.036)	NA	NA	322* (178)	255* (142)
Sales outside the village	NA	NA	NA	0.662* (0.346)	NA	NA	NA
Sales of aromatic varieties	NA	NA	NA	NA	0.100*** (0.054)	NA	NA
Household control variables	Yes	Yes	Yes	Yes	Yes	Yes	Yes
Block fixed effect	Yes	Yes	Yes	Yes	Yes	Yes	Yes
# of observation	718	718	718	718	718	718	718

Source Authors' own
Note Standard errors clustered at the village level are in parentheses
$*p < 0.1$, $**p < 0.05$, $***p < 0.01$

marketing were induced by our information provision, demonstrating that there are opportunities to sell "better quality paddy" at higher prices outside the village. It is confirmed by the finding that either sales outside the village or sales of aromatic varieties increase paddy sales prices.

With respect to quality-enhancing technologies, the impact of our intervention was not substantial, and farmers need to learn more about good practices. Since our intervention was only a short training that provided farmers with information about technologies and did not show how to implement them in practice, this low and modest impact is not unexpected. Rather, we can say that our low-cost intervention can generate sufficiently good outcomes. If more technical training is done simultaneously, the impact could be much greater. It is therefore advisable that any technical training should include quality and marketing aspects for the participants to get a larger benefit.

Table 13.6 Comparison of baseline characteristics between remained and attrited

	(1) Remained	(2) Attrited	P-value
A: Household (respondent's) and village characteristics			
Gender (Male = 1, Female = 0)	0.96	0.95	0.89
Age (years)	41.2	40.6	0.72
English literacy (speak, read, and write) (1 or 0)	0.08	0.05	0.42
Household head (1 or 0)	0.67	0.64	0.73
Born in the current village (1 or 0)	0.78	0.57	0.00***
Polygamously married (1 or 0)	0.44	0.50	0.44
Good health status (1 or 0)	0.98	0.93	0.02**
Rice experience (years)	12.9	15.5	0.08*
Household value of asset holdings (1000 GHS)	14.6	39.5	0.04**
Ethnic group (Dagomba = 1, others = 0)	0.87	0.88	0.77
Distance to the central market from the village (km)	20.02	19.03	0.11
Number of observations	1023	42	
B: Plot characteristics			
Household's own plot (1 or 0)	0.82	0.67	0.00***
Steep sloping (1 or 0)	0.058	0.065	0.75
Fertile soil (1 or 0)	0.37	0.40	0.49
Rainfed plot (1 or 0)	0.96	0.94	0.47
High frequent flood (1 or 0)	0.016	0.049	0.02**
Rice plot size (ha)	1.32	1.45	0.50
Number of observations	849	123	

Source Authors' own
*$p < 0.1$, **$p < 0.05$, ***$p < 0.01$

References

Abate GT, Bernard T, de Janvry A, Sadoulet E, Trachtman C (2021) Introducing quality certification in staple food markets in Sub-Saharan Africa: four conditions for successful implementation. Food Policy 105:102173

Africa Rice Center (2017) Africa Rice Center Annual Report 2016: towards rice self-sufficiency in Africa, Abidjan, Côte d'Ivoire

Bergquist LF, Dinerstein M (2020) Competition and entry in agricultural markets: experimental evidence from Kenya. Am Econ Rev 110(12):3705–3747

Bernard T, de Janvry A, Mbaye S, Sadoulet E (2017) Expected product market reforms and technology adoption by Senegalese onion producers. Am J Agric Econ 99(4):1096–1115

Bold T, Ghisolfi S, Nsonzi F, Svensson J (2022) Market access and quality upgrading: evidence from three field experiments. Am Econ Rev, Conditional Acceptance

CARD (Coalition for African Rice Development) (2018) Coalition for African Rice Development (CARD) initiative Final Review Assessment, Final Report, May 2018. https://www.riceforafrica.net/. Accessed 24 July 2020

de Janvry A, Sadoulet E (2020) Using agriculture for development: supply-and demand-side approaches. World Dev 133:105003

Demont M (2013) Reversing urban bias in African rice markets: a review of 19 national rice development strategies. Glob Food Sec 2(3):172–181

Demont M, Neven D (2013) Tailoring African rice value chains to consumers. In: Wopereis MCS, Johnson DE, Ahmadi N, Tollens E, Jalloh A (eds) Realizing Africa's rice promise. CABI Publishing, Wallingford, pp 303–310

Demont M, Rutsaert P, Ndour M, Verbeke W, Seck PA, Tollens E (2013) Experimental auctions, collective induction and choice shift: willingness-to-pay for rice quality in Senegal. Eur Rev Agric Econ 40(2):261–286

Demont M, Ndour M (2015) Upgrading rice value chains: experimental evidence from 11 African markets. Glob Food Sec 5:70–76

Demont M, Fiamore R, Kinkpe AT (2017) Comparative advantage in demand and the development of rice value chains in West Africa. World Dev 96:578–590

Deutschmann JW, Bernard T, Yameogo O (2021) Contracting and quality upgrading: evidence from an experiment in Senegal. Working Paper

Do Nascimento Miguel J (2022) Return to quality in rural agricultural markets: evidence from wheat markets in Ethiopia. IFPRI Discussion Paper 2101. Washington DC. https://doi.org/10.2499/p15 738coll2.134984

Diagne M, Demont M, Ndour M (2017) What is the value of rice fragrance? Consumer evidence from Senegal. Afric J Agric Resour Econ 12(2):99–110

Dillon B, Dambro C (2017) How competitive are crop markets in sub-Saharan Africa? Am J Agric Econ 99(5):1344–1361

Fiamore R, Demont M, Saito K, Roy-Macauley H, Tollens E (2017) How can West African rice compete in urban markets? A demand perspective for policymakers. Euro Choices 17(2):51–57

Futakuchi K, Manful J, Sakurai T (2013) Improving grain quality of locally produced rice in Africa. In: Wopereis MCS, Johnson DE, Ahmadi N, Tollens E, Jalloh A (eds) Realizing Africa's rice promise. CABI Publishing, Wallingford, pp 311–323

He X, Sakurai T (2019) Transferability of green revolution in Sub-Saharan Africa: impact assessment of rice production technology training in Northern Ghana. Japanese J Agric Econ 21:74–79

Ibrahim L, Sakurai T, Tachibana T (2020) Local rice market development in Ghana: experimental sales of standardized premium quality rice to retailers Japanese. J Agric Econ 22:118–122

Magnan N, Hoffmann V, Opoku N, Garrido GG, Kanyam DA (2021) Information, technology, and market rewards: incentivizing aflatoxin control in Ghana. J Dev Econ 151:102620

Magruder JR (2018) An assessment of experimental evidence on agricultural technology adoption in developing countries. Ann Rev Resour Econ 10:299–316

Mano Y, Njagi T, Otsuka K (2022) An inquiry into the process of upgrading rice milling services: the case of the Mwea Irrigation Scheme in Kenya. Food Policy 106:102195

McKenzie D (2012) Beyond baseline and follow-up: the case for more T in experiments. J Dev Econ 99(2):210–221

Ogura T, Sakurai T (2019) Does a market-oriented variety change rice farming in SSA? Evidence from a new aromatic rice variety in northern Ghana. Japanese J Agric Econ 21:68–73

Ogura T, Awuni J, Sakurai T (2020) The impact of quality-based pricing scheme on local paddy transactions in the northern region of Ghana. Japanese J Agric Econ 22:147–151

Park S (2021) State certified lemons: a randomized intervention on agricultural quality certification. Working Paper

Park S, Yuan Z, Zhang H (2021) Technology, Training, buyer-supplier linkage, and quality upgrading in an agricultural supply chain. Working Paper

Rao M, Shenoy A (2021) Got (Clean) Milk? Transparency, governance, and incentives for cleanliness in Indian dairy cooperatives. Working Paper, Agricultural Technology Adoption Initiative. https://www.atai-research.org/got-clean-milk-transparency-governance-and-incentives-for-cleanliness-in-indian-dairy-cooperatives-2/

Rutsaert P, Demont M, Verbeke W (2013) Consumer preferences for rice in Africa. In: Wopereis MCS, Johnson DE, Ahmadi N, Tollens E, Jalloh A (eds) Realizing Africa's rice promise. CABI Publishing, Wallingford, pp 294–302

Saenger C, Qaim M, Torero M, Viceisza A (2013) Contract farming and smallholder incentives to produce high quality: experimental evidence from the Vietnamese dairy sector. Agric Econ 44(3):297–308

Soullier G, Demont M, Arouna A, Lançon F, Del Villar PM (2020) The state of rice value chain upgrading in West Africa. Glob Food Secur 25:100365

Takahashi K, Muraoka R, Otsuka K (2020) Technology adoption, impact, and extension in developing countries' agriculture: a review of the recent literature. Agric Econ 51(1):31–45

Treurniet M (2021) The potency of quality incentives: evidence from the Indonesian dairy value chain. Am J Agric Econ 103(5):1661–1678

Tatsuya Ogura is a post-doctoral fellow at Graduate School of Agricultural and Life Sciences, the University of Tokyo, Japan. He received a Ph.D. in agricultural and resource economics from the University of Tokyo in 2021. His area of research is agricultural technology adoption and its impact assessment.

Joseph A. Awuni is a senior lecturer at University of Development Studies, Tamale, Ghana. He obtained a Ph. D in management science and engineering from Jiangsu University, China in 2015. His area of research is agricultural economics.

Takeshi Sakurai is a professor of the University of Tokyo since 2014. He obtained a Ph.D. in agricultural economics from Michigan State University in 1995. His expertise is in agricultural economics and development economics.

Part IV
Conclusion

Chapter 14
Toward a Full-Fledged Rice Green Revolution in Sub-Saharan Africa

Keijiro Otsuka, Yukichi Mano, and Kazushi Takahashi

Abstract The main purpose of this book is to provide an appropriate strategy for realizing a rice Green Revolution in SSA with a special focus on the role of training and other complementary strategies, comprising the diffusion of power tillers, the expansion of irrigated areas, and the quality improvement of milled rice. Through the research presented here, we found that the central strategy for achieving this goal ought to be strengthening the extension system to promote rice farming intensification by adopting improved rice management practices. We also identified the important roles played by complementary strategies and policies to further strengthen those roles. We conclude that a full-fledged rice Green Revolution is possible if adequate supporting policies are implemented.

14.1 Introduction

It remains a puzzle why a rice Green Revolution has failed to take place in most areas of sub-Saharan Africa (SSA), even though it transpired in tropical Asia a half-century ago. It is still more puzzling if we recall the fact that Asian rice Green Revolution technology is highly transferable from tropical Asia to SSA. Historically, improved rice farming technology was transferred from Japan to Taiwan in the 1920s, to the Philippines in the 1960s, and further to other Southeast Asian countries and South

K. Otsuka (✉)
Graduate School of Economics, Kobe University, 2-1 Rokkodai-Cho, Nada-ku, Fourth Academic Building, 5th Floor, Room 504, Kobe 657-8501, Hyogo, Japan
e-mail: otsuka@econ.kobe-u.ac.jp

Y. Mano
Graduate School of Economics, Hitotsubashi University, 2-1 Naka, Isono Building Room 24, Kunitachi-shi, Tokyo 186-8601, Japan
e-mail: yukichi.mano@r.hit-u.ac.jp

K. Takahashi
Graduate School of Policy Studies, National Graduate Institute for Policy Studies (GRIPS), 7-22-1, Roppongi, Room 1211, Minato-ku, Tokyo 106-8677, Japan
e-mail: kaz-takahashi@grips.ac.jp

© JICA Ogata Sadako Research Institute for Peace and Development 2023
K. Otsuka et al. (eds.), *Rice Green Revolution in Sub-Saharan Africa*, Natural Resource Management and Policy 56, https://doi.org/10.1007/978-981-19-8046-6_14

Asia in the 1970 and 1980s (Barker and Herdt 1985; Hayami and Ruttan 1985; Otsuka and Zhang 2021). Otsuka and Larson (2013, 2016) argue that Asian rice Green Revolution technology is also highly transferable even to sub-Saharan Africa (SSA), given the high yield performance of Asian rice varieties in various parts of SSA. Therefore, the key question is: what are the major constraints that prevent the full-fledged success of the rice Green Revolution in SSA?

As pointed out in Chap. 1, policymakers, and even leading agricultural economists, have assumed that the Green Revolution requires only the use of modern inputs, such as modern varieties (MV) and chemical fertilizers (Gollin et al. 2021; Carter et al. 2021), ignoring the role of improved cultivation practices, such as transplanting in rows, land leveling, and bunding (Abe and Wakatsuki 2011). We argue strongly in this book that the widely held view that Green Revolution can be called a "seed-fertilizer revolution" is misleading. After reviewing the literature on the role of extension in Chap. 2, case studies of rice cultivation training programs in Cote d'Ivoire, Tanzania, Uganda, and Mozambique were examined in Chaps. 3–6. These case studies demonstrated that the key to disseminating such management-intensive technologies—consisting of not only the use of improved varieties and inorganic fertilizer but also improved cultivation practices—is the training of farmers by extension agents. It is truly remarkable to observe that rice yield, income, and profit increase significantly after rice production training programs are offered, even without changing the marketing system, input subsidies, irrigation conditions, or credit provision. We cannot over-emphasize this exceedingly important finding because, for the rice Green Revolution to be successful in SSA, the notion that the rice Green Revolution is management intensive must be widely shared among policymakers, researchers, and extension workers.

While the rice cultivation training program is a crucial entry point to increase the intensity and productivity of rice farming in SSA, there are important complementary strategies that are indispensable for realizing a full-fledged rice Green Revolution. After reviewing the literature on the role of mechanization and irrigation in the intensification of rice farming in Chap. 7, case studies from Cote d'Ivoire and Tanzania examining the impact of agricultural mechanization (particularly tractorization) on the intensity of rice farming are reported in Chaps. 8 and 9, respectively. The rate of return to a large-scale irrigation scheme in Kenya is discussed in Chap. 10, whereas the efficiency of large scale relative to small and medium-scale irrigation schemes in Senegal is assessed in Chap. 11. Chapter 12 reports on the critical role played by rice millers in improving the quality of milled rice in Kenya, while the effect of the information provision about quality-based pricing to farmers on paddy quality is discussed in Chap. 13.

In this concluding chapter, we summarize the significant findings of each case study to draw clear policy recommendations that will contribute toward a full-fledged rice Green Revolution in SSA.

14.2 Training as an Indispensable Entry Point

This section summarizes the results of five new case studies presented in this volume, three of which are concerned with longer-term impacts and the extent of information spillovers from participants to non-participants in training in irrigated areas (Chap. 3 on Cote d'Ivoire and Chap. 4 on Tanzania) and rainfed areas (Chap. 5 on Uganda). The remaining two case studies are concerned with the short-run impacts of rice cultivation training in rainfed areas (Chap. 4 on Tanzania and Chap. 6 on Mozambique).

Concrete evidence on the sustainability of training impacts and the spillover of technological knowledge is provided by the experimental study of irrigated areas in Cote d'Ivoire in Chap. 3. Eligibility to participate in training was randomly allocated to avoid any imbalance in pre-training characteristics between potential participants and non-participants. Thus, there were no significant differences between treated (eligible) and non-treated (ineligible) households in paddy yield, fertilizer use, and the adoption rates of leveling and transplanting in the pre-training year of 2014. When training was offered in 2015, treated and non-treated households were initially requested not to communicate with one another in order to identify the pure impact of the training. It was revealed that the yield of treated households, their fertilizer application, and the adoption of leveling and transplanting in rows significantly increased from 2014 to 2015 compared with non-treated households. After the 2015 season, treated and non-treated households were advised to communicate and spread the new practices. Consequently, the adoption rates of improved management practices of non-treated households increased from 2015 to 2016, indicating spillover effects. As a result, the yield gap between the two groups of farmers disappeared.[1]

Longer-term and spillover effects of rice cultivation training can also be confirmed by a study of the diffusion of improved rice production practices in irrigated areas in Tanzania, as reported in Chap. 4. In the training program, competent and motivated farmers, called key farmers, were initially selected and directly trained by extension agents at nearby training institutes for 12 days before the main crop season in 2009. Each key farmer was requested to choose five intermediary farmers and train them in the improved rice production methods. Intermediary farmers were then expected to train other ordinary farmers. As expected, key farmers' performance was better than the intermediary and other farmers in 2008 before the training program commenced (Table 4.3). The performance of key farmers substantially improved immediately after they took the training program in 2009, including an increased adoption rate of MVs. A critically important observation is that this high performance of key farmers was sustained for the next three years, indicating that the impact of the rice production training program is sustainable. Also noteworthy is that the performance of intermediary farmers improved gradually, followed by improved performance of other farmers in later years. As shown in Table 4.3, paddy yields of four to five tons per hectare were achieved, which are very high by any standard (see Fig. 1.5 for yield

---

[1] In the drought year of 2016, the yield of non-participants slightly exceeded that of participants.

trends of SSA and India). Thus, there is no question that a rice Green Revolution occurred in irrigated areas in Tanzania due to the rice cultivation training program.

A spectacular example of high paddy yield triggered by training in rainfed conditions is found in the Kilombero Valley in Tanzania (see Chap. 4). Rice cultivation management training was offered by a large private rice plantation to nearby farmers in 2012 and 2013. The production management approach was called a "system of rice intensification (SRI)." However, unlike its original definition,[2] the use of MVs and chemical fertilizer was recommended, the use of irrigation was not assumed, and straight-row dibbling was promoted.[3] This is why Chap. 4 referred to this approach as a "modified SRI" or MSRI. As can be seen from Table 4.1, paddy yield was as high as 4.7 tons per hectare on plots where the trainees adopted MSRI technologies. To the best of our knowledge, this is higher than the highest paddy yield under rainfed conditions in tropical Asia. There were no changes in technology adoption on other plots. Thus, the significant difference in paddy yield between MSRI plots and others can be attributed not only to the difference in the adoption of MVs and the use of chemical fertilizer but also to the difference in cultivation practices. While substantial yield gains were observed, trained farmers did not adopt MSRI practices on all plots they had access to. The authors also did not find systematic evidence of information spillover from the trained to non-trained farmers, even though their ongoing study anecdotally shows some signals of spillover.[4] Thus, in this study site, it remains unclear how sustainable and widespread the impact of MSRI training in rainfed conditions will be.

According to Fig. 5.1 in Chap. 5, the average yield per hectare of participants in the rice cultivation training in rainfed areas in Uganda increased by roughly 50% from 2008/09 to 2011/12. This high yield was maintained for five to six years, suggesting that the impact of rice cultivation training could be substantial and sustainable. In contrast, the average yield of non-participants did not increase as much as that of participants in 2011/12 but caught up in 2015/16. In this study site, however, participants in the training program were not randomly selected: participants were those who expressed interest in the training program, whereas non-participants did not show interest or were not informed about the training by village leaders. Thus, rigorous comparisons were made based on the propensity score matching method between participants in training villages and farmers with similar characteristics in non-training villages, as well as between non-participants in training villages and farmers with similar characteristics in non-training villages. According to this analysis, while

---

[2] Original SRI principles include shallow and widely spaced transplanting of young seedlings, intermittent irrigation, and application of organic matter. The original SRI generally does not require additional purchased inputs, such as MVs or chemical fertilizer (Takahashi and Barrett 2014).

[3] Dibbling is a method of crop establishment wherein seeds are planted in holes prepared by simple tools, such as sticks.

[4] The authors reported that non-trained farmers started to adopt some recommended management practices five years after the training was provided. This result was presented in a research meeting held at the JICA Ogata Research Institute on February 24, 2022. However, the draft report has not yet been formally circulated, and further accumulation and evaluation of the evidence will be required.

participants are found to improve yields and adopt transplanting in 2015/16 relative to their counterparts in non-training villages, there is no evidence that non-participants in the training villages improved their performance relative to farmers in non-training villages. The results thus do not provide supporting evidence on information spillovers from participants to non-participants in the training villages, even though non-participants improved the planting method from broadcasting to transplanting, which indicates that non-participants partially learned technology from participants.

Rice farming in Mozambique is underdeveloped and rainfed, with direct seeding of local varieties without any fertilizer. Chapter 6 evaluated a randomized controlled trial (RCT) of rice cultivation training in the Central Region implemented by the Japan International Cooperation Agency (JICA). After the baseline survey was conducted in the 2016–17 season, rice cultivation training programs were offered to six farmer's associations in 2017 and another six in 2018, referred to as Demo groups 1 and 2, respectively. According to Table 6.2, paddy yield was lowest among the control farmers (i.e., no assignment of training) in the post-training year of 2018–19, even though it was highest in the pre-training year of 2016–17. Since paddy yield depends on rainfall in the study sites as well as household and plot characteristics, it is challenging to identify the impact of rice cultivation training from the descriptive data. According to the regression analysis, which controls for relevant characteristics, yield increased significantly by 450–550 kg per hectare among farmers in the treated groups compared with the control group. This result is consistent with higher adoption of improved practices, such as plot leveling and straight-row transplanting, among the treated groups more than in the control groups. Yield gain of 450–550 kg may appear modest, but this accounts for an approximately 30% improvement compared to the control group. This was achieved without applying any additional modern inputs, such as improved varieties or chemical fertilizer. These results are consistent with earlier studies of the impact of rice cultivation training on the rice production performance in rainfed areas in northern Ghana by de Graft-Johnson et al. (2014) and eastern Uganda by Kijima et al. (2012). Note also that rice farmers learn new cultivation practices from participating in the training program at the demonstration plot or from extension workers. In contrast, there is no clear evidence of "social learning" or information spillover from participants to non-participants, at least in the short run in the Mozambique sites.

To sum up, the evidence reviewed in this section indicates that the impacts of cultivation training are significant and sustainable in both rainfed and irrigated areas, as well as being transmissible from participants to non-participants, especially in irrigated areas. It must be emphasized that such impacts were realized without any improvement in irrigation, marketing, or credit programs, among others. Thus, our findings can be taken to imply that rice cultivation training is a crucial entry point to the rice Green Revolution in SSA.

14.3   Complementary Development Strategies

14.3.1   Impacts of Tractorization

Power tillers were introduced in Asia intensively in the 1980s to reduce the use of draught animals and labor (David and Otsuka 1994). However, SSA differs in this respect because manual labor has primarily been used for land preparation due to the unavailability of draught animals in most areas (Chap. 7). In other words, power tillers are substitutes for draught animals as well as labor in Asia, whereas they are substitutes for human labor in SSA.

According to a study on the use of power tillers in Cote d'Ivoire reported in Chap. 8, the average paddy yield is significantly higher for power tiller users (4.7 tons/ha) than non-users (3.6 tons/ha). Furthermore, both family and hired labor were more intensively applied on plots plowed and harrowed by power tillers. The use of power tillers also increased fertilizer application and enhanced the implementation of improved cultivation practices. Cultivation size is also significantly larger for power tiller users (0.9 ha) than non-users (0.7 ha), indicating that the use of power tillers contributes to both intensification and extensification, which is consistent with the conceptual framework shown in Fig. 1.8.

The case of Tanzania is unique because not only power tillers and hand hoes but also four-wheel tractors and draught animals are used in land preparation. While four-wheel tractors were most common in 2018, the use of power tillers has been increasing more sharply. Several important observations can be made from the analysis of Chap. 9. First, the adoption of power tillers is associated with the higher yield and the higher adoption of modern inputs and straight-row transplanting compared with draft animals. Second, neither the adoption of power tillers nor four-wheel tractors is associated with a larger rice cultivation area per household. Third, there is no evidence that using four-wheel tractors contributes to intensification compared to using draught animals. It may well be that it is difficult to maneuver heavy four-wheel tractors, which bog down in muddy small paddy fields. Judging from this analysis, the introduction of power tillers seems to be highly conducive to intensifying rice farming and improving rice yield in Tanzania.

14.3.2   Impacts of Irrigation

There is no question that irrigation has significant impacts on rice farming performance because rice plants rely on a steady water supply. According to Balasubramanian et al. (2007), the average paddy yield in 16 countries with less than a 10% irrigated ratio in SSA was 1.6 tons per hectare in 2004. By contrast, the average of four countries with more than a 90% irrigation ratio (i.e., Cameroon, Kenya, Mauritania, and Swaziland) was 3.9 tons per hectare in the same year. A similar tendency was found in tropical Asia in the late 1980s (David and Otsuka 1994).

One of the critical questions is whether the rate of return to investment in large-scale irrigation schemes is high enough to justify the investment. Chapter 10 estimated the rate of return to irrigation investment in the Mwea Irrigation Scheme in Kenya by asking the hypothetical rate of return if the Mwea Irrigation Scheme was constructed as a new scheme now. The estimated internal rates of return are reasonably high (10.7–14.9%) if the value-added ratio of 0.8 is assumed but lower than 7% if the value-added ratio of 0.5 and low rice prices are assumed.[5] Note that the Mwea Irrigation Scheme is considered one of the most successful irrigation schemes in SSA because of its extremely high yield. The apparent conclusion is that the rate of return to investment for such a successful large-scale irrigation scheme as Mwea is not necessarily very high, mainly because world rice prices have remained low after the success of the Asian Green Revolution.[6]

We want to make a couple of additional comments on the rate of return to investment in irrigation in SSA. First, because of the possible complementarity between improved cultivation practices and the availability of irrigation water, well-designed training on appropriate rice cultivation may significantly enhance the rate of return to large-scale irrigation investments. Second, rates of return would be higher if we assume that the benefits of irrigation schemes are accrued not only to producers, rice millers, and traders but also to various economic sectors, including input supplies and other related businesses, through economic linkages and transactions. If such multiplier or market-wide "general-equilibrium" effects are taken into account, the net benefits for the entire economy could be enlarged to justify investment in irrigation.

While rates of return to investment in large-scale irrigation schemes in SSA are considered to be generally low, as discussed in Chap. 10, there is no evidence that larger-scale irrigation schemes are less efficient than smaller ones in Senegal River Valley, as reported in Chap. 11. It is also identified that government-invested irrigation schemes perform worse than private ones. Thus, it seems that not only the scale of irrigation schemes but also the type of investors and management capacities of users critically affect the efficiency of irrigated rice farming.

14.3.3 Role of Upgrading Milled Rice and Grading Paddy Quality

Aside from the proper timing of paddy harvesting and its impeccable drying, the quality of milled rice depends on the quality of milling machines, particularly the use of destoners and color sorters. Rice millers are in a good position to provide

---

[5] Since value added is defined as the value of production minus paid-out cost, it corresponds to income accrued to household-owned resources, such as family labor and land. Thus, if the ratio of value added to the value of production is high, a high proportion of the increased value of production becomes farmers' income, thereby leading to a high benefit from irrigation investment.

[6] Another potential reason raised by Kikuchi et al. (2021) is that, because the coverage and the number of beneficiaries in Mwea is relatively small among large-scale irrigation schemes, it cannot fully exploit scale economies, resulting in relatively high average construction costs.

information about appropriate harvesting and drying to farmers and local traders and provide information about milled rice quality to urban traders and consumers through branding and marketing for supermarkets in both Asia (Reardon et al. 2014) and SSA (Ogura et al. 2020).

Chapter 12 observes significant improvements in rice milling machines in the Mwea Irrigation Scheme in Kenya. As in many other places in SSA, a major factor impairing the quality of milled rice is the inclusion of small stones and other impurities. However, these can be removed by installing destoners. Three rice millers adopted destoners in 2011 and 34 millers had done so by 2019.[7] The estimated market share of non-adopters was 80% in 2011 but decreased to less than 10% in 2019. Rice millers adopting destoners have a greater milling capacity, can charge higher milling fees, and fetch 10% higher prices for milled rice sold than non-adopters. Furthermore, 50% of the early adopters and 25% of the late adopters had brand names as of 2019, and only these millers sold milled rice to supermarkets in large cities.

Chapter 13 is concerned with the improvement of the quality of paddy produced in northern Ghana. It was hypothesized that the lack of knowledge of paddy quality and its relationship with price causes farmers to continue producing low-quality paddy. The authors conducted a field experiment in northern Ghana to verify this hypothesis. They randomly selected 108 villages and ten rice producers from each sample village. From this group, they randomly chose 54 treatment villages and provided the farmers with information about paddy quality-enhancing technologies and quality parameters appreciated by the market. It was found that the intervention significantly influenced farmers to adopt some quality-enhancing practices. Moreover, the intervention induced significant behavioral changes among the treated farmers: they sold more aromatic varieties of paddy outside the village than the control farmers and received a higher sales price. Thus, Chap. 13 concludes that providing information about paddy quality and quality-based pricing improved farmers' paddy quality management and market sales.

14.4   Concluding Remarks

To conclude, our major findings and policy implications toward realizing a full-fledged Rice Green Revolution in SSA are summarized as follows. First, we found that the impact of rice cultivation training programs is significant not only in the short run but the results indicate that it is likely to be long-lasting. It is also being disseminated through information spillovers from training participants, particularly in irrigated areas. Thus, we advocate the rice cultivation training program as a critical entry point. This strategy is different from tropical Asia because Asian farmers largely adopted basic rice cultivation practices at the dawn of the Green Revolution, most likely due to the long tradition of rice farming. However, whether and to

---

[7] Note that most adopters of destoners also adopted pre-cleaners and graders, whereas some of them adopted color sorters, in contrast to non-adopters of destoners who did not adopt these devices.

what extent new technology is disseminated from participants to non-participants in the training program in rainfed areas must be further analyzed. Such participant-to-non-participant information dissemination may be more difficult in rainfed areas than in irrigated areas, partly because of the greater heterogeneity of agro-ecological farming conditions and partly because of the weaker social interaction among farmers in rainfed villages without a water user association.

Second, we found that the use of power tillers promotes the intensification of rice cultivation by thorough plowing, harrowing, and leveling and by inducing increased labor use for subsequent care-intensive activities in both Cote d'Ivoire and Tanzania. To overcome the handicap arising from the unavailability of draft animals in many parts of SSA, the use of power tillers should play a significant role in the intensification of rice farming. However, when soil is very hard, powerful four-wheel tractors may be more suited for plowing than power tillers. The defect of four-wheel tractors is their heavy weight: they bog down in muddy paddy fields when harrowing is performed. To what extent power tillers can be successfully disseminated in SSA must be further examined.

Third, we found that the rice farming intensification, including the application of modern inputs and adoption of recommended rice management practices induced by rice cultivation training—coupled with intensive land preparation with oxen and four-wheel tractors—improved the return to large-scale irrigation investment in Mwea. However, large-scale irrigation investment is often not economically viable because of the low rice price resulting from the Asian rice Green Revolution. Yet, relatively large irrigation schemes perform better than smaller ones in Senegal River Valley, which does not indicate that large-scale irrigation schemes are inherently inefficient. Although evidence is not decisive, it seems to us that large- and medium-scale irrigation schemes in SSA significantly contribute to the improvement of productivity of rice farming in SSA, judging from the extremely high paddy yield in our study sites in Kenya and Senegal. An under-explored issue is whether small-scale irrigation schemes are more cost-effective than large- and medium-scale schemes.

Fourth, we also found evidence that the introduction of improved milling machines has a significant impact on the quality of milled rice. Rice millers must be trained or provided information about improved milling machines to produce high-quality milled rice that can compete with imported rice from Asia. Similarly, to produce high-quality paddy, rice farmers must be trained to learn paddy quality-enhancing technologies and informed about the relationship between paddy quality and market prices.

In conclusion, we argue firstly that strengthening the public extension system for improved rice cultivation must be a central strategy to realize the rice Green Revolution in SSA. Secondly, we argue that the promotion of power tillers must play a complementary role in supporting the rice Green Revolution. Thirdly, since the availability of irrigation water is a decisive factor affecting the performance of rice cultivation, the benefits and costs of large-scale irrigation projects in SSA should be carefully reconsidered. Finally, we recommend the training of rice millers in the use of improved rice milling machines to enhance the quality of African-milled rice. There is little doubt that a full-fledged rice Green Revolution can take place in SSA if the rice extension system is adequately strengthened, power tillers are widely diffused, irrigated areas expand significantly, and the quality of milled rice is improved.

References

Abe SS, Wakatsuki T (2011) Sawah ecotechnology—a trigger for rice Green Revolution in sub-Saharan Africa. Outlook on Agric 40(3):221–227

Balasubramanian V, Sie M, Hijmans RJ, Otsuka K (2007) Increasing rice production in sub-Saharan Africa: challenges and opportunities. Adv Agron 94(1):55–133

Barker R, Herdt RW (1985) The rice economy of Asia. Resources for the Future, Washington, DC

Carter M, Laajaj R, Yang D (2021) Subsidies and the African Green Revolution: direct effects and social network spillovers of randomized input subsidies in Mozambique. Am Econ J: Appl Econ 13(2):206–229

David CC, Otsuka K (1994) Modern rice technology and income distribution in Asia. Lynne Rienner, Boulder

de Graft-Johnson M, Suzuki A, Sakurai T, Otsuka K (2014) On the transferability of the Asian rice Green Revolution to rainfed areas in sub-Saharan Africa: an assessment of technology intervention in Northern Ghana. Agric Econ 45(5):555–570

Gollin D, Hansen CW, Wingender A (2021) Two blades of grass: the impact of the Green Revolution. J of Political Econ 129(8):2344–2384

Hayami Y, Ruttan VW (1985) Agricultural development: an international perspective. Johns Hopkins University Press, Baltimore

Kijima Y, Ito Y, Otsuka K (2012) Assessing the impact of training on lowland rice productivity in an African setting: evidence from Uganda. World Dev 40(8):1610–1618

Kikuchi M, Mano Y, Njagi T, Merrey D, Otsuka K (2021) Economic viability of large-scale irrigation construction in sub-Saharan Africa: what if Mwea irrigation scheme were constructed as a brand-new scheme? J Dev Stud 57(5):772–789

Ogura T, Awuni JA, Sakurai T (2020) The impact of quality-based pricing scheme on local paddy transactions in the northern region of Ghana. Japanese J Agric Econ 22:147–151

Otsuka K, Larson DF (2013) An African Green Revolution: finding ways to boost productivity on small farms. Springer, Dordrecht

Otsuka K, Larson DF (2016) In pursuit of an African Green Revolution: views from rice and maize farmers' fields. Springer, Dordrecht

Otsuka K, Zhang, X (2021) Transformation of the rural economy. In: Otsuka K, Fan S (eds) Agricultural development: new perspectives in a changing world. International Food Policy Research Institute, Washington, DC

Reardon T, Chen KZ, Minten B, Adriano L, Dao TA, Wang J, Gupta SD (2014) The quiet revolution in Asia's rice value chains. Ann N Y Acad Sci 1331:106–118

Takahashi K, Barrett CB (2014) The system of rice intensification and its impacts on household income and child schooling: evidence from rural Indonesia. Am J Agric Econ 96(1):269–289

Keijiro Otsuka is a professor of development economics at the Graduate School of Economics, Kobe University and a chief senior researcher at the Institute of Developing Economies in Chiba, Japan since 2016. He received Ph.D. in economics from the University of Chicago in 1979. He majors in Green Revolution, land tenure and land tenancy, natural resource management, poverty reduction, and industrial development in Asia and sub-Saharan Africa.

Yukichi Mano is a professor at Hitotsubashi University, Japan, and is a fellow at Tokyo Center for Economic Research (TCER). He received Ph.D. in Economics from the University of Chicago in 2007. His scholarly interests include agricultural technology adoption, horticulture and high-value crop production, business and management training (KAIZEN), human capital investment, migration and remittance, and universal health coverage in Asia and sub-Saharan Africa.

Kazushi Takahashi is a professor at the National Graduate Institute for Policy Studies (GRIPS), and is a director of the Global Governance Program at GRIPS, Japan. He received Ph.D. in Development Economics from GRIPS. His scholarly interests include agricultural technology adoption, rural poverty dynamics, microfinance, human capital investment, and aid effectiveness in Asia and sub-Saharan African countries.